Bibliometrics and Citation Analysis

From the *Science Citation Index* to Cybermetrics

Nicola De Bellis

The Scarecrow Press, Inc.
Lanham, Maryland • Toronto • Plymouth, UK
2009

SCARECROW PRESS, INC.

Published in the United States of America
by Scarecrow Press, Inc.
A wholly owned subsidiary of
The Rowman & Littlefield Publishing Group, Inc.
4501 Forbes Boulevard, Suite 200, Lanham, Maryland 20706
www.scarecrowpress.com

Estover Road
Plymouth PL6 7PY
United Kingdom

British Library Cataloguing in Publication Information Available

De Bellis, Nicola, 1969–
Bibliometrics and citation analysis : from the Science citation index to cybermetrics / Nicola De Bellis.
p. cm.
Includes bibliographical references and index.
ISBN-13: 978-0-8108-6713-0 (pbk. : alk. paper)
ISBN-10: 0-8108-6713-3 (pbk. : alk. paper)
ISBN-13: 978-0-8108-6714-7 (ebook)
ISBN-10: 0-8108-6714-1 (ebook)
1. Bibliometrics. 2. Information science–Statistical methods. 3. Scientific literature–Statistical methods. 4. Technical literature–Statistical methods. 5. Bibliographical citations. I. Title.
Z669.8D43 2009
020.72'7–dc22 2008043216

∞™ The paper used in this publication meets the minimum requirements of American National Standard for Information Sciences—Permanence of Paper for Printed Library Materials, ANSI/NISO Z39.48-1992.
Manufactured in the United States of America.

For Viktor
and his clique of science ministers without portfolio

Now let us come to those references to authors which other books have, and you want for yours. The remedy for this is very simple: You have only to look out for some book that quotes them all, from A to Z as you say yourself, and then insert the very same alphabet in your book, and though the imposition may be plain to see, because you have so little need to borrow from them, that is no matter; there will probably be some simple enough to believe that you have made use of them all in this plain, artless story of yours. At any rate, if it answers no other purpose, this long catalogue of authors will serve to give a surprising look of authority to your book. Besides, no one will trouble himself to verify whether you have followed them or whether you have not.

—Miguel De Cervantes, *Don Quixote* (the author's preface)

Contents

Preface

The present work grew out of a research project started eight years ago, soon after the completion of a Ph.D. thesis on the history of Renaissance scientific encyclopedias, and culminated in 2005 with the submission of the final manuscript to the competition for the Biblioteche Oggi Prize announced by the Italian Editrice Bibliografica. The Italian version, entitled *La citazione bibliografica nell'epoca della sua riproducibilità tecnica: Bibliometria e analisi delle citazioni dallo Science Citation Index alla cybermetrica*, was inspired by a twofold purpose: to trace the residues of the encyclopedic ideal in the current uses of bibliographic citations and to make citation analysis a less unfriendly subject for Italian librarians by placing its "hard" quantitative core in a broader historical and philosophical context. The library-oriented nature of the competition, however, forced me to place more and more emphasis on the latter aspect at the expense of the former, thus turning the initial project into a fairly comprehensive introductory treatment of bibliometric concepts and techniques. After winning the prize, it was my conviction that the manuscript's duties had been completely fulfilled until, having sent it to Eugene Garfield, he encouraged an English translation, which would allow a better assessment of its value by exposing its content to the attention (and criticisms) of a wider audience. Whatever the final outcome, for which I'm solely responsible, I'm grateful to him for the input. The English edition, however, is quite different from the original version because of a massive work of revision and updating.

I wish to thank Vincenza Catalano, Dipartimento di Storia Antica (University of Bologna), Angela Pacillo, Biblioteca Civica Antonio Delfini (Modena), and Antonio Petrone, Centro Servizi Informatici (University of Bari) for their invaluable technical support in the gathering of the huge amount of documentation necessary for the completion of the work throughout the past eight years. I am also grateful to Peter Harriman for revising the English in the book's outline at an early stage, and to my friends John Eko, Franco Pizzardi, Anthony De Curtis, Desmond Hume, Luigi Matarrese, Fabrizio Facchini, Alain Baumsong, Salvatore Pilato, Luca Gennaro Piccolo, Gioacchino il Timido, and Joseph De Filippo for their insightful comments on draft versions of single chapters.

Bibliographic Notice

The following criteria were adopted for the compilation of bibliographic notes:

1. Free online availability has been an important criterion for source selection throughout the book: whenever two or more sources have been considered of equal or comparable informative value to the reader, the preference has been given to the one freely accessible in whatever form (preprint, postprint, etc.). That rule holds true also for materials that do not entirely comply with the open access requirements, such as the titles occasionally retrieved from the Google Books digital archive (when the full text is entirely available; books with limited preview are excluded).
2. If a paper published in a non-open access source has a freely available version (preprint or postprint) that came to my notice, the URL of the latter is given next to the reference to the former (after, of course, a preliminary check for the substantial equivalence between the two versions at the content level). The same treatment is reserved to Eugene Garfield's papers collected in the fifteen volumes of the *Essays of an Information Scientist* and freely accessible on his personal website at www.garfield.library.upenn.edu/.
3. For the sake of clarity, the expression "ISI citation indexes" is used throughout the book to indicate products developed after the acquisition of ISI by Thomson Scientific, Inc. (now part of Thomson Reuters). Finally, trademark and registration symbols have been removed from database and software names.
4. All URLs were updated and checked for availability on 23 August 2008.

Introduction

This book is intended to provide readers from a wide range of cultural backgrounds with a simple and accessible survey of the main concepts, techniques, theoretical premises, and historical developments in the subfield of information science that deals with the quantitative analysis of scientific and technological literature. For now, before the terminological digression presented in section 1.1, let's call "bibliometrics" this area of investigation and use "scientometrics" to emphasize its concern with scientific information. Obviously bibliometric techniques, exactly as any other set of mathematical tools, do not need be restricted to scholarly products, let alone scientific reports. Investigating the formal properties of the scholarly publication system, however, is undeniably their primary target and the area where their application prompts us to ask fascinating questions about how it could happen that one of the most highly regarded human achievements—scientific knowledge—turned into the object of itself.

It is assumed that the reader has some intuitive grasp of notions as amazingly complex as "science" and "information," whose ubiquity is nothing but the best guarantee of the multiplicity of technical and non-technical meanings assigned to them. Far from disentangling the skein, the following pages confine themselves to showing how those concepts are handled within the intellectual tradition of quantitative science studies. The exposition deploys quite systematically, though not exclusively, historical documents and figures, but it is by no means an exhaustive or erudite history of bibliometrics. Early breakthroughs and theories are recalled, partly because they keep on living in current research agendas, and partly as a way for putting recent developments in the right perspective.

In placing science and scientific research at the center of bibliometric discourse, moreover, the focus is on areas traditionally more sensitive to this kind of investigation, namely physical and life sciences, medicine, applied and engineering sciences, and limited portions of the social sciences. For the most part, indeed, social sciences and humanities conform to communication habits that don't fit the analytic capabilities of current bibliometric facilities: the flair for book (instead of journal article) publishing and referencing; the frequent bias toward themes of local/regional interest prevailingly dealt with using a national, non-English language; and the tendency to rely on a pool of sources older than those employed by natural scientists and much more loosely knitted to the ongoing research activity.[1] The main character of our story, in a sense, the Ariadne's thread of the entire overview, is the bibliographic citation: a choice dictated by the selection of a tightly defined perspective on what kind of information is being measured and for what purpose.

Information can be sized in various ways and for different ends. Adopting a utilitarian stance, one could develop a set of methods and algorithms for gauging the degree of informativeness of a document, that is, the extent to which it provides valuable information to an individual, the user, for the purpose of testing and possibly improving upon the performance of an information service.[2] Such an insight would certainly project on the measurement activity an aura of usefulness a far cry from any intellectualistic meddling with the ultimate enigmas of scientific knowledge, but here a somewhat inverse perspective is assumed. A myriad of valuable records is already out there: scientific documents. They are already valuable, insofar as they result from the well-established process by which organized communities of experts in distinct research areas produce and validate, against common sets of methodological criteria, a wide range of knowledge claims about natural and social phenomena. Bibliographic citations form one visible and traceable channel linking scientific documents; hence, they have been asked to tell us, via increasingly sophisticated analytic toolkits, something about the way those knowledge claims are generated, connected to each other, and validated. This concern with the epistemological status of bibliometric phenomena, and specifically of bibliographic citations, runs throughout the rest of the book at the expense of the equally important issue of the quantitative patterns of document–user interaction in a library service perspective.

Another limiting choice, closely intertwined with the previous point, refers to the particular subset of the science and technology system under investigation. As extensively documented by Eliezer Geisler, measurements of science and technology can take a wide variety of forms and use several kinds of metrics, each addressing a particular facet of the system's complexity, such as economic/financial inputs and outputs, organizational aspects related to specific research and development projects, documents (the stuff of bibliometrics), patented inventions, and peer-reviewing processes.[3] Bibliometrics, therefore, is but one of several options, and clearly differentiates itself from other metrics in that it revolves around the measurable properties of the formal communication system of science and technology, that is, the network of published documents, above all journal papers.

Admittedly, all metrics fall short of an ultimate understanding of scientific and technological change. Constructs as elusive as "scientific knowledge" and "technological innovation," indeed, escape any pretension to capture their deep nature in universal statements, let alone mathematical formulas. Compared to other units of analysis, nonetheless, bibliographic citations are quite unique in this respect, because the connections they establish between documents are the operation of the scientists themselves in the process of exposing (and propagandizing) their findings to the community of their peers. Hence it is reasonable to assume that any regularity revealed by citation patterns is, to a certain extent, the faded reflection of parallel regularities occurring in the process of generation, validation, and communication of scientific knowledge. The limits of such an assumption will become clear later on. For the moment, in further introducing the book, let's adopt an unusual point of departure, resorting to a cursory flash-forward to its completion, the moral of the story.

When the research effort of a scholar comes to an end, a lexical dictionary may be used to play a sort of game: taking the word or words that best represent the research subject under investigation and checking if and to what degree, owing to an unpredictable cunning of the language, the conclusions drawn from the entire discussion can be stretched to the point of fitting the listed meanings. Let's try this game. For the entry "citation," an English dictionary usually lists four main definitions:[4]

1. "The action of citing or quoting any words or written passage, quotation." The English language neatly differentiates a citation/quotation in

this sense from the "reference," that is, the more technical indication of a source document's bibliographic data (author, title, page, etc.). A similar distinction exists, for example, in German, between "Zitat" and "Referenz," and in French between "citation" and "référence."

2. By extension, a citation/quotation is also "a short passage or tune taken from one piece of music to another or quoted elsewhere." There is no reason, of course, to restrict this definition to music alone: movies, paintings, sculptures, and any other form of artistic expression may (or must necessarily) incorporate more or less explicit allusions to earlier models.
3. "A mention in an official dispatch" for a particularly honorable act or career, as in the sentence "The officer was awarded a citation for bravery."
4. "A citing or summoning to a court of justice, a summons": an acceptation applied, as mentioned in Tomlins's *Law Dictionary,* particularly to a process in the spiritual court.

The four definitions share an etymological root in the Latin *citāre*, an iterative form of *ciēre*, which means to move, to set in motion, to stir, and also to call, to invite, to invoke, to summon (a person, a text, a divinity). The last definition hints at a power relationship between an individual and an institution in charge of judging personal behavior; the summons is motivated by the suspicion that a preestablished set of norms has been infringed. The third definition also entails a value judgment about an individual by a collective entity, but now with an unequivocally positive orientation; a soldier may be cited for having demonstrated bravery and loyalty, and the citation is likely to foster a promotion or a more prestigious assignment. The first definition is confined to the linguistic domain, wherein it signifies the act and the result of an extraction-transfer-reallocation routine. What is set in motion, here, is a text or speech fragment culled from a written or oral source and "grafted" onto another text or speech. Scholarly quotations and bibliographic references do not differ so much in this respect, save that in a quotation the textual transfer is visible (the quoted text is repeated verbatim), while in a reference it is symbolized by the recording, in a conventional format, of the cited source's bibliographic data: "Read this document"—the bibliographic note seems to suggest—"and you'll find the sentence that warrants my argument." Both operations undeniably provide the cited fragment with a new context, hence with a whole array of potentially new meanings. Besides, the two operations often coexist. Quoting a passage along with the corresponding bibliographic coordinates in a note, indeed, is a basic rule of the game in modern scholarly writing, and its punctual violation marks indelibly the huge share of medieval and Renaissance scholarship wherein citations/quotations/paraphrases are such stuff as texts are made of. But if one simply leafs through a present-day scholarly journal, it takes just a few minutes to realize that scanty lists of bibliographic references dominate the stage, or rather the backstage, of the citation arena, while in-text quotations and discursive foot- or endnotes are sacrificed in the name of fluency and the crystal clear writing advocated by style manuals.

In the most subtle manifestation, stretching far beyond the boundaries of the scientific publishing system, the quotation takes its full revenge as a tool of artistic expression. Now the second dictionary definition is applied: the uniqueness of an artistic creation is always shaped against the background of existing models, stereotypes, *tòpoi*, whose modes and forms of survival across space and time are addressed by literary, music, film, or art criticism. Classical music, for instance, has long been a fertile soil for experimenting with ideas, themes, fragments, and combinations of timbres of extremely diverse origins. Even more so, jazz music has affirmed its identity through the joyful celebration of the musical quotation and the repetition/variation on preexisting riffs and tunes, whereas pop music has to such an extent exasperated the recycling of melodic combinations made up of standardized sequences of chords and harmonic progressions that the originality of individual pieces is often a matter of controversy in intellectual property infringement litigations.

Beyond the transfer of ideas and motifs as an integral part of the cultural transmission process, the cult for repetition, quotation, and allusion to the déjà vu and déjà entendu is ubiquitous in current literary and artistic discourses. Yet, numbered among the most evident symptoms of postmodern sensitivity, this citation/quotation mania has disclosed all of its reactionary side effects. In trying to restore a synchronic image of every possible history and every conceivable memory, citation is not only a chance for the experimentation with new poetic and narrative forms, but also a renunciation of originality, a condemnation of authors and readers to an endless intertextual game consisting of the obsessive repetition and updating of past codes and materials. Such awareness can be tantamount to admitting that nothing substantially new can arise beyond what has already been said, but it contains also the seed for more fruitful re-readings of cultural history. In a much debated book published in 1973, the literary theorist Harold Bloom put forth a theory of poetry pivoting on the idea that, because texts originate necessarily in contact with other texts, and new poets are irremediably influenced by a small group of masterful forebears, poetic history is indistinguishable from poetic influence. Writing, accordingly, is a struggle of the young poet against the old masters, a sort of anxiety-laden oedipal conflict of the artist precisely with those precursors whom he or she most admires, motivated by the desire to "clear imaginative space" for himself or herself, the final outcome being a systematic creative misinterpretation ("misprision") of previous texts. Great writing—as Bloom pointed out—"always proceeds by a misreading of the prior poet, an act of creative correction that is actually and necessarily a misinterpretation."[5] As we shall see below, a quite similar dependence on the burden of existing paradigmatic theories grown out of the revolutionary insights of a few past masters qualifies the below-the-water-line influence argument taken up by the bibliometric tradition in the wake of Thomas Kuhn's theory of scientific change.

The dictionary game played so far, to put it bluntly, goes a long way toward philosophical issues that do not in any way constitute the subject of the present work. To the extent that the central thread of much scientometric literature is the quantitative analysis of bibliographic citation links, only the first of the above definitions is strictly pertinent. But the dictionary game is forward looking in that it suggests one of the most fundamental points to come: as the chapters in this book will extensively show, the vicissitudes of bibliographic citations in the bibliometric tradition would make no sense without the constant reference to the extra-bibliographic dimensions implicit in the other dictionary definitions, namely,

1. the juridical-prescriptive dimension (to cite following some rules of scientific conduct or deviating from them);
2. the social dimension of rewarding individual merits (to cite recognizing the value of a document while raising the status of its author); and
3. the postmodern dimension of citations as literary devices capable of dynamically rewriting the past according to the present author's own

agenda (the citation as a rhetorical tool in the social legitimation of scientific discourse).

Transversal to each dimension is the omnipresent, multipurpose, and somewhat equivocal concept of influence: the influence of people on other people's lives through transactions firmly embedded in the social network of power relations (an echo of the primeval rootedness of the concept in the astrological theory of the influence exercised by the heavens on the course of human actions); and the influence of texts on other texts through the uninterrupted chains of endorsements and rebuttals linking past ideas with new ideas. Both facets are deeply intermingled with the creative process of every scholarly and artistic work, but whereas Bloom's theory of poetry pointed to the dismissal of any "external" history of ideas produced by the "wearisome industry of source hunting, of allusion counting," in the realm of scientific communication surveyed by the present book, the stakes revolve precisely around the tracing of influences. Now the hero (the scientist) is still struggling, as in Bloom's universe, to clear cognitive space for himself or herself, but the chances for the struggle to be successful depend as much on the adherence of the final product (the scientific report) to a set of preestablished tenets and methodological criteria (a paradigm) as on the adoption of the reported findings by other researchers. However imperfect, bibliographic citations are the most accessible and visible tracks of this double movement of complying with the past while projecting into the future. Hence, even though scarcely anything of the oedipal conflict behind their production is visible in the finite product, it is perfectly legitimate to use them as proxies for deeper connections in the attempt to pick the lock on the door of scientific communication. This strategic position assigned to bibliographic citations throughout the bibliometric tradition stems directly from the way science works differently from other types of scholarly and artistic expression.

Scientists do research and, at least in natural and life sciences, publish their results, mostly in the form of peer-reviewed scientific articles. Research activity is never isolated in space and time. Rather, it implements techniques and concepts previously established by other scientists and colleagues in the same research field. A silent code of honor requires the acknowledgment of such dependency through bibliographic references. In a scientific paper, contrary to a novel or a newspaper article, citations are made and must be made—or, in a worst-case scenario, appear to be made—to acknowledge intellectual debts. So, if used appropriately, a citation makes visible an intellectual link in the process of transmitting and re-elaborating scientific knowledge, thus working as a peculiar form of currency in the market of official scientific communications. The banknote denomination, actually, is minimal (it costs very little to cite), but its symbolic buying power is higher than one might expect. Through citations, in fact, the author debriefs and calls for the support of a host of colleagues and predecessors working in the same research area. Their presence is integral to the credibility gained by the text both for "what" they say and for "who" they are. Some are summoned for having supplied basic methods and concepts; others, more spuriously, are called in to reinforce, to challenge (in a battle that is won even before it is fought), or simply to broaden the author's point of view.

In the humanities, and particularly in historical and philological disciplines, where footnote tracing from current and past literature is still the most popular retrieval technique, the accuracy of bibliographic references and the mastering of an appropriate citation style certify the author's full membership in a professional community. Footnotes, for a long time the natural habitat of bibliographic citations and the filter imposed by the author on the interpretation of their meaning, are to the historian nearly the equivalent of what factual and experimental evidence are to the scientist, and perform an inseparably cognitive and social function as well.[6] They make tangible and verifiable the empirical foundation of the proposed reconstruction; at the same time, they lend authoritativeness and respectability to the text while furnishing evidence to the effect that, having inspected primary (the data to be interpreted) and secondary (other historians' interpretations of the same data) sources, the rules of the game were obeyed, the author is consequently ready to reproduce the mode of knowledge production upon which the institutional survival of the discipline is founded. Checking whether a bibliographic note was really needed for the completion of a study, however, is quite a difficult task. Only those who are familiar with the same archival records and handle the same documents can, at least in principle, venture to say. And even so, the opportunity to cheat is always at hand. To take just one extreme example, Leopold von Ranke, a key figure in the eighteenth-century development of modern scientific history, professed the cult for archival research and the direct inspection of primary sources. He restlessly countered the questionable philosophical vocation of Enlightenment historiography with the famous statement—actually a quotation from Thucydides—that history should limit itself to just reporting "how it really was" (*wie es eigentlich gewesen ist*). The manuscript of his *History of the Latin and Teutonic Nations from 1494 to 1514* (1824), however, raises the suspicion that he first composed the treatise as a whole, without worrying about philological matters, and then looked for bibliographic evidence to support its arguments. In other words, as Anthony Grafton puts it, "he used a salt-shaker to add references to an already complete stew."[7]

In natural sciences, where the unit of analysis is not a sequence of unique historical events but a set of variables related to phenomena that virtually anyone can observe or reproduce under similar experimental conditions, the emphasis on the accuracy and appropriateness of bibliographic citations is no less important. The ability to foreground one's claims, to place locally generated knowledge into an intertextual framework by citing the right sources in the right places, helps strengthen the credibility of the author as an expert in the field, whereas the inability to do so may lead to professional failure. Such a discriminant role has a comparatively recent history, which unfolds in strict connection with the steadily increasingly critical tasks assigned in contemporary physical (and subsequently also biological and social) sciences to the experimental paper. Even if its origins as a literary genre date back to the late Renaissance, it was not until the second half of the nineteenth century that the scientific paper underwent a series of internal transformations that gradually shaped its current structure, turning it into the most effective means of disclosing new ideas while simultaneously securing the author's intellectual property rights over the reported findings. Critical changes involved also the organization and treatment of bibliographic data. Initially scattered throughout the text and devoid as much of an unambiguous reference to the cited authors' works as of a precise correlation with the statements they were meant to support, their compilation gradually became more accurate, their format standardized, and the physical position aligned with the conceptual one (the right source cited in the right place). But here, too, as in the example about Ranke, the evolution generated a two-faced Janus, perfectly functional in the double story told by bibliographic references henceforward. One is the bright story of the good guys, those who use references as they should be used in the "perfect world," namely to help the reader trace a certain statement back to its source while acknowledging and partially rewarding the contributions of colleagues and predecessors at any level, whether they be methods, concepts, or experimental results. The other is the counter-story of the bad guys, the bluffers who overcite, undercite, and miscite previous work so as to reinforce the persuasiveness of the argument or to accomplish some other nondocumentary purpose.

As will soon become clear, a key objection raised against the value of bibliographic citations in the quantitative study of scientific communication is that, more often than not, good and bad guys cohabit in the same author, preventing external observers from answering the following questions: How much of the cited stuff has actually nurtured the scientist's mind? Where is the borderline between documenting and cheating? Science, after all, is also a kind of writing, a set of socially recognized and repeated textual strategies enacted by authors belonging to the same community to achieve similar goals in situations perceived as similar. The creation and refinement of the modern literary genre of the experimental article, consequently, is also the creation and refinement of our modern concept of the experiment, and of the laboratory as the chosen place where scientific truths are built, by definitely linking text to a nonliterary, "external" reality.[8] So, when science came under the purview of constructivist sociologists in the 1970s and 1980s, any simplistic definition of the communicative role of bibliographic citations was jeopardized by the fading away of the apparently straightforward distinction between a scientific writing hung on the impartial reporting of laboratory experiences and a nonscientific (humanistic, historical, juridical, poetic) writing focused more on itself than on a supposedly separate piece of reality.

Beginning in the 1960s, the apparently ordinary mission of bibliographic citations to repay intellectual debts became the focus of unprecedented attention from administrators, politicians, sociologists, and historians, and the citation started to take on a life of its own, independent of the documents forming its original habitat. The change took place when, within the tradition of information retrieval studies, increasing importance was ascribed to the cognitive potential of citation indexes, or at least of the most important one ever compiled, the *Science Citation Index* (*SCI*), conceived by Eugene Garfield, the founder of the Institute for Scientific Information (ISI) in Philadelphia.[9] A citation index lists the documents cited in the reference section of articles from a selected group of scientific journals next to the source of each citation, that is, the article or articles in which those documents appear among the bibliographic references. This seemingly straightforward operation doesn't alter minimally the above dilemma of whether, in citing previous literature, the individual author is documenting or cheating, but affords a completely new perspective on the quantity and quality of information that can be mined out of the network of citations woven around individual authors and documents as they enter the publication (and citation) game. Turning the lists of references upside down, indeed, a citation index allows three types of operations implicit in the network structure of collective referencing patterns:

1. using a document relevant for me in the retrieving of new documents by authors who, like me, have judged the same contribution relevant and, presumably for the same or similar reasons, decided to cite it (the bibliographic citation as an information retrieval tool);
2. exploring, in qualitative and quantitative fashion, the intellectual links between citing and cited authors, to identify key documents in a research field, to describe the structure and the dynamic modifications of research fronts at any given moment, and to draft a map of scientific specialties (the bibliographic citation as a tool of historical and sociological analysis); and
3. measuring, in terms of citation frequencies, the cognitive impact of individual documents, journals, and authors (the bibliographic citation as a tool for research quality control).

The present book follows these three operations throughout recent bibliometric history, relying on a necessarily limited but representative sample of bibliometric literature.

Chapter 1 leads off with a terminological discussion of the rich nomenclature associated with bibliometric studies in the era of digital networking (section 1.1). It continues with an overview of the pioneering measurements of scientific literature carried out, either on speculative grounds or for library management purposes, within the tradition of statistical bibliography (section 1.2). The extension of bibliometric techniques to science policy issues was largely an effect of the rise of citation indexes during the 1960s and 1970s. The explosive growth of scientific literature, along with the increasing attention of local governments to the economic potential of basic research, called for new investments in specialized information services. Simultaneously, the growth in the expense and complexity of experimental equipment in many fields, and the increasing constraints on research budgets urged technocrats to pursue a quality control of the research output by means of tools more refined than simple publication counts. Citations promised to accomplish both these objectives to the extent that they symbolized the degree of endorsement of the cited texts by the citing authors; hence, their statistical analysis could be employed to build indicators of scientific performance (section 1.3).

Chapter 2 reviews the inspiring principles, planning stages, structure, and searching potential of the *SCI*. In sections 2.2 and 2.3 the focus is, respectively, on the historical development and the structure of the database, whereas section 2.1 dwells on the quite ironic circumstance that what was going to become the main tool of the trade for scientometric appraisals developed from a completely different set of problems and theoretical concerns, namely the American tradition of information retrieval studies. In a period during which the efforts of most librarians and information scientists revolved around the choice of selective lists of subject headings and the perfection of semiautomatic systems of word indexing, Garfield designed a radically different bibliographic tool, a highly comprehensive, interdisciplinary, and "subject-less" index to journal literature, whose entries were bibliographic citations instead of keywords or subject headings, and whose authorial provenance was scientists themselves instead of professional indexers. Apparently diverting from the mainstream of sociologically oriented and evaluative bibliometrics, section 2.1 features a synthetic overview of basic indexing and retrieval concepts in full-text databases. The rationale for such a digression is that information retrieval cannot be considered a merely accidental fellow traveler of science measures, in at least three senses:

1. Any large-scale research assessment exercise, as well as any mapping of research fields for sociological or science policy purposes, is heavily conditioned by the quality of the information retrieval and data-cleansing routines applied to the source database.
2. The search for quantitative patterns in the content and uses of information systems, along with the practical implications of bibliometric

laws for information systems design and management, have been the focus of growing attention from information and computer scientists over the last few decades in view of the design and perfection of online retrieval systems.[10]

3. The very possibility of using cited references as search entries in a database hangs on a premise that strikes at the heart of bibliometric work, namely the relative stability of meaning and patterns of use of citations within a community; conversely, the validity of citations as indicators of intellectual and social links between documents and authors hinges on a trust in the same kind of conceptual stability that guarantees the success of citation-based retrieval systems.

What makes particularly intriguing such a threefold coalition of information retrieval and sociopolitical issues in the role-play enacted by citations on the stage of scientific communication is that an apparent weakness in one part turns into a strength of the other part: not all documents citing a previous document pertain to the same subject or field of knowledge; thus a potential weakness on the information retrieval side has to be recognized if subject relatedness is the coveted goal. At the same time, the considerable overlap in the citation net among documents on different topics is perhaps the most visible trace of that cross-fertilization of fields and topics that is of paramount importance in the advancement of scientific knowledge.

The disciplinary foundation of scientometrics runs parallel with the increasing political importance gained by citation analysis, but it could hardly take place without the two underpinnings unveiled, respectively, in chapters 3 and 4. First is a strong theoretical commitment to emerging trends in the history and sociology of science, above all the normative sociology of Robert K. Merton and the quantitative history of science of Derek John de Solla Price. Second is the incorporation of citation regularities into the broader mathematical framework set up, from the 1960s onward, in connection with the discovery of hyperbolic empirical distributions in information production processes.

Chapter 3 outlines the theoretical foundations of bibliometrics through the writings of John Desmond Bernal, Robert K. Merton, Derek Price, Eugene Garfield, and Henry Small. Bernal, as did Price after him, advocated the application of scientific methods to the study of science itself.

Even more important, he legitimized the idea, currently reassessed by the partisans of the open access movement, that not only does the formal communication system of science reflect, to a great extent, its cognitive and social structure, but the latter can be improved by a drastic transformation in the way the former organizes and disseminates recorded knowledge (section 3.1). The place of bibliographic citations in this undertaking depended, in turn, on the legitimization of its sociocognitive potential. In fact, a set of bibliographic citations accrued to a document is a key for both bibliographic control and scientometric analysis only if one admits that, on the one hand, scientists do not cite arbitrarily, but follow a "tacit rule" that imposes the recognition of colleagues' and predecessors' work (Merton's theory of citation as a kind of peer recognition, outlined in section 3.2); and, on the other hand, references point, directly or indirectly, to scientific concepts for which the citer recognizes the right of total or partial intellectual property to the cited (Garfield and Small's theory of citations as concept symbols, discussed in section 3.3). The next section outlines Derek Price's research program for the foundation of a new research field that could supply an empirical basis for science policy (section 3.4).

Chapter 4 surveys the mathematical foundations of bibliometrics. Between the 1920s and 1930s, three milestone studies in the history of the discipline were published, respectively, by Alfred Lotka on the distribution of scientific productivity, by Samuel Bradford on the scattering of papers among disciplinary journals, and by George Zipf on the statistical properties of text corpora. From different starting points and analytic perspectives, the three authors formalized a set of regularities—the "bibliometric laws"—behind the processes by which a certain number of items (scientific papers, text words) are related to the sources generating them (authors, journals, text corpora). Their common feature is an amazingly steady tendency to the concentration of items on a relatively small stratum of sources. That few very productive scientists exist compared with a much higher number of one-shot authors; that most of the literature relevant to a research area is issued by a small number of journals; and that few words occur much more frequently than others in written (and spoken) language, were actually no secret to sociologists, librarians, historians of science, prosopographers, and lexicographers during the late 1800s and early 1900s. Lotka's, Bradford's, and Zipf's laws did not explain why this happened. Nonetheless, the chance they offered to bring a gleam of mathematical clarity into traditionally elusive communication patterns encouraged a good deal of empirical and theoretical work on the strengths and limitations of statistical measures of scientific products. On a practical basis, moreover, bibliometric laws put into rough mathematical terms a principle that is at the core of the extra-bibliographic uses of citation data, both in information systems construction and in science management: in modeling an information production process, few elements are sufficient to account for the structural properties of a complex of many elements. In sections 4.2, 4.3, and 4.4, the significance of the three laws for the subsequent development of bibliometrics is explicitly addressed. Bradford's Law, in particular, spurred information scientists' desire for a rigorous mathematical treatment of the complex of informetric regularities (subsection 4.3.1) and met with an important application by Garfield, who relied on citation analysis to extend its validity to the global network of scientific journals (subsection 4.3.2). The conceptualizations of scientific literature's growth dynamics and aging processes grown out of Derek Price's work are reviewed, respectively, in sections 4.5 and 4.6. Section 4.7, then, is fully devoted to the search for a mathematical foundation of bibliometric laws performed, from the late 1950s onward, in the writings of several authors.

Chapters 5 and 6 delve into the extra-bibliographic uses of bibliographic citations in sociology, history, and science policy and management. Here citations are taken as indicators of something more than a sheer subject relationship: cognitive and social structures, intellectual kinship or influence, and research quality. An indicator, in general, is anything but a plain object: it is a construct derived from a mathematical operation (usually arithmetical) on the data aimed at capturing, in a convenient and simplified fashion, an aspect or dimension of a system that, because of its original complexity and multidimensionality, escapes any clear-cut global representation. Citation-based indicators, specifically, are supposed to reflect important aspects of the science communication and reward system but, in addition to the multiplicity of ways they can be constructed from within different reference frameworks, their building blocks, bibliographic citations, are by themselves multifaceted data, whose ability to represent sociocognitive transactions cannot be taken for granted.

Chapter 5 highlights the main applications of citation indexes to the history and sociology of science. An analytic tool alternative to citation analysis for mapping the dynamics of science and technology is outlined: co-word analysis, developed during the 1980s at the École Nationale Supérieure des Mines of Paris. Particular emphasis is placed on the non-Mertonian rationale of the mapping technique, which builds on the assumption that facts reported in scientific articles are constructed and shaped according to the interests and strategic positions advocated by researchers and institutions in the endless rearrangement of their sociocognitive networks (section 5.1). The next section lingers over Garfield's algorithmic historiography and the conceptual issues arising from any attempt to capture, via bibliographic links, the influence exercised by a paper (or an author) over another (section 5.2). At a deeper layer, citations have been asked to perform an even more difficult job. It can safely be assumed, in fact, that an article citing two documents in its bibliography establishes some kind of association between them. But if, as usually happens in a tightly defined scientific or technical field, other articles use the same association, co-citing the same documents in a statistically significant way, then it might also be expected that the association is something more than a coincidence, hinting at the conceptual structure of the field. If the frequency of co-citation measures the degree of affinity (conceptual, methodological) between documents as it is perceived by the group of citing authors, and if frequently cited documents are the equivalent of key concepts or methods of a discipline, then co-citation analysis may be used to trace the map of relationships among these documents/key concepts, to outline and graphically visualize the structure of a research field, its connections with other fields, and its articulation into subfields and new research fronts. The basic technical steps required to build a co-citation map are outlined in subsection 5.3.1, along with some crucial issues raised by a pioneering model of co-citation mapping developed by Small and Griffith at the ISI. Next, the possible utility of bibliometric maps in providing a concrete, operational meaning to Kuhn's construct of scientific paradigms is tackled (subsection 5.3.2). Kuhn's theory of scientific development strongly influenced bibliometricians, in that it prompted them to seek empirical evidence of paradigm changes either by matching the idea of drastic, conceptual switches in the history of science against mathematical and empirically testable models of scientific growth, or by tracing the documentary roots of a paradigm, the "exemplary documents," by means of bibliometric techniques.

Chapter 6 surveys the uses and abuses of citation indexes in the evaluation of scientific research. Faced with the crisis in traditional peer reviewing, the number of citations accrued to a document has often been welcomed enthusiastically as an unobtrusive indicator of the quality of its author's work, an antidote against funding and academic rewarding criteria governed by obscure logic and personal favoritism. As such, though, it has caused misunderstandings and misuses, above all the habit, widespread in European universities, of associating the number that expresses the citation impact of the journal wherein an article was originally published to the article (and the author) itself. The debate on the improper uses of the journal impact factor, about which Garfield himself complained on more than one occasion, is documented in section 6.1. Severe methodological issues, however, also come up when, moving from the journal to the individual author, scientific quality is carelessly identified with the number of citations settled on the publications of a single scientist (subsection 6.2.1). The elusiveness of concepts as diverse as quality, impact, and influence imposes extreme caution in the use of bibliometric data to assess individual performance. That's why avant-garde bibliometricians have promoted more sophisticated conceptual and analytic tools for the handling and interpretation of citation data. The overview of such tools, conducted in subsection 6.2.2, revolves around four milestone research programs in evaluative bibliometrics, developed, respectively, by Francis Narin's team at CHI Research (now ipIQ); Ben Martin and John Irvine at the Science Policy Research Unit (SPRU), Brighton; the Information Science and Scientometric Research Unit (ISSRU) at the Library of the Hungarian Academy of Sciences, Budapest; and the Leiden Group at the Centre for Science and Technology Studies (CWTS). A common denominator of advanced bibliometric projects is the tendency to overcome some of the technical and conceptual problems of citation analysis by selecting a collective instead of individual unit of evaluation (university, faculty, department, research group, country) and by integrating statistical analysis with a set of methodological remarks aimed at specifying the context and conditions of applicability of bibliometric measures. The case study of patents, introduced in section 6.3, illustrates exemplarily how the strategic value of citations in research evaluation and business intelligence is inseparable from the ambiguity inherent in the process of their generation.

Chapter 7 delves into the basic issues raised by the use of citation statistics for sociological and political purposes: Is the Mertonian normative theory of citing still defensible if the actual role played by citations in carefully defined samples of scientific literature is examined (section 7.1)? What do citations have to say about the stratification system of science? Do higher citation rates amount to (or prepare) outstanding intellectual achievements in any meaningful way (section 7.2)? To what extent does the status of the contemporary scientific author, vexed by the explosion of multiauthorship in many disciplines, undermine the full intellectual responsibility of a scientific contribution, thereby destabilizing the final destination of every citation count for political-administrative purposes (section 7.3)? The attention constantly devoted by sociologists and information scientists to the citing behavior of scientific authors, in the hope of gaining a deeper insight into the complex mechanism of the citation process, is documented from four distinct perspectives: the reasons to cite propagandized by the authors themselves in questionnaires or interviews (subsection 7.1.1); the analysis of the content and context of citations in relation to the argument they are supposed to underpin (subsection 7.1.2); the discrepancies between what authors and texts say on citations and the actual behavior of scientists as emerging from "ethnographic" accounts of laboratory life (subsection 7.1.3); and the effort, manifested by many published mathematical and empirical models of citation behavior, to "count the uncountable" anyway and in spite of all evidence against its very possibility "constructed" by "constructivist" theories (subsection 7.1.4).

Chapter 8 addresses the status of bibliographic citations in the universe of the World Wide Web, wherein they accomplish two main tasks. First, they keep connecting scientific papers just as the traditional lists of references did in the paper world, save that now papers are in digital format and a bibliographic link between any two of them is also, for the most part, a physical link between their respective full-text versions. Second, to the extent that hyperlinks between web pages may be considered high-tech versions of bibliographic references, they form the connective structure of the medium itself, the World Wide Web. This dual personality accounts for the two paths taken by citation analysis in the new environment: the design of automatic indexing systems for citation data scattered in e-journals and open archives (the story of *CiteSeer* and *Citebase* is told in subsection 8.1.2); and the application of bibliometric methods to the study of the hyperlink network of the World Wide Web. The former path has opened the way to an entirely new array of issues concerning the extension of impact metrics on scientific documents freely available via open access journals and disciplinary or institutional repositories; one related question, for instance, is whether the open access to scientific literature increases its citation impact (subsection 8.1.1). The latter path is, in turn, split into three main directions of inquiry, each embedded in a specific disciplinary tradition: complex network analysis in the wake of statistical physics (subsection 8.2.1), hyperlink network analysis in the tradition of social network analysis, and webometrics in the mainstream of information science (both reviewed in subsection 8.2.2).

The reader should be advised that, just as much work in evaluative bibliometrics is empirical in nature, there is also a bibliometrics for bibliometricians, which wends its way through complex statistical reasoning and in-depth mathematical modeling, but any introductory treatment can't do justice to either the empirical or the mathematical conduits. On the empirical side, it should be remembered that statements such as "evidence exists that . . ." and "studying the literature of . . . the author empirically demonstrated that . . ." are never to be understood to be absolute; empirical evidence, even more so in the social sciences, is the final product of carefully planned studies exhibiting a high variability in research design and methodology. Hence it is always "controlled" evidence, shored up by a specific research question, by a set of (theoretical, methodological) assumptions about the phenomena under investigation, and by the resolution of the observation instruments (bibliographic databases, citation indexes, mapping software, surveys, and the like). An interesting and somewhat extreme example of this point is given in subsection 6.2.2, which outlines the debate on the alleged decline of British science taking place throughout the 1980s and the 1990s. On the abstract modeling side, it is worth stressing that, even though mathematical reasoning is an essential part of current bibliometric research, a book pretending to reproduce even the smallest part of it would inevitably run the risk of losing itself in the details of the formalism. Hence, save for the oversimplified algebraic expressions of the historical bibliometric laws, any mathematics has been removed also from the places where it would constitute the only possible way to synchronize one's mind with the game actually played by militant scientometricians. Concepts like those of probability distribution and stochastic models, nonetheless, occur so frequently in bibliometric parlance that a purely qualitative grasp of their meaning is highly recommended. Having this in mind, one might wish to give section 4.1 a cursory glance before reading the other chapters, because it discursively reviews some basic constructs preliminary to the understanding of the significance of bibliometric laws for the quantitative studies of science.

The volume and thematic variety of bibliometric literature piled up over the past forty years would make absurd and perhaps even useless any item-by-item approach. Exclusion and inclusion criteria have been consequently adopted quite systematically. One notable exclusion, motivated by the infancy of the subject, concerns the recent flourishing literature on the structure, coverage, and scientometric potential of the two main competitors of the *SCI* in multidisciplinary citation indexing: *Scopus* and *Google Scholar*. A second, more radical omission, partially justified by the conviction that the most important trends and contributions to the discipline leak out through a limited number of well-established international journals, is the bibliometric literature published in non-English-language sources. As to the inclusions, besides the milestone papers and books marking an undisputable turning point in the history of the discipline, only those references are cited that either were deemed representative of original research lines or turned out especially pertinent to the points made from time to time. The approach to the subject is partly descriptive, sometimes perhaps even didactic if basic concepts and techniques have to be introduced, partly critical when the relative strengths and weaknesses of those concepts and techniques are discussed from a broader perspective. A preferential treatment, however, is reserved for monographs and review papers enabling the reader to quickly locate previous work relevant to each topic being discussed. This choice is in tune with the book's threefold rationale: 1) introducing nonspecialist readers to a variety of issues and resource materials, which they can then pursue in greater depth if they wish; 2) promoting a polyphonic perspective on quantitative evaluations of scientific research so as to avoid the equally biased extremes of total rebuttal and uncritical acceptance; and 3) injecting an even minimal dose of bibliometric-enhanced reflection on the complexity of science evaluation criteria into areas, like the Italian university system, still largely affected by nepotism and the arrogance of self-regenerating "old boy" and "young boy" networks.

NOTES

1. A recent overview of the bibliometric peculiarities marking the research and publication practices in the social sciences and humanities can be found in Mu-hsuan Huang and Yu-wei Chang, "Characteristics of Research Output in Social Sciences and Humanities: From a Research Evaluation Perspective," *Journal of the American Society for Information Science and Technology* 59, no. 11 (2008): 1819–28.

2. This is, for instance, the approach of Jean Tague-Sutcliffe in *Measuring Information: An Information Services Perspective* (San Diego, Calif.: Academic Press, 1995).

3. Geisler, *The Metrics of Science and Technology* (Westport, Conn.; London: Quorum Books, 2000).

4. The dictionary quotations are from "Citation," "Reference," and "Quotation" in *The Oxford English Dictionary on CD-ROM*, 2nd ed. (Oxford, Oxford University Press, 1999). 5. Bloom, *The Anxiety of Influence*, 2nd ed. (Oxford: Oxford University Press, 1997), 30.

6. On the uses and misuses of footnotes in modern historical scholarship, see Anthony Grafton, *The Footnote: A Curious History* (London: Faber and Faber, 1997).

7. Grafton, *The Footnote*, 65.

8. A classic study on the development of the experimental report as a literary genre, mainly based on articles published in the *Philosophical Transactions of the Royal Society of London* from its founding in 1665 until 1800, is Charles Bazerman, *Shaping Written Knowledge: The Genre and Activity of the Experimental Article in Science* (Madison: The University of Wisconsin Press, 1988), wac.colostate.edu/books/bazerman_shaping/.

9. The first annual edition of the index was Garfield, *Science Citation Index. An International Interdisciplinary Index to the Literature of Science* (Philadelphia: Institute for Scientific Information, 1963).

10. The research topics developed at the intersection between informetrics and information retrieval studies have been recently reviewed by Dietmar Wolfram in *Applied Informetrics for Information Retrieval Research* (Westport, Conn.: Libraries Unlimited, 2003).

Chapter One

Biblio/Sciento/Infor-metrics: Terminological Issues and Early Historical Developments

A scientist is not simply a gifted person engaged in the production of new knowledge claims through scientific publications. More prosaically, he or she is an individual growing out of a singular, substantially irreproducible sequence of biological, biographical, and historical circumstances. It might be reasonably contended, therefore, that a comprehensive measurement of science should be performed at different layers, applying mathematical tools not only to the final output, the stylish and irreproachable book or journal paper, but to any type of quantitative data somewhat referable to scientific achievements. And such a claim is even more reasonable because an extra-bibliographic concern with science measures appeared in history long before the bibliometric zooming in on publications and citations, being the secret hope of many scholars to grasp in precise, mathematical terms the material conditions for the occurrence of genius and creativity in view of their artificial reproduction for the sake of progress.

The belief that social activities, including science, could be reduced to quantitative laws, just as the trajectory of a cannonball and the revolutions of the heavenly bodies, traces back to the positivistic sociology of Auguste Comte, William Ogburn, and Herbert Spencer. The same idea informed several practical and conceptual achievements throughout the second half of the nineteenth century, including the historiographic studies by Georges Sarton and Pitrim Sorokin on the distribution of scientific genius in various epochs and the debate, involving Alphonse de Candolle and Francis Galton, over the environmental and social conditions of intellectual prominence as opposed to the biological constraints dictated by heredity laws. De Candolle's *Histoire des sciences et des savants depuis deux siècles*, published in 1885 (second augmented edition), contained perhaps the first systematic attempt to investigate in exact, mathematical fashion some crucial indicators of scientific prominence, the most highly regarded by the author being the national share of scientists affiliated to international scientific societies. His work, however, rapidly sank into oblivion while, as recently pointed out by Benoît Godin, the British scientist Francis Galton and the American psychologist James McKeen Cattell were especially influential in developing a sound quantitative approach for the reduction of science to a measurable historical agency. In Galton's project, measuring the distribution of excellence was a necessary move toward the eugenic way of re-creating the conditions for its artificial reproduction; similarly, the first edition of Cattell's *American Men of Science* (1906), with its astoundingly simplistic rating system of asterisks attached to individual entries in proportion to the estimated eminence of the starred scholar, provided the author with raw materials for testing the psychological differences among individuals in view of shedding light on the factors behind scientific prominence.[1]

Eighteenth-century scientists' concern with the statistical distribution of scientific merit, being primarily driven by the search for the true, material causes (biological, psychological) of its manifestations, leaned on a ready-made definition of scientific value that revolved around the sealing of individual excellence by past achievements, such as the affiliation with a prestigious academy, the inclusion in a dictionary, or the opinion of qualified peers. Bibliometrics, on the contrary, evolved from the analysis of quantitative patterns pertaining to the network of scientific documents produced by the scientists themselves. When it searched for explanations, it didn't appeal to external agents or material causes, but to the empirical laws of Lotka, Bradford, and Zipf. And when it met citation indexes, its ability to cooperate or compete with peer ratings for the appraisal of scientific value opened up an entirely new set of opportunities.

1.1. HOW MANY METRICS?

"Bibliometrics," "scientometrics," "informetrics," "webometrics," "netometrics," "cybermetrics": the metrology of scientific communication is rich in terms hinting at various and often indistinguishable research areas. An obvious temptation would be to establish a direct genealogical relationship, bibliometrics leading to cybermetrics by passing through scientometrics and informetrics. Actually, the question is more complicated. The objective of each research area is to analyze, quantify, and measure communication phenomena to build accurate formal representations of their behavior for explanatory, evaluative, and administrative purposes. Differences lie in the order of the factors and the boundaries of the object being measured.

Winking at the tradition of library studies, the term "bibliometrics," coined by Alan Pritchard in the late 1960s,[2] stresses the material aspect of the undertaking: counting books, articles, publications, citations, in general any statistically significant manifestation of recorded information, regardless of disciplinary bounds. "Scientometrics," instead, emphasizes the measurement of a specific type of information, upon which a certain kind of value judgment—relative to the status of "scientific"—has already been made by someone put in charge of (and trusted for) giving it. In the widest possible sense, scientometrics encompasses all quantitative aspects and models related to the production and dissemination of scientific and technological knowledge. Provided some preliminary assumptions about what science actually is and how a true scientific achievement is to be recognized, scientometrics ultimately addresses the quantitative and comparative evaluation of scientists', groups', institutions', and countries' contribution to the advancement of knowledge. As noted above, published documents are but one of the several possible units of analysis; manpower, instrumentation, facilities, and economic and financial inputs and outputs are worth considering as well. All the same, insofar as scientometric investigation is carried out through publications and citations, or, stated alternatively, insofar as bibliometric techniques are applied to scientific and technical literature, the two areas of scientometrics and bibliometrics overlap to a considerable degree.

In conformity with one of its authoritative definitions, "informetrics" is "the study of quantitative aspects of information in any form, not just records or bibliographies, and in any social group, not just scientists."[3] Today information is a key concept in several contexts, with a somewhat mystical propensity to ubiquity. In the wake of computers' and digital networks' revolution, indeed, scholars have progressively used computers and networks both as a tool and as a metaphor for modeling information flows at whatever scale and level of complexity, from submolecular systems to black holes.

Of course, pretending to measure information in all of its natural and social manifestations would be senseless. But each time such manifestations are recorded in any form suitable to present or future communication, they enter de facto the domain of information science, and the issues inherent in the collection, storage, retrieval, and transmission of their symbolic expressions can be addressed quantitatively. At such a level of generality, bibliometric techniques take leave definitely of any kinship with the universe of books and library studies and gain full inclusion in the realm of information science, by which it is signified that also the subset of information exchanges occurring in a library environment is nothing but a particular case of an information production process amenable to general mathematical treatment. And if the generalization is to be taken seriously, then informetrics may be regarded as a super-set comprising all the other metrics sets insofar as they count some type of information. Yet a too broad definition of this kind also has some drawbacks, notably its scarce resolution power in neatly distinguishing informetric studies from other well-established research areas dealing with formal treatments of information processes, above all information retrieval.

In a digitally networked universe, "webometrics" and sister terms like "netometrics" and "cybermetrics" signify the extension of informetric methods and concepts to information transactions taking place on the Internet. As long as such transactions are recorded somewhere, permanently or temporarily, webometrics intersects the domain of bibliometrics and, as long as the analysis focuses on scientific or technological cyber-traces, it obviously intersects scientometrics' path too. Lennart Björneborn and Peter Ingwersen, two pioneers of webometrics, propose also to distinguish between the quantitative study of web resources (webometrics in a strict sense) and the more general quantitative analysis of all Internet applications (cybermetrics).[4]

In 1994, at the dawn of the World Wide Web revolution in scholarly communication, Wolfgang Glänzel and Urs Schoepflin set forth a provocative analysis of the terminological chaos reigning in the quantitative studies of science, putting it down to a deep identity crisis of the whole field.[5] Stagnation in basic and methodological research, drifting apart of the various subdisciplines and research groups, lack of integrating personalities, passive subjection to the immediate interests of science policy, and consequent reduction of scientific contents to the mere presentation of data sets were, in the authors' opinion, the main symptoms of the crisis. Many scholars reacted to this apocalyptic scenario, arguing that, despite its being right on target in some respects, after all, things were not that bad. Fragmentation of research interests was not necessarily a sign of theoretical weakness, as it hinted at the coming-of-age of the discipline through its progressive ramifications into specialized subdomains; the internationalization of the field was supported by a growing institutional framework that reinforced the sense of professional identity while fostering inter- and intradisciplinary communication and joint research projects; and furthermore, the steadily increasing availability of full-text scientific documents in electronic form and their progressive merging into what was going to become a universally "webbed" world of scholarship brought a whole set of new, interesting problems along. With hindsight, it could be contended that one of the alleged signs of weakness, the fragmentation of research objects and methodologies across subfields and the lack of an overarching consensus on fundamental issues, would eventually turn into a plus point for bibliometric studies, fostering a multidisciplinary perspective on science communication processes that is perfectly attuned to the steadily growing awareness of the complexity and multidimensionality of science itself. On the other hand, a serious and still persisting hindrance to a wider recognition of bibliometrics' potential in science studies, seemingly understated in Glänzel and Schoepflin's pamphlet, was readily traced back by Quentin Burrell to the ever-expanding gap between the theoretical work of mathematicians (the increasingly sophisticated models of informetric processes) and the "dirty job" of the practitioners (the practical implementation of the models).[6]

As regards the present overview, save for the terminology unequivocally related to web measures in chapter 8, the terms "bibliometrics," "scientometrics," and "informetrics" are used somewhat interchangeably. In so doing, the fuzziness of boundaries between the above definitions is simply bypassed by the pragmatic stance that reviewing or discussing a particular knowledge claim in the quantitative tradition of science studies compels the reviewer to specify the (or one possible) historical context of its emergence, an operation largely unaffected by the way that claim is pigeonholed with one term or another.

1.2. EARLY APPROACHES TO BIBLIOMETRICS: STATISTICAL BIBLIOGRAPHY

Notwithstanding its spiritual aura, the cultural legacy of books and manuscripts has a long-established partnership with common people and landed property on the score of its measurability for bureaucratic, fiscal, and administrative purposes. To become objects of investigation in their own right, however, quantitative phenomena related to books and other cultural artifacts had to wait for the domino effect on multiple knowledge domains triggered by a revolution that started silently in the seventeenth century, when the popularity of gambling and the financial interests related to life insurance gave impetus to the development of probabilistic techniques that helped to capitalize on the uncertainty reigning in physical and socioeconomic settings. Down the path toward modern scientific attitudes, the renowned correspondence between Blaise Pascal and Pierre de Fermat gave rise, toward the mid-1600s, to the mathematical theory of probability, whose methods formed the nucleus of inferential statistics. At the end of the nineteenth century, statistics and the theory of probability started invading areas previously immune to their appeal, including the universe of library and documentation. Books and written records of scholarly activity were increasingly subjected to quantitative assessment, in the hope of gaining either a tangible idea of scientific progress or, more modestly, useful indications for library acquisition policies. Initially, the product of this activity was labeled "statistical bibliography" and was made possible by the great availability of abstracting and indexing services developed, from the eighteenth century onward, to counter the documentary explosion brought about by scientific specialization.[7]

In 1917, Francis J. Cole and Nellie B. Eales applied quantitative analysis to the comparative anatomy literature from 1543 through 1860. Their declared objectives were both of a descriptive and an evaluative nature: to represent by a curve the documentary growth rate over the three centuries; to plot separately what they called "the performance" of each European country; to determine which aspects of the subject had most attracted the scholars' efforts from time to time; and to correlate the evolution and recession phases of research activity with human, economic, and social factors, such as the foundation of new scientific societies and the influence of prominent individuals. Their study, often credited as one of the first full-fledged bibliometric investigations, reveals a conceptual framework firmly backed up by three assumptions central to every future project of science measurement.

First of all, the object: publications. Science and scientific progress can be sized through statistical analysis of publications because a publication "is an isolated and definite piece of work, it is permanent, accessible, and may be judged, and in most cases it is not difficult to ascertain when, where, and by whom it was done, and to plot the results on squared paper."[8] Every aspect of scientific activity not amenable to quantitative treatment was thereby ruled out in favor of the finite product, the disembodied ensemble of discursive or numerical assertions enveloped in a book or journal article.

Second, the purposes: performance evaluation and mapping of scientific areas. Science has a basic, transcultural unity in space and time, at least from the mid-1600s onward, so it is possible "to reduce to geometrical form the activities of the corporate body of anatomical research, and the relative importance from time to time of each country and division of the subject."[9] In addition, statistical analysis offers a tool for mapping scientific research and locating the most prominent actors on the stage: "a sure indication of contemporary interests and activities . . . the branches of the subject that were attracting attention at a particular time, and what influence, if any, was exercised by the more important workers."[10]

Third, the limits: quantitative analysis is necessarily shortsighted, as it imposes conventional and somewhat arbitrary choices in the dataset construction. This is particularly evident when choices have to be made about questions such as, Was Cuvier's work to be credited to Germany or France? Had the year to be assigned from the publication date or from the date a work was actually accomplished? Furthermore, numbers alone do not suffice to tell the whole story about science, nor do they necessarily tell a true one. On a purely quantitative basis, "the author of fifty small ephemeral papers is, judged by figures, of greater importance than William Harvey, represented only by two entries, both of great significance."[11] Therefore, any conclusion drawn from the analysis must be supplemented by an independent, qualitative appraisal of scientific literature's value.

On a scale smaller than Coles's and Eales's, literature growth patterns were investigated during the early 1930s by the Japanese botanist H. Tamiya in the field of fungi *Aspergillus* and by the American agricultural chemists Perry Wilson and Edwin Fred in the area of nitrogen fixation by plants. Both studies attempted to capture in mathematical fashion what Wilson and Fred called the "biological properties" of scientific literature, and both found a plausible solution in a logistic curve quite similar to the one made popular three decades later by Derek Price. It expressed in precise terms the simple observation that the growth rate at any given time is roughly proportional to the extant total production, until an upper limit or saturation point is attained.[12]

In the first half of the twentieth century, Western science was perceived by many scholars as a markedly cumulative enterprise, an uninterrupted sequence of steps toward progress and civilization, to the point that a physical arrangement of landmark books in a library, or, alternatively, the careful chronological organization of their records in several subject bibliographies, could well be deemed to represent the evolution of scientific knowledge. Thus, having gauged the annual growth of published scientific papers in a catalog of the Royal Society of London and the distribution of scientists by discipline in thirteen years of the *International Catalogue of Scientific Literature*, E. Wyndham Hulme, librarian at the London Patent Office, claimed a deeper cognitive value for his effort than that of a simple scanning of catalog entries: "If civilization," he argued in 1922, "is but the product of the human mind operating upon a shifting platform of its environment, we may claim for bibliography that it is not only a pillar in the structure of the edifice, but that it can function as a measure of the varying forces to which this structure is continuously subjected."[13] To Hulme, the process of civilization kept pace with scientific specialization. Therefore, by classifying all the books in the world according to some universal criteria and ordering them chronologically within each class, a reliable bibliographic picture of the human mind's evolution could be worked out.

In the hands of librarians and information scientists, the idea that the macrocosm of human knowledge is mirrored in the microcosm of the library had to cope with the practical constraints imposed by tightening budgets and lack of physical space against an ever-increasing volume of potentially relevant purchasable documentation. That is why someone started to consider the possible advantages afforded by a quantitative insight into library collections, catalogs, and bibliographies. In 1927, wishing to improve upon the library acquisition policies at Pomona College in

Southern California, P. Gross and E. M. Gross felt that it was no longer sufficient to "sit down and compile a list of those journals which one considers indispensable" because "often the result would be seasoned too much by the needs, likes and dislikes of the compiler." So they tabulated 3,633 bibliographic references from one year of the prestigious *Journal of the American Chemical Society* and ranked the cited journals by the number of citations received "in such a way that the relative importance of any single periodical for any five year period can be seen."[14] In addition, they emphasized that the sheer total citation count was not the only reasonable criterion to be taken into account for journal selection, because the age distribution of references furnished a similarly important index of utility: the number of citations being equal, journals receiving citations to the most recently published articles ranked higher because "the 'present trend' rather than the 'past performance' of a journal should be considered first."[15] In so doing, the Grosses established an equation between quality, citation rates, and time distribution of citations that would revolutionize the way information scientists sized information for library management and research evaluation purposes. Similar studies were performed during the late 1930s and the 1940s, among others, by Husley Cason and Marcella Lubotsky for psychology, Hermann Henkle for biochemistry, and Herman Fussler for physics and chemistry,[16] while Charles Gosnell tried to express mathematically the "life expectancy and mortality" of library materials implicit in the Grosses' criteria. Having examined the publication date of references in three standard lists of books for college libraries, he came to an exponential formula for which "some of the simplest and most satisfactory analogies may be found in the field of radioactivity and the decay of radioactive substances."[17]

Meanwhile, a new window on information measurement was opened by Paul Otlet, father of European documentation and cofounder with Henry La Fontaine of the International Institute of Bibliography (1895). Otlet clearly distinguished bibliometrics from other forms of statistical enquiries applied to the records of human knowledge. In a section of the *Traitée de documentation* (1934) entitled *Le livre et la mesure—Bibliometrie*, he celebrated "the measure" as a superior form of knowledge and envisaged the development of a subfield of "bibliology" entirely devoted to the organic collection of measures related to documents of all kinds.[18] Whereas statistics dealt chiefly with external measures of output—i.e., number of published and printed books; number of editions; distribution of libraries, booksellers, prices, and so on—bibliometrics focused on the internal aspects of a text and on its influence on people and society, one of its purposes being to "determine the place, time and, insofar as the readers are concerned, the probability for texts to be read, hence for exerting their action on the society."[19] This is, in current terms, a problem of impact, and Otlet addressed it by the resolution of texts into atomic elements, or ideas, which he located in the single paragraphs (*alinéa*, *verset*, *articulet*) composing a book. Each paragraph conveys a simple but complete concept, so the ultimate question was, How can the transmission of elementary ideas from authors to readers be measured in the light of this one-to-one relationship between the molecular structure of texts and that of ideas? Had citation practices been more developed at that time, and had he admitted the possibility that elementary ideas embedded in single paragraphs could be conveyed, in a synthetic fashion, through a bibliographic reference, we would probably talk of him as the main inspirer of later theories on the cognitive potential of citation indexes.

The interest of early statistical bibliographers in the valuable amount of extra-bibliographic information hidden in bibliographies, as well as Paul Otlet's visionary outline of a wide-ranging bibliometric project, were neglected for many years. It was not until the post–World War II period, indeed, that the suitable historical and cultural conditions for making science measures a desirable, if not necessary, job came about.

1.3. THE BIRTH OF EVALUATIVE BIBLIOMETRICS

"We can say," wrote Derek Price in 1963, "that 80 to 90 percent of all the scientists that have ever lived are alive now."[20] In those years, science appeared an essentially cumulative, collaborative, and above all contemporary enterprise, wherein the essential had happened from Einstein onward and the entire process had been gradual yet inexorable. The fatalism inherent in this perception made progress an almost necessary and predictable effect, to the point of asserting that "if Michelangelo or Beethoven had not existed, their works would have been replaced by quite different contributions. If Copernicus or Fermi had never existed, essentially the same contributions would have had to come from other people. There is, in fact, only one world to discover."[21] The occurrence of independent multiple discoveries, that is, similar discoveries made quite simultaneously but independently by different scientists in different places, confirmed the inexorability of scientific advance. Given certain environmental conditions, some theoretical acquisitions seemed bound to occur sooner or later, regardless of the credit being awarded to a particularly brilliant individual or to a multitude of less prominent researches functionally equivalent to the individual genius.

World War II had shown that science and technology could be controlled, manipulated, and directed toward specific goals. An efficient organization of the scientific and technological information system was thus expected to play a decisive role also in peacetime, when it could provide additional value to any national policy striving to promote economic growth. Science, by that time, had climbed over the ivy-covered walls of academia, joining its destiny to that of industrial capitalism and national economic policies. Postwar governments and industries fostered the creation of private laboratories and reserved larger portions of their budgets for contracts and grants to universities and research institutes. A standard interpretation of this transformation holds that, with wartime expansion of industry and government-funded research, the era of "big science" officially began. Physicists and life scientists were the prime ministers of the new regime, thanks to a series of long-range projects supported by large experimental facilities employing hundreds of scientists and costing hundreds of millions of dollars. From the Manhattan Project and the massively funded research on penicillin at the dawn of the new antibiotics industry to the Collider Detector experiments at Fermilab, the Hubble space telescope, and the Human Genome Project, all the way to the recent National Nanotechnology Initiative injecting vital funding into nanoscience and engineering research, breakthrough science has since relied steadily on expensive equipment operated by cross-national teams of highly specialized personnel. Fragmentation of competences, new loci of scientific production, and increased funding, however, entailed also a higher volume of information to access as quickly and effectively as possible. Hyperspecialization was one immediate response to the information overload, but it did not relieve researchers and information scientists of the burden to cope with an exponentially growing body of increasingly complex literature. After all, though apparently inexorable, progress had to be supported and, if possible, accelerated by means of adequate tools for the storing, diffusion, and retrieval of information, a concern already evident, in the immediate postwar period, in the strong commitment to information management issues of the Royal Society Conferences held in 1946 and 1948.

Following the 1957 launching of *Sputnik* by the Soviet Union, the political interest in scientometric indicators grew rapidly. The Soviet breakthrough had thrown into relief the undeniable circumstance that the liberal U.S. model was failing to adapt to the new scenario with the same rapidity as the nonliberal one of the opposing block. The Soviet Union, unlike the United States, had for some time mastered a bureaucratic apparatus for the control and orientation of scientific research; in its economic and fiscal planning regime, measuring scientific and technological research in terms of documentation produced and used by scientists was a strategic goal, one of the many ways to consolidate a centralized control over intellectual and material activities. Scientometrics, from this point of view, can be regarded mostly as a Soviet invention. In the late 1950s, two influential scientometric research programs were pioneered, by Gennady M. Dobrov at Kiev and by Vassily V. Nalimov at Moscow. Whereas the former had solid institutional roots and a clearly policy-oriented imprint, the latter revolved around Nalimov's cosmic-eye view of science as a "self-organizing information system, ruled by its information flows,"[22] and structurally embedded in uncertainty, a system that could hardly be mechanistically oriented toward specific goals but nonetheless was amenable to strict quantitative analysis. Such a view, while fostering the development of what he came to define as a "cybernetic" attitude toward science studies, kept him from blindly retiring into the shell of a purely technical approach, as it underpinned the conviction, unfortunately dismissed by most later scientometricians, that quantitative analysis ought not to be separated from a philosophical investigation of science's logical development. Following Derek Price's lead, Nalimov used simple mathematical tools to model international scientific growth and suggested the term *naukometriya* ("scientometrics") for this kind of inquiry in 1966 and 1969.[23] At around the same time, he also enthusiastically welcomed the start-up of the *SCI* as the advent of a new age in the history of science, because it filled the traditional gap between historical facts and abstract speculations, a step comparable, in his opinion, to the appearance of the first chronicles that had made possible the documented history of ancient human societies.[24]

Reaction in the United States to *Sputnik* materialized in intensified efforts to enhance national scientific information systems' quality and efficiency, as testified by the 1958 International Conference on Scientific Information, sponsored by the National Science Foundation, the National Academy of the Sciences, and the American Documentation Institute (the future American Society for Information Science). On the institutional level, that effort resulted in the creation of the National Aeronautics and Space Administration (NASA), the Advanced Research Project Agency (ARPA), and the Organization for Economic Cooperation and Development (OECD). The OECD, born in 1961 from the ashes of the Marshall Plan, was the main catalyst for the development of national science and technology policies in industrialized countries, supplying, under the leadership of the United States, Canada and Great Britain, standards and widely accepted methodologies for local and comparative international statistics. Nonetheless, at least initially, governments didn't pursue the social control of research activities. Their attention focused on the input, the human and financial resources devoted to research and development (R&D), rather than the output, the achieved results. Increasing the national expenditure for research and development remained the primary goal of politicians and administrators, who continued to treat science by and large as an independent, self-governing machinery inexorably running toward progress. This attitude, already perceptible in the renowned report *The Endless Frontier* (1945) by Vannevar Bush,[25] is well reflected in compilations like the *Frascati Manual* and later the *Oslo Manual* and the *Canberra Manual*, which shared with Bush's report the basic assumption that scientific progress originates from the free interplay of free intellects, and that the institutional independence of science had to be preserved if its socioeconomic promises were to be fulfilled. Governments, accordingly, should limit themselves to funding basic research without worrying about issues of productivity and impact. Scientists would do the rest alone, entrusting internal quality control to the well-tested peer-reviewing system. As a result, the early R&D measurements disdained bibliometric indicators on the ground that the complex interrelationships among science, technology, economy, and society, for which those indicators allegedly provided a shortcut, were actually too complex to be captured in a simplified fashion, given also the lack of pertinent international guidelines.

It is precisely here, at the crossroads where a deeper concern of policymakers with science and the increasing complexity of the documentation process intersect, that the unusual political story of Eugene Garfield's *SCI* starts. As we shall see in section 2.2, after 1961 the project for an interdisciplinary, citation-based index of scientific literature, besides becoming technically feasible, acquired political desirability and cultural credibility. Bibliographic references linked documents and authors in accordance with the commonly perceived dynamics of knowledge production: scientists pursuing new knowledge cite previous documents from which they drew significant concepts, thereby raising the cognitive and social status of cited authors. Hence, in addition to facilitating literature research, citations promised an unobtrusive measure of the impact of cited documents (and authors) on citing documents (and authors). Needless to say, for how science was perceived at that time, impact always implied a push forward, rather than a sudden stop or hesitation. Thus, by virtue of their supposed ability to detect influential authors and papers in the advancement of knowledge, citation indexes turned into the keystone to the realm of research output measurement and evaluation.

During the 1960s and 1970s, the use of bibliometric indicators to measure outcome and impact of R&D spending gained ground first in the United States, then in Europe, coming very soon under the spell of citation indexes. In 1973, the first of a series of *Science Indicators Reports* by the National Science Board was published, with the objective of portraying the status of scientific research in the United States. The *Reports*, which inspired many subsequent similar compilations outside the United States, comprised citation rankings among quality indicators and took data directly from Garfield's *SCI*. The trend they inaugurated was strong enough to urge scholars of various extractions to make a closer inspection of the theoretical background surrounding scientometric indicators in a conference held at Stanford in June 1974. The conference proceedings, published in 1979 under the title *Toward a Metric of Science*, provided one of the most important sources of inspiration for the scientometricians to follow.[26] Meanwhile, the first Western center of excellence for scientometric studies, the Information Science and Scientometric Research Unit at the Library of the Hungarian Academy of Sciences, Budapest, was established under the direction of Tibor Braun, founder and editor-in-chief of the journal *Scientometrics*, in 1978. Braun's institution played a central part in shaping the field of scientometrics as we currently know it, not only for the journal's gatekeeping role in separating the sheep from the goats, but also for the ISSRU members' commitment to the definition of international standards for research evaluation and, more generally, for its ability to connect symbolically and materially the two main tributaries flowing into the newborn scientometric paradigm: the Russian tradition of Nalimov and Dobrov and the Anglo-American tradition of Bernal, Price, Merton, and Garfield.

In the early 1980s, new insights into the design and application of scientometric indicators to sensitive targets of research management, especially at the level of the research group and the academic institution, came by way of Martin and Irvine's work at the SPRU, University of Sussex, and the Research Policy and Science Studies Unit at the University of Leiden, which would eventually turn into the Centre for Science and Technology Studies under van Raan's headship. Ever since, quantitative studies have evolved into a full-fledged discipline, with its own apparatus of research facilities and hallmarks of excellence, which encompass the categories of materials collected under the following headings:

Journals. After the 1978 launching of *Scientometrics*, a joint publication of Elsevier, Amsterdam, and Akadémiai Kiadó, Bucarest, the disciplinary research front relied steadily on journal papers as the primary means of communication. Nonetheless, given the interdisciplinary character of much bibliometric research and the inherent fuzziness of journals' subject coverage, relevant materials were increasingly scattered in an imprecisely defined set of sources. In the Western tradition, as far as English-language publications are concerned, they range from strictly discipline-oriented titles, such as *Journal of Informetrics* and *Cybermetrics*, to notable portions of leading information science journals, including *Journal of the American Society for Information Science and Technology*, *Journal of Documentation*, and *Journal of Information Science*, all the way to sources specializing in particular aspects of information processes (*Information Visualization*, *Webology*) and journals covering adjacent areas of scholarship that take often advantage of bibliometric methods, such as sociology (*Social Studies of Science*, *Science Studies*,

American Sociological Review), science policy and management (*Research Evaluation*, *Research Policy*), information retrieval (*Information Processing & Management*), library science (*Library Trends*, *Library Quarterly*), and communication studies (*Journal of Computer-Mediated Communication*).

Handbooks and authoritative monographs. The basic statistical techniques employed in informetric studies were reviewed by Ravichandra Rao in 1983 and by Leo Egghe and Ronald Rousseau in 1990. On the applicative side, two handbooks with a marked propensity toward science policy issues appeared, respectively, in 1988 and 2004.[27] Both are collections of essays from specialists of various extraction that present, in a polyphonic fashion, the state of the art in the main areas of bibliometrics at two critical stages of its historical development. In 2005, Henk Moed's *Citation Analysis in Research Evaluation* supplied the most up-to-date organic treatment of the theoretical and technical aspects underpinning the application of citation-based indicators to science policy and management.[28] At almost the same time, Mike Thelwall's *Link Analysis* (2004) accomplished a comparable task for the cybertech equivalent of citation analysis, making accessible concepts, tools, and techniques currently used to investigate, from an information science perspective, the quantitative properties of academic web spaces through statistical analysis of hyperlink patterns.[29]

Reviews, bibliographies, dictionaries. In 1976, Narin's *Evaluative Bibliometrics*, published under contract with the National Science Foundation, supplied an early comprehensive review of citation studies that laid the groundwork for some of the most fruitful research lines in bibliometrics for generations to follow.[30] Then, starting in 1977 and more systematically from 1989 onward, the *Annual Review of Information Science and Technology* (*ARIST*) devoted increasing attention to the selective review of studies in bibliometrics and sister areas. In addition, after a comprehensive 1980 bibliography of over 2,000 bibliometric studies, other occasional bibliographic compilations attempted to keep under control the explosion of increasingly specialized contributions.[31] On the popularization front, an English-language dictionary of bibliometric terminology was issued in 1994.[32]

Scientific societies and international conferences. The International Conference on Scientometrics and Informetrics series, launched in 1987 by Egghe and Rousseau with a first meeting entitled *International Conference on Bibliometrics and Theoretical Aspects of Information Retrieval*,[33] continued to be held on a biennial basis, coming later under the auspices of the International Society for Scientometrics and Informetrics, founded in Berlin in 1993. It has been complemented, since 1988, by another (almost biennial) series of Conferences on Science & Technology Indicators, organized by the CWTS, and, since 1998, by the Berlin Workshop on Scientometrics and Informetrics, which led in 2000 to the establishment of the COLLNET international research network on collaboration in science and technology, and subsequently to an International Conference (formerly Workshop) on Webometrics, Informetrics and Scientometrics. A spillover of the COLLNET experience is the COLLNET *Journal of Scientometrics and Information Management*, started in 2007.

Discussion lists. The American Society for Information Science and Technology's *SIGMETRICS* listserv is a discussion group officially devoted to all theoretical aspects and practical implementations of information metrics. Its archive (listserv.utk.edu/archives/sigmetrics.html), spanning back to June 1999, is a valuable resource for professional and aspiring scientometricians who wish to keep abreast of current "hot" topics and new resources in the field of quantitative science studies.

Prizes. In 1983, scientometrics started to canonize its heroes with the Derek de Solla Price Medal, awarded to the most prominent scholars in the field. The first awardee was Eugene Garfield in 1984.

NOTES

1. Godin, *From Eugenics to Scientometrics: Galton, Cattell and Men of Science,* Project on the History and Sociology of S&T Statistics. Working Paper No. 32 (Montreal: Canadian Science and Innovation Indicators Consortium, 2006), www.csiic.ca/PDF/Godin_32.pdf; Godin, *On the Origins of Bibliometrics,* Project on the History and Sociology of S&T Statistics. Working Paper No. 33 (Montreal: Canadian Science and Innovation Indicators Consortium, 2006), www.csiic.ca/PDF/Godin_33.pdf. For an early reevaluation of de Candolle's work in a scientometric perspective, see A. T. Szabó, "Alphonse De Candolle's Early Scientometrics (1883, 1885) with References to Recent Trends in the Field," *Scientometrics* 8, nos. 1–2 (1985): 13–33. On Cattell's role in shaping the

American reward system throughout the first half of the twentieth century, see Michael Sokal, "Stargazing: James McKeen Cattell, American Men of Science, and the Reward Structure of the American Scientific Community, 1906–1944," in *Psychology, Science, and Human Affairs: Essays in Honor of William Bevan*, ed. Frank Kessel, 64–86 (Boulder, Colo.: Westview Press, 1995).

2. "BIBLIOMETRICS, i.e., the application of mathematics and statistical methods to books and other media of communication," is the definition given in Pritchard, "Statistical Bibliography or Bibliometrics?" *Journal of Documentation* 25, no. 4 (1969): 349. In the same year, Pritchard's statement was rephrased in "quantitative treatment of the properties of recorded discourse and behaviour appertaining to it" by Robert A. Fairthorne, "Empirical Hyperbolic Distributions (Bradford-Zipf-Mandelbrot) for Bibliometric Description and Prediction," *Journal of Documentation* 25, no. 4 (1969): 319.

3. Jean Tague-Sutcliffe, "An Introduction to Informetrics," *Information Processing & Management* 28, no. 1 (1992): 1. A clue to terminological issues can also be found in Robert N. Broadus, "Toward a Definition of 'Bibliometrics'," *Scientometrics* 12, nos. 5–6 (1987): 373–79; Bertram C. Brookes, "Biblio-, Sciento-, Infor-Metrics??? What Are We Talking About?" in *Informetrics 89/90: Selection of Papers Submitted for the Second International Conference on Bibliometrics, Scientometrics and Informetrics*, ed. Egghe and Rousseau, 31–43 (Amsterdam: Elsevier Science, 1990), doclib.uhasselt.be/dspace/handle/1942/857; I. N. Sengupta, "Bibliometrics, Informetrics, Scientometrics and Librametrics: An Overview," *Libri: International Library Review* 42, no. 2 (1992): 75–98; William W. Hood and Concepción S. Wilson, "The Literature of Bibliometrics, Scientometrics, and Informetrics," *Scientometrics* 52, no. 2 (2001): 291–314. For a seminal definition of "Webometrics," see Tomas C. Almind and Peter Ingwersen, "Informetric Analysis of the World Wide Web: Methodological Approaches to 'Webometrics'," *Journal of Documentation* 53, no. 4 (1997): 404–26.

4. Björneborn and Ingwersen, "Toward a Basic Framework for Webometrics," *Journal of the American Society for Information Science and Technology* 55, no. 14 (2004): 1217–18, www.db.dk/binaries/PerspectivesWebometrics-Jasist.pdf. To get just a broad idea of the shift in content and theoretical focus that has occurred in the field of biblio/sciento/informetrics over the past three decades, contrast the theme issue of *Library Trends,* "Bibliometrics," which appeared in summer 1981 (30, no. 1), hdl.handle.net/2142/7189, with the special issue of *Information Processing & Management* on informetrics, which appeared in 2005 (41, no. 6).

5. Glänzel and Schoepflin, "Little Scientometrics, Big Scientometrics . . . and Beyond?" *Scientometrics* 30, nos. 2–3 (1994): 375–84. The comments and replies to this paper are contained in the same issue of the journal.

6. Burrell, "Scientostochastics?," *Scientometrics* 30, nos. 2–3 (1994): 505–9.

7. A comprehensive historical overview of bibliometric studies and research traditions can be found in Dorothy H. Hertzel, "Bibliometrics History," in *Encyclopedia of Library and Information Science,* 2nd ed., 288–328 (New York: M. Dekker, 2003), books.google.it/books?id=tPnkxsklgXoC. This is a reprint of a 1987 essay, and may be complemented with Farideh Osareh, "Bibliometrics, Citation Analysis and Co-Citation Analysis: A Review of Literature I," *Libri: International Library Review* 46, no. 3 (1996): 149–58, and Osareh, "Bibliometrics, Citation Analysis and Co-Citation Analysis: A Review of Literature II," *Libri: International Library Review* 46, no. 4 (1996): 217–25.

8. Cole and Eales, "The History of Comparative Anatomy, Part I: A Statistical Analysis of the Literature," *Science Progress* 11, no. 43 (1917): 578.

9. Cole, "The History," 578.

10. Cole, "The History," 579.

11. Cole, "The History," 579.

12. Wilson and Fred, "The Growth Curve of a Scientific Literature: Nitrogen Fixation by Plants," *Scientific Monthly* 41, no. 3 (1935): 240–50. The two authors acknowledge Tamiya for having reported a similar exponential pattern in 1931.

13. Hulme, *Statistical Bibliography in Relation to the Growth of Modern Civilization* (London: Grafton, 1923), 43.

14. Gross and Gross, "College Libraries and Chemical Education," *Science* 66, no. 1713 (1927): 386.

15. Gross and Gross, "College Libraries," 387.

16. Cason and Lubotsky, "The Influence and Dependence of Psychological Journals on Each Other," *Psychological Bulletin* 33, no. 1 (1936): 95–103; Henkle, "The Periodical Literature of Biochemistry," *Bulletin of the Medical Library Association* 27 (1938): 19–35; Fussler, "Characteristics of the Research Literature Used by Chemists and Physicists in the United States," *Library Quarterly* 19, no. 2 (1949): 139–47.

17. Gosnell, "Obsolescence of Books in College Libraries," *College & Research Libraries* 5, no. 2 (1944): 121.

18. "En tout ordre de connaissance, la mesure est une forme supérieure que prend la connaissance. Il y a lieu de constituer en un ensamble coordonné les mesures relatives au livre et au document, la Bibliométrie." Otlet, *Traitée De Documentation. Le Livre Sur Le Livre. Theorie Et Pratique* (Bruxelles: Van Keerberghen, 1934), section 124.1, 13, https://archive.ugent.be/handle/1854/5612?mode=full.

19. Otlet, *Traitée*, section 124.2, 16.

20. Price, *Little Science, Big Science . . . and Beyond* (New York: Columbia University Press, 1986), 1. The first edition of this fundamental book appeared in

1963 as a result of the Pegram Lectures delivered by the author at the Brookhaven National Laboratory.

21. Price, *Little Science*, 62.

22. Quoted in Lyubov G. Gurjeva and Paul Wouters, "Scientometrics in the Context of Probabilistic Philosophy," *Scientometrics* 52, no. 2 (2001): 122.

23. A machine-aided translation of Nalimov's 1969 book on science-metrics is available on Garfield's website: V. V. Nalimov and B. M. Mulchenko, *Measurement of Science: Study of the Development of Science as an Information Process* (Washington, DC: Foreign Technology Division, US Air Force System Command, 1971), www.garfield.library.upenn.edu/nalimovmeasurementofscience/book.pdf. For a detailed account of Soviet scientometrics in the 1960s and 1970s, see Gurjeva, "Early Soviet Scientometrics and Scientometricians" (thesis for the degree of MSc in Science Dynamics, Universiteit van Amsterdam, 1992), www.chstm.man.ac.uk/people/gurjeva/thesis-msc.pdf. Nalimov's foundational work was the object of a 2001 memorial issue of *Scientometrics* (52, no. 2); for a synthesis of his main contributions to the field see, in particular, Yuri V. Granovsky, "Is It Possible to Measure Science? V. V. Nalimov's Research in Scientometrics," *Scientometrics* 52, no. 2 (2001): 127–50.

24. "[W]ith the advent of SCI the documented history of science began in the same way as with the advent of chronicles the documented history of society began." Nalimov and Mulchenko, *Measurement of Science*, 96. Garfield himself, on the fiftieth anniversary of VINITI (the Russian national agency for technical and scientific information), pointed out the multiple lines of connection between ISI and VINITI, depicting the Soviet interdisciplinary approach to the indexing of scientific literature as a model for ISI's philosophy: Garfield, "The ISI-VINITI Connection—Remarks on the 50th Anniversary of VINITI" (paper presented at the 6th International Conference of Information Society, Intelligent Information Processing and Information Technologies, Moscow, 16–18 October, 2002), www.garfield.library.upenn.edu/papers/viniti101502.html.

25. Bush, *Science: The Endless Frontier* (Washington, D.C.: U.S. Government Printing Office, 1945), chapter 1, www.nsf.gov/about/history/vbush1945.htm.

26. Yehuda Elkana et al., eds., *Toward a Metric of Science: The Advent of Science Indicators* (New York: John Wiley & Sons, 1979).

27. Inna K. Ravichandra Rao, *Quantitative Methods for Library and Information Science* (New Dehli: Wiley Eastern Ltd., 1983); Egghe and Rousseau, *Introduction to Informetrics: Quantitative Methods in Library, Documentation and Information Science* (Amsterdam: Elsevier, 1990), http://hdl.handle.net/1942/587; Anthony F. J. van Raan, ed., *Handbook of Quantitative Studies of Science and Technology* (Amsterdam: Elsevier, 1988); Henk F. Moed, Wolfgang Glänzel,

and Ulrich Schmoch, eds., *Handbook of Quantitative Science and Technology Research: The Use of Publication and Patent Statistics in Studies of S&T Systems* (Dordrecht: Kluwer Academic, 2004).

28. Moed, *Citation Analysis in Research Evaluation* (Dordrecht: Springer, 2005).

29. Thelwall, *Link Analysis: An Information Science Approach* (Amsterdam: Elsevier, 2004). For an update see Thelwall, "Bibliometrics to Webometrics," *Journal of Information Science* 34, no. 4 (2008): 605–21, www.scit.wlv.ac.uk/~cm1993/papers/JIS-0642-v4–Bibliometrics-to-Webometrics.pdf.

30. Narin, *Evaluative Bibliometrics: The Use of Publication and Citation Analysis in the Evaluation of Scientific Activity* (Cherry Hill, N.J.: Computer Horizons, 1976), www.conceptsymbols.com/narin/narin_1975_eval-bibliometrics_images.pdf.

31. Roland Hjerppe, *A Bibliography of Bibliometrics and Citation Indexing & Analysis* (Stockholm: Royal Institute of Technology, 1980). A supplement appeared in *Scientometrics* 4, no. 3 (1982): 241–73. Other bibliographies occasionally covered the subject during the 1980s and 1990s: Roger Hahn, *A Bibliography of Quantitative Studies on Science and Its History* (Berkeley: Office for History of Science and Technology, University of California, 1980); Robert C. Stowe, *Annotated Bibliography of Publications Dealing with Qualitative and Quantitative Indicators of the Quality of Science* (Cambridge, Mass.: Jefferson Physical Lab., Harvard University, 1986), eric.ed.gov/ERICDocs/data/ericdocs2sql/content_storage_01/0000019b/80/2f/77/bf.pdf; Mary K. Sellen, *Bibliometrics: An Annotated Bibliography, 1970–1990* (New York: G.K. Hall, 1993); and Ronald N. Kostoff, *The Handbook of Research Impact Assessment* (Arlington, Va.: Office of Naval Research, 1997), handle.dtic.mil/100.2/ADA296021. Early ARIST reviews are Narin and Moll, "Bibliometrics," *Annual Review of Information Science and Technology* 12 (1977): 35–58; and Howard D. White and Katherine W. McCain, "Bibliometrics," *Annual Review of Information Science and Technology* 24 (1989): 119–65. For recent overviews see Wilson, "Informetrics," *Annual Review of Information Science and Technology* 34 (1999): 107–247; and Judit Bar-Ilan, "Informetrics at the Beginning of the 21st Century—A Review," *Journal of Informetrics* 2, no. 1 (2008): 1–52.

32. Virgil Pasquale Diodato, *Dictionary of Bibliometrics* (New York: Haworth Press, 1994). A critical review of the dictionary is in Burrell, "Review of Virgil Diodato, Dictionary of Bibliometrics," *Journal of Documentation* 51, no. 4 (1995): 448–50.

33. The proceedings of the first two conferences were published in, respectively, Egghe and Rousseau, eds., *Informetrics 87/88: Select Proceedings of the First International Conference on Bibliometrics and Theoretical Aspects of*

Information Retrieval (Amsterdam: Elsevier, 1988), and Egghe and Rousseau, eds., *Informetrics 89/90: Selection of Papers Submitted for the Second International Conference on Bibliometrics, Scientometrics and Informetrics* (Amsterdam: Elsevier Science, 1990). The individual papers from both proceedings are freely downloadable from the open archive of the Hasselt University at doclib.uhasselt .be/dspace/.

Chapter Two

The Empirical Foundations of Bibliometrics: The *Science Citation Index*

The genesis of the *SCI* conceals a paradox. Projected into the future, designed to promote research and to advance scientific communication, Garfield's index is, from a technical point of view, the direct descendant of a cultural model—the Anglo-Saxon legal system—firmly rooted in the past, in the formalism of "reasoning from precedent" (the core of legal training) and the conservative doctrine of *stare decisis*: a court judging a case has to follow precedents laid down by higher courts; thus, in citing authorities to back up new arguments, a lawyer must check if they're still valid and have not been overturned by later sentences. Although tables of previous cases appended to legal reviews were already available in the eighteenth century, and a citation index of Californian sentences appeared in 1860, the best known application of this model is *Shepard's Citations*, a reference tool started in 1873 that soon became very popular among American lawyers.[1] Given a case debated in one of the American state or federal courts, the index lists all the subsequent cases reconfirming, reversing, or overruling the original sentence, each unambiguously identified by the volume and page number of the court report wherein it appeared.

Shepard's Citations, it is worth stressing, differs from the *SCI* in some crucial respects. Aside from the object, case law instead of literature, it can be said to perform something less and something more than the *SCI*. Something less, because it doesn't give the clue to a network of citations, hence of scientific papers tied up with basic sociocognitive links, but simply connects sentences associated with a plus or minus sign; something more, because in a legal index each link is unambiguously qualified by the mark of its validity over time, so that some sort of context analysis is embedded within the tool (a case is confirmed or reversed by subsequent cases; the same is not necessarily true for scientific literature). An underlying legal component is still preserved in the *SCI*, but with a unique meaning: the modern scientist cites to pay an intellectual debt in such a way that anyone can realize he or she also has a credit (cognitive, social) to gain, namely the priority claim of a discovery, the intellectual property on a new or purportedly new idea disclosed to the community of qualified peers. Before turning out to be a tool for sociological and scientometric analysis, however, Garfield's index grew out of a tangle of issues, conceptualizations, and technical breakthroughs firmly embedded within the intellectual tradition of information retrieval studies. To a large extent, in fact, the story of the birth and development of the *SCI* is the story of an obsession with the idea of finding a fresh solution to the old problem of retrieving relevant and useful information from a huge amount of recorded data.

2.1. CONTEXTUALIZING THE *SCIENCE CITATION INDEX*: INFORMATION RETRIEVAL AND THE "LINGUISTIC CONNECTION"

One of the key problems faced by library and information scientists is how to store and organize large collections of documents so as to make them retrievable from as many points of view as possible in a timely fashion. Let's take the concept of "document" in the broadest meaning of carrier of some type of information and, at the risk of circularity, let's pragmatically stick to a loose definition of "information" including any object regarded as informative, that is, any physical expression, description, or representation of knowledge being communicated and somehow recorded. To simplify, let's focus on a single subset of the all-inclusive class of documents, namely documents containing written text.

If one assumes that documents are the books stored in a library, then the library card catalog is the primary tool for the retrieval of a particular book, or a certain number of books, matching a specified set of search criteria, such as being written by a certain author or dealing with a given subject. Similarly, if documents are the pages of a book and the purpose is to find all pages on which a particular subject is dealt with, the alphabetical subject index is the ordinary tool for quickly finding out what page number(s) correspond(s) to that subject. The two examples, though at different scales, show one basic feature of information retrieval systems: quick access to the desired information unit, whether a book on the shelf or a specific page in a book, entails the creation of document surrogates, or index terms, stored in a separate archive, serving as pointers to specific documents. The entries in a library card catalog, such as author names, subject headings, and so on, are document surrogates, each pointing to the physical copies of all the books sharing the particular property specified by the entry: all the books written by the same author, all the books on a given subject, and the like. Similarly, the words, or combinations of words, listed in the subject index of a book next to the page numbers where the corresponding subjects occur are the surrogates of the single pages of the book dealing with those subjects.

Indexes are a basic component of manual or computerized information retrieval, because in most systems the user queries do not match directly the collection of documents, but rather an index previously prepared by manual or automatic operations. In the two foregoing examples, the assignment to the source documents of indexing terms, whether carried out manually or with the aid of a computer, entails two basic, albeit not mutually exclusive, choices: the indexing terms can be extracted from the document itself (the full text, the abstract, the title), or the indexer supplies them through an interpretation of the text. The first choice is obvious for indexes of author names and book titles, which do not require any kind of interpretation as to their meaning. But if one is willing to index the conceptual units embedded in the text, then things get dramatically complicated. Indeed, if the index words are extracted from the text itself, they might be of little use because the author and the user are likely to use different words to address the same or similar concepts. The intermediation of human indexers, on the other hand, improves the uniformity and consistency of indexing language, because it allows the assignment to each document of predefined terms or combinations of terms, called "descriptors" or "subject headings," whose formulation obeys strict semantic and syntactic rules. They typically draw upon a controlled vocabulary or a disciplinary thesaurus and, whenever they comprise more than one word, the combination of words follows a fixed scheme. Human indexing, nevertheless, although ensuring that all content-related documents enjoy a consistent treatment from the same indexer, does not guarantee consistency among different indexers and is seriously time-consuming on the part of the user, who has to figure out in advance the entries of the controlled vocabulary.

With the advent of modern, high-speed digital computers, the idea that indexing, that is, the assignment of content identifiers to documents, could be carried out automatically so as to build intelligent and cost-effective retrieval systems, gained ground rapidly. Given a document collection in machine-readable form, the computer would be able to analyze its linguistic content, generate appropriate content identifiers, and attach them to each stored document. User needs, concomitantly, would be identified and expressed in machine-readable language by Boolean combinations of index terms. Finally, representations of document content and user needs would be compared, leading to the retrieval of the matching items. This is what typically happens, at a higher level of technological sophistication, in modern full-text retrieval systems, where automatic indexing amounts essentially to the automatic creation of an "inverted file," which contains a complete index of words for all the searchable text, each associated with a pointer, that is, an entry indicating the exact position of all its occurrences in the database of stored documents.

Automatic indexing was pursued long before the advent of digital computers and the inverted file structure. One might even argue that the current disciplinary status of information retrieval would be unimaginable without the key conceptual and practical developments occasioned, before and during the same period that witnessed the birth of the *SCI*, by the search for a mechanical solution to the problem of indexing.[2] Throughout the 1950s and 1960s, commercial organizations pioneered the application of automatic indexing to small collections of internal research reports, thereby defining objectives and capabilities that later systems would have accomplished and consistently improved upon. At that time, key punching was the favored method for converting text into machine-readable form, and the mechanization of indexing profited by such cumbersome tools as punched cards and tabulating machines combined with various kinds of coding systems. The research front of information retrieval in the United States was not joint or homogeneous. Rather, it appeared as a good example of Merton's construct of multiple discoveries: many people worked simultaneously on similar problems using quite comparable techniques, so that certain fruits were ripe and ready to be picked up by many minds conjuring up the same ideas within months of one another.

In the mainstream, the idea that any automatic content analysis should rest on the statistical properties of the vocabulary contained in the text of individual documents had been pursued by many researchers since the mid-1950s. One of the most prominent figures was the IBM engineer Hans Peter Luhn, who delineated the paradigm of much subsequent work in the field by two simple assumptions: "communication of ideas by means of words is carried out on the basis of statistical probability,"[3] so the more two representations of the same document agree in terms of elements and their distribution, the higher the probability that they represent similar information; and "the frequency of word occurrence in an article furnishes a useful measurement of word significance," while "the relative position within a sentence of words having given values of significance furnishes a useful measurement for determining the significance of sentences."[4] The subjects of documents, therefore, can be explicitly linked to the pattern of occurrences of words, and each word can be assigned a weight estimating its relative value as subject identifier. Excessively frequent as well as too rare words are likely poor discriminators of document content, so Luhn used Zipf's Law to define cut-off levels for filtering out the nontopical words falling above or below a certain threshold frequency. Later studies by Gerard Salton and Karen Sparck Jones formalized this regularity, demonstrating that the ability of an index term to discriminate between different documents is somewhat inversely correlated with the number of documents in which it occurs, the best terms being those that occur frequently in individual documents but rarely in the rest of the collection.

Seminal experiments dating back to the 1960s, such as the SMART (Salton's Magic Automatic Retriever of Text) Project and the Cranfield experiments, confirmed the fruitfulness of Luhn's paradigm, definitely showing that, while human language is undeniably a complex system of communication based on the human mind's ability to form meaningful associations of words in conformity with a number of fixed structures, "the phrase languages are not substantially superior to single terms as indexing devices, and . . . sophisticated analysis tools are less effective than had been expected."[5] Thus, in the wake of Luhn's ideas, the term-weighting problem and the distributional properties of language units came progressively to the forefront of information retrieval research and experimentation.

During the 1960s and 1970s, the first large-scale information systems were established and the technologies for the systematic, laboratory-based evaluation of the merits of different automatic indexing methods consolidated, thereby anticipating the big collaborative evaluation programs envisaged by the DARPA/NIST Text Retrieval Conferences (TRECs) in the 1990s. Formal models were devised, accordingly, that emphasized the multidimensional and probabilistic character of information retrieval. Two stand out as most successful: the vector-space model, originally developed by Gerard Salton, and the multiple offspring of Luhn's probabilistic model. In the vector-space model, each document (and each user query as well) is represented as a vector in a multidimensional space, the dimension being determined by the number of unique terms in the document collection. Each term, in turn, is weighted proportionally to its degree of representativeness of a document's content. Algebraic operations on vectors thus allow computing similarity coefficients between any two documents as well as between a query and a document. In the probabilistic approach, term weights and similarity coefficients are computed alike, but the likelihood that a document will be relevant to a given query is grounded in assumptions of probability theory, and so is the decision whether a word should be assigned as an index term given its distributional behavior in the document collection.[6]

Despite the predominance of Luhn's paradigm, early information scientists were perfectly aware of the limits inherent in the statistical approach. Text, after all, is more than a simple sequence of words. It is written in natural language, and the meaning and weight of single words in natural language use are context-dependent, reflecting its involvement in cultural and social issues. To work well in the indexing process, then, a machine should manage to reproduce the human indexer's understanding and processing of natural language. That's why several information scientists, including Garfield before the *SCI*'s breakthrough, tackled the problem of automatic indexing with the conviction that a substantive help to its resolution could derive from the methods of linguistics, hence from the study of language as a system for communicating information. Since the very beginning of information retrieval studies, in fact, linguistics had a say in information retrieval as long as it helped system designers reduce the ambiguity of language use through lexical, syntactic, and semantic analysis.

Lexical analysis. Lexical and morphological variants of terms pointing to the same concept have to be conflated in some way. In a well designed system, for instance, "encyclopedia" and "encyclopaedia" map to the same lexical unit, and the singular form "neutron" in a user query retrieves also documents containing the plural "neutrons." The simplest way to achieve such conflation is suffix stripping and detection of equivalent stems, an otherwise ambiguous solution insofar as it doesn't take into consideration the context: the root "neutr," for instance, is as good to retrieve "neutron," as it is for "neutrin" and "neutral." Suffix stripping, moreover, doesn't work in the case of multiword terms subject to syntactic variation.

Syntactic analysis. Text words in natural language are connected to form phrases, clauses, and sentences in accordance with syntactic rules. An effective automatic indexing, consequently, should implement a module, called "parser," capable of going beyond mere word frequency counts by assigning a grammatical structure to meaningful segments of text. Stated alternatively, an intelligent retrieval system has to learn to place brackets in the right places so as to drive the selection of better indexing entries. Statistical analysis alone can hardly accomplish this task. Computational linguists, on their part, have designed transformational parsers capable of identifying syntactical variations of the same conceptual unit, as would be necessary, for example, to establish the conceptual equivalence in the following two phrases: "Nicola writes the book" and "The book is written by Nicola." The transformational approach starts from the assumption that each sentence has a deep underlying structure that comes afloat in written or spoken language, and a set of transformational rules operates from time to time on the deep structure representation to produce the tangible output. The work of Zelig Harris during the 1950s, in particular, formed the basis for some practical applications of transformational grammar to information retrieval systems. It is worth remembering here because, as we shall see in the next section, Garfield's road to citation indexing went through a keen reconceptualization of the linguistic difficulties in automatic indexing, mainly in the light of Harris's discourse analysis.

Semantic analysis. Words organized into syntactically ordered sentences carry concepts, but concepts are not, as it were, univocally "embedded" in words and sentences; there are words that share the same spelling while carrying distinct meanings (homonyms), words that point to the same concept despite different spellings (think of the chemical and brand names for the same drug), and sentences whose underlying structure can be understood in different ways, as in "I'm going to write the book" ("going" can indicate that I'm moving toward the destination "to write the book" or it can be an auxiliary indicating near future). To cope with semantic ambiguities at the word level, an information retrieval system usually resorts to field-specific dictionaries and thesauri, which supply semantic specifications and variant forms for single entries. In the most advanced applications, artificial intelligence studies introduce complex knowledge bases to represent the conceptual structure of a subject area, its main concepts, and the network of their interrelationships.

In the past few years, natural language processing techniques have progressed considerably in the areas of speech recognition, question answering, and information extraction, slipping concomitantly into the world of commercially available databases and search engines, as testified by the currently widespread facilities for searching structured databases in natural language. Nevertheless, though recurrently invoked, designed, and tested by researches over the past forty years, automatic indexing systems based on linguistic competence have often seen their promises to improve information retrieval broken in the impact with experience. In this regard, little evidence has been produced that, taking advantage of linguistic methods, automatic routines can equal or outperform human skills in the comprehension of language for indexing purposes. Statistical analysis, on the contrary, has unremittingly proved its ability to perform such an impossible mission in a fairly acceptable, albeit approximate, fashion.

Despite the undeniable merits of statistical modeling, as the storing and processing capabilities of digital computers grew by orders of magnitude, the limits of traditional, document-oriented approaches became evident, and information scientists felt it was increasingly necessary to implement techniques for handling the uncertainty fed into the system by the user. Traditional measures of effectiveness in information retrieval centered upon two well-known ratios, originally proposed by James W. Perry and Allen Kent in 1957 and later canonized by Cyril Cleverdon, who had argued for the existence of a generalized inverse relationship between them: recall, or the fraction of relevant documents retrieved out of a given collection, and precision, or the fraction of retrieved documents that are relevant. These two criteria, useful as they might be in laboratory-based assessments, appeared nonetheless more and more constrained by the inability to capture the actual relevance of a search output, that is, the extent to which the retrieved set of documents meets the information needs of a user. User-seeking behavior and cognitive structures, consequently, entered the laboratory and became a variable of the utmost importance to study and model.

Carried to the extreme, the idea that any processing of information is mediated by a preexisting system of categories or concepts, hence by a model of the world (something sharing some properties with the world but also differing from it in many substantial respects), has been extended, by the supporters of the "cognitive" view of information retrieval, to all the actors participating in the seeking and retrieval operations: the authors, the system architecture and interface designers, the searchers with their knowledge and emotional status, the indexers, and the selectors (publishers, editors, etc.). As a result, information seeking and retrieval is increasingly being characterized and modeled in terms of a multilayer, historically situated, and context-dependent process in which each actor "interacts with other actors at various levels under influence of social contexts and work task or interest situations over time."[7]

Citation indexes supplied the above tradition of information retrieval studies not only with an additional, conceptually innovative tool for organizing the mess of extant scientific literature, but also with a new idea of relevance, in which the retrieving of a document perceived as relevant by a user is inextricably interwoven with all the previous judgments of relevance embedded in the network of papers that referred to it by means of a bibliographic reference. And if, in the mid-1980s, someone endeavored to reconnect that idea with canonic probabilistic indexing through Bayes decision theory,[8] about fifteen years later, when the hypertextual machine of the World Wide Web began to work at its fullest, the citation-driven concept of relevance applied to the network of hyperlinks between web pages would revolutionize the way Web search engines let users quickly pick useful materials out of the anarchical universe of digital information.

2.2. DEVISING AND REALIZATION OF THE *SCIENCE CITATION INDEX*

The story of the "bits and pieces of experience and insight" that culminated in the foundation of the *SCI* has been told many times, in many good books and papers, including Garfield's own autobiographical accounts.[9] In 1951, the chemist Eugene Garfield entered the field of information science by joining the Johns Hopkins University's Welch Medical Library Indexing Project, sponsored by the Army Medical Library, the future National Library of Medicine. Here he worked on a team with Sanford Larkey, a pioneer in the application of machine methods to bibliographic control. The main scope of the project, centered on the *Current List of Medical Literature* (ancestor of the *Index Medicus* and *MEDLINE*), was to examine if and how computers could be used to automate the indexing of medical literature. In those years, given the technological limitations of the available calculators, their applicability to text analysis seemed successful only for carrying out mechanical tasks, such as compiling concordances. During the mid-1940s, for instance, the Italian Jesuit Father Roberto Busa had started to produce, with punched cards fed into an IBM calculator, the index of St. Thomas Aquinas's complete works (*Index Thomisticus*). By contrast, the goal of computerizing an intellectual task similar to the one performed by human indexers appeared far from being attainable. The main difficulty arose from the inability of computer algorithms to reproduce the intellectual process of document analysis and to simulate human judgment about the relative importance of the units composing a text.

At about the same time, computational linguistic experts were experiencing a similar frustration in the attempt to construct an automatic translation system. Weighing on them was the admonition of Yehoshua Bar-Hillel, that a completely automated translation of a text could be achieved only at the expense of accuracy. To translate quickly and thoroughly, indeed, a computer should incorporate not only a dictionary, but an entire encyclopedia of concepts along with their multilayered relationships.[10]

Apparently, indexing was simpler than translating because it amounted to the extraction, from a specialized body of literature, of a carefully selected pool of terms to be implemented in a retrieval system by virtue of their representativeness of the full-text content. Yet experience raised obstacles of various kinds.

One possible solution, favored by the members of the Welch Project, was the automatic extraction of index terms from the most accessible place where, at least statistically, they were located: the title and abstract section of scientific articles. The degree of selectivity thus attainable had many advantages in terms of processing time and costs, but titles and abstracts, as soon became manifest, are often poorly representative of the document content, so the alleged economization was achievable only at the cost of losing information potentially crucial in the retrieval stage. Garfield felt that the full text itself had to be taken into consideration and that the clue for mechanical indexing should be some sort of linguistic analysis of the text combined with a careful consideration of the mental process of indexing. His acquaintance with structural linguistics, especially with Zelig Harris's theories and Casimir Borkowski's pioneering work at the border between linguistics and computer science, prompted him to the idea that, if natural language is made up of a small number of basic, irreducible components, out of which all sentences can be constructed by proper transformations, then the grammatical structure of the natural language embedded in scientific texts, disclosed by automatic syntactical analysis, could supply the building blocks of a formal indexing language. This strand of research, however, ultimately met with the same constraints of automatic translation, above all the inability of computers to judge the relative importance of words and to exploit conceptual relationships lying outside the mere sequence of text words. Even so, the linguistic commitment in the search for a new solution to the problem of automatic indexing affected Garfield's subsequent work in at least three ways:[11]

1. *Language and structure*. Garfield's chemico-linguistic doctoral dissertation contained an early attempt at reducing the complexity of scientific language to elementary components via a reproducible sequence of operations. Garfield used Harris's analytical methods to devise an algorithm for the mechanical translation of chemical nomenclature into molecular formulas and the graphical display of structural diagrams.

The outcome confirmed that chemical nomenclature contains enough information for inferring structural properties: "[T]he basic language of all naming systems in organic chemistry is essentially the same . . . the use of different systems corresponds to the problem of handling dialects rather than treatment of separate distinct languages."[12] The search for a similar basic unity in the language of science would ultimately persuade Garfield that, insofar as the manifold themes embedded in a scientific paper depend on the other documents it cites, an effective algorithmic indexing of scientific literature can be pursued by replacing the hidden and fundamentally unpredictable process of text generation with the comparatively more predictable rules governing the generation of citations.

2. *Integration of citation and text analysis.* Later on in the history of the *SCI*, linguistic models of semantic and syntactic analysis supported the development of automatic keyword indexing systems, notably the ISI *Permuterm Subject Index* (*PSI*) and the *Key Word/Phrase Subject Index* (*KWPSI*), aimed at supplementing the search potentialities of the *SCI* by means of subject entries extracted from articles' titles. The *PSI*, introduced in 1964 to overcome the shortcomings of IBM's Key-Word-In-Context (KWIC) system, exploited the complete permutation of all significant title words to produce all possible and potentially searchable pairs.
3. *Metatext.* A third input of Garfield's structural apprenticeship relates to the importance of metatext in discourse analysis. Metatext is text whose function is not to communicate a conceptual content but rather to introduce and locate it, as in phrases such as "it was found that . . . ," and "these results suggest that . . ." A particular kind of metatextual relationship is also the one established by a text with concepts not directly exposed but hinted at or summarized through the bibliographic references to other documents in which they are embedded. Along this line, inspired by the reflections of John Desmond Bernal and further encouraged by the polymath Chauncey D. Leake, Garfield devoted a good deal of attention to the structure and communicative functions of review articles, namely articles that do not communicate original research findings but merely synthesize the paradigmatic core of concepts and predominantly shared assumptions whereby the state of the art in a particular subject area is established. Here the relation-

> ship between text and bibliographic references is metatextual to the nth power: the author adds a bibliographic reference to almost every sentence, and the sentence's main function is to introduce and partly anticipate the cited document's conceptual content with "an unusually definitive indexing statement."[13]

The frustration with the state of the art in automatic indexing ultimately led Garfield to switch lanes abruptly. Rather than pursuing the dream of an automatic text word or subject indexing system capable of competing with human judgments, he tackled the entire question on a new ground by radically transforming the unit of analysis: no longer words, by themselves ambiguous due to the dependence of their meaning on ever variable contexts, but bibliographic citations, whose semantic stability is ensured, at least within a specialized research area, by the tendency of the authors to use them in relation to a comparatively fixed set of concepts. In March 1953, William C. Adair, former vice president of the Frank Shepard Company, had sent the Welch Project's staff a letter in which he suggested a well-tried solution to the problem of automatic indexing, namely to "shepardize" biomedical literature, to untangle the skein of its content by following the thread of citation links in the same way the legal citator did with court sentences. The proposal was ignored at the institutional level, but it turned out to be the missing link in the chain of events preceding the *SCI* breakthrough. The structure of *Shepard's Citations* demonstrated definitely that, from a technical point of view, the automatic indexing of scientific literature could be performed outside the traditional paradigm of subject headings.

The 1955 article "Citation Indexes for Science,"[14] whose importance for the foundation of scientometrics will be further emphasized in section 3.3, disclosed the benefits and theoretical underpinnings of a citation-based retrieval system modeled on *Shepard's Citations* and outlined the plan of the new system along with some technical troubles to overcome. Contextually, by emphasizing the ability of citation indexes to reduce the complexity of scientific language to a manageable set of atomic units and to let the user run back and forth through the citation network (from the cited to the citing and vice versa), it linked citation indexing both with the below-the-water-line paradigm of structural linguistics and with the forthcoming extra-bibliographic uses of citation indexes in the sociology and politics of science.

In the meanwhile, Garfield had successfully entered the commercial publishing arena, setting up Garfield's Associates, renamed Institute for Scientific Information (ISI) in 1960. The company's main business revolved around the recovery and practical implementation, beyond the biomedical field, of an idea circulating within the Welch Project: the production of a semiautomatic information system based on the photoreproduction of journals' tables of contents. This idea is the origin of *Current Contents* (*CC*), a bibliographic service that gradually extended its coverage to the areas of library and information science (1953), the management and social sciences (1955), pharmacy and biomedicine (1957), and the physical sciences (1960). *Current Contents* enjoyed great popularity among researchers because, removing the congenital delay of traditional bibliographies, it allowed an increasingly wider audience to access, on a weekly basis and often ahead of print, the most up-to-date information on scientific journals' forthcoming content. The popularity grew over the years to the point of prefiguring a condition of reciprocal dependence among the bibliographic tool, the publishing industry, and the scientific community that the *SCI* would ratify once and for all: listing in *CC* (and subsequently among the *SCI* sources) could noticeably enhance the symbolic power and financial status of a journal, whose editorial standards should nonetheless meet ISI quality requirements for the inclusion to take place. Scientists, on their part, started submitting their works preferably to journals processed by the ISI.

A citation index had a far greater appeal for science administrators than journals' tables of contents, but it also entailed a stronger financial effort. A series of pilot tests were performed during the late 1950s and early 1960s, including a citation index to the Old Testament and one to 5,000 chemical-pharmaceutical patents. All confirmed the feasibility and fruitfulness of the enterprise, but large-scale production required money and the right amount of support from the cultural and political establishment. The best chance came when the section of genetic studies of the National Institutes of Health (NIH) was requested to account for the impact of its own research projects. A citation index seemed tailored specifically to such a purpose. Thus, having noticed Garfield's 1955 article, Stanford University geneticist and Nobel Prize winner Joshua Lederberg endorsed the plan of building a citation index for the literature of genetics and strongly advocated its potential benefits before the sensitive audience of a presidential advisory committee on scientific information (the Weinberg Committee). Lederberg played a crucial role in assigning citation indexes the status of politically desirable objects by promoting them as an effective remedy to the flood of scientific information that was thought to submerge American scientists in the Cold War era. In so doing, he passed the *SCI* off as a matter of practical and political necessity, an antidote to the crisis that threatened the cultural identity of science itself.

In 1963, thanks to the NIH's grant obtained through the mediation of Lederberg, Garfield managed to publish the *Genetics Citation Index*, quickly followed by the first volume of the *SCI*. The explosion of what should originally have been a disciplinary tool into a multidisciplinary index was quite inevitable if one thinks of how concretely science works. Genetics, all the more at that time, was a rapidly evolving field, spurred by the recent discovery of the double-helical structure of DNA, and with firmly rooted interdisciplinary connections to mathematics, statistics, and other domains of the physical and life sciences. Papers of genetic interest, accordingly, were not concentrated in a limited number of specialized sources, but scattered throughout several specialized and general journals. In trying to develop a set of rules for automatically extracting all the genetics-related records from an initial, omni-comprehensive citation index to a multidisciplinary set of journals, Garfield realized that the by-product, the multidisciplinary index, was even better than the initial target because, by avoiding artificial boundaries between research areas, it more effectively coped with the practical issue of retrieving relevant documents across disciplinary barriers. Furthermore, the comprehensive index disclosed new and promising areas of application, such as the mapping of cognitive networks connecting papers and authors, and the assessment of the impact of individual work on the scientific community, exactly what science policymakers and people involved in science management at any level were looking for.

Notwithstanding the initial support and the success of the advertising campaign for the broad range of uses to which citation indexes could be put, the funding governmental agencies didn't accept Garfield's proposal to take on the publishing enterprise, nor did they give the ISI additional money, so the work could continue only on a commercial basis. Starting in 1964, the ISI *SCI* was regularly issued every three months, followed in 1972 by the *Social Sciences Citation Index* (*SSCI*) and in 1978 by the

Arts & Humanities Citation Index (*A&HCI*). Ever since, despite a basic continuity in structure, ISI indexes have undergone a complete redefinition of relationships and strategic alliances with the other nodes of the scientific communication network, shifting away from the theoretical background of information retrieval and entering definitively the toolbox of scientometrics. In the new setting, they have become as essential to the disciplinary survival of the host field as the telescope was to Galileo Galilei's early exploration of the moon or, to give a more appropriate idea of its theory-ladeness, as the synchrocyclotron has been to the discovery of new particles in contemporary nuclear physics. The similarity, of course, involves also the drawback that the degree of validity and reliability of the results obtained through the instrument (telescope, synchrocyclotron, citation index) is strictly dependent on its conceptual and technical limitations.

Garfield himself actively concurred in setting the scene for the marriage between citation indexes and scientometrics through the systematic exploration of the statistical regularities governing the citation patterns of authors, papers, and journals in the annual *SCI* files. One important by-product of these efforts is the Law of Concentration (discussed in section 4.3.2). A second, practical achievement is the huge amount of quantitative and qualitative evidence in support of the thesis that high impact science, namely science contained in highly cited papers or citation classics, is good science. This is well exemplified as much by the annual citation analyses of the Nobel prize winners' work started in 1980, as by the 4,000 *Citation Classic Commentaries* published in *Current Contents* since 1977, in which the authors themselves, while suggesting their own reasons for the flattering citation score of a specific book or paper, also had the opportunity to broaden the picture with further details on their personal and intellectual background. Most-cited author rankings were straightforward in equating excellence with citation impact, to the point that even the prediction of "Nobel class authors," notably those highly cited scientists who later went on to win the prize, appeared within the reach of citation analysis. On the other hand, in focusing on the sparkling world of top class scientists, those rankings supplied many superficial readers with a distorted picture of the analytical power of citation analysis in the detection of scientific performance at lower levels of productivity and citation impact. That's why, in the years following, Garfield himself had to spend a lot of time repeating, over and over again, the refrain that "great care must be taken when using citation data to evaluate the impact of the *average* individual. Those evaluations can be both revealing and reliable, but only when performed properly—with expert interpretation, peer assessment, and recognition of potential artifacts and limitation."[15]

In 1992, the ISI was acquired by Thomson Business Information, a subsidiary of the Thomson Corporation, which ferried Garfield's index into cyberspace. The transfer enhanced its interdisciplinary structure and its capability to integrate contents and analytical instruments of diverse origins and purposes within a common framework steadily propped up by the citation network. Hypertext technology eventually eliminated the physical separation between citation and keyword search strategies, turning them into interchangeable options for retrieving bibliographic information from a vast array of sources hosted on *ISI Web of Knowledge*, the Web platform merging Garfield's two creatures, the *Web of Science* (the citation indexes) and *Current Contents*, with the scientometric evaluation tools sponsored by ISI-Thomson.

2.3. "SUBJECT-LESS" RESEARCH: THE *SCIENCE CITATION INDEX* AS AN INFORMATION RETRIEVAL TOOL

In the operative stage, indexing citations on a large scale required a basic, cost-effective choice about disciplinary coverage and inclusion criteria, in short, how many journals to index and how to select them. The selection criteria were partly inspired (and justified) by Bradford's Law of scattering, specifically by Garfield's reinterpretation of it. This fundamental law of information science is discussed in section 4.3. For now, suffice it to note its true practical significance: the core of really important scientific literature, in whatever discipline, is published by a small number of journals; hence, it would be useless, if not financially wasteful, to strive for total inclusion. The problem was not so much in identifying the core journals in a discipline—an operation that any specialist would be able to perform quickly and better than anyone else—as in the definition of qualitative and quantitative criteria for expanding the coverage beyond the nucleus of "obviously" outstanding titles.

The qualitative criteria were inspired by simple and well-established rules: peer judgment; publishers' reputation; editorial staff members' geographic distribution; authors' curricula; and adherence to some elementary editorial standards, such as regularity and punctuality of issues, accuracy, completeness and easy-to-process format of bibliographic references, and inclusion of authors' names and addresses for personal contacts. The quantitative criteria, quite predictably, centered on the neonate citation indexes, which Garfield and Irving Sher tailored to journal management by introducing a new citation measure doomed to set off controversy and misunderstandings: the impact factor (IF). The new index (further examined in chapter 6) counted citations accrued to a journal over a limited time span (two years) and weighted the result, dividing it by the number of items published during the same period.[16] The short time window and the use of a ratio instead of a total citation count were intended to normalize the measure of impact for age and size, and allowed capturing, at least in part, the journals' current usefulness to the scientists. Significantly, the ranking of journals by IF score pushed toward the top positions those primarily devoted to review articles, thus indirectly corroborating Garfield's perception of the importance of reviews in scientific communication and their elective affinities with citation indexes. Though at the beginning, on the admission of Garfield himself, the pressure and friendliness of a publisher could decidedly affect the selection of a title, once set in motion, the citation machine became self-feeding and, thanks to the IF, could proceed automatically with the inclusion of new, high-quality sources. "It is difficult," he pointed out, "to imagine an important journal escaping this citation net."[17]

In the original print version, whose basic structure has been substantially preserved after the transition to electronic and online editions, the *SCI* combined three different files: the *Source Index*, the *Citation Index*, and the *Permuterm Subject Index*, later augmented by the *KeyWords Plus* system. Separate files were also devoted to a *Patent Citation Index* (for patents cited in source journals) and authors' affiliations (*Corporate Index*). The *Source Index* comprised the full bibliographic records of all the articles issued by the core journals whose reference section had been scanned for citation processing and indexing. The *Citation Index* listed in alphabetical order, by name of the first author only, all the documents cited in the source articles, and linked each one to the corresponding set of citing articles.

It is an obvious, though often overlooked, fact that, whereas the *Source Index*'s content is limited both in format and chronological depth, because it comprises only papers published in the core journals fulfilling ISI journal selection policies, the *Citation Index* is universal, because it doesn't discriminate among documents by format or publication date. That is to say, it includes not only journal papers, but also technical reports, theses, conference proceedings, and personal communications, and not only contemporary authors and scientists on the research front, but also Leonardo da Vinci, Copernicus, and Galileo Galilei, if someone ever thought it right to cite them. An annual cumulation of the *SCI*, therefore, makes retrievable a significant percentage of papers published in that year by the small set of ISI core journals (*Source Index*), along with a much larger percentage of documents published in previous years and connected to the former via citation links (*Citation Index*).

The *Permuterm Subject Index* exploited all possible permutations of title words from source articles so as to build a searchable list of natural language keywords. In 1990, it was enhanced by a kind of algorithmic indexing, which further exploited the idea of subject affinity between citing and cited papers. The *KeyWords Plus* system, developed by Irving Sher out of a series of previous experiments and insights, enriched the indexed record of a paper with additional terms (words and phrases) appearing in the titles of its cited references.[18]

In the architecture of the *SCI*, keyword searches are, at best, a simple starting point, a shortcut to the retrieval of a relevant, seminal document useful to spark off the citation cycling routine. Once the seminal document is identified, locating its record in the *Citation Index* grants immediate access to a list of articles citing it; then, by means of the *Source Index*, the complete bibliographic records of the citing articles are retrieved and, if one of them turns out especially relevant, either it or one of the entries in its reference section can be reused as a fresh search entry in the *Citation Index*. The relevance of the results clearly depends largely on the relevance of the seminal document, but also on the citation practices prevalent in the research field to which the cited document belongs, particularly on the authors' attitude to citing predominantly pertinent references in their bibliographies.

Though unfamiliar and unconventional when compared to traditional bibliographic tools, citation indexes make easier some search strategies that usually require a considerable amount of work in keyword or subject-oriented retrieval systems:[19]

1. *Methodology searches.* Often, scientific authors cite documents to address specific methodological techniques that hardly come to the surface of traditional subject indexes, whose primary focus is on the main theme of a document.
2. *Follow-up and "state-of-the-art" searches.* The citation thread is especially effective when one sets out to follow the transformation of a concept or technique throughout the literature, provided the document in which that concept or technique was described for the first time is used as a starting point.
3. *Multidisciplinary and interdisciplinary searches.* When the subject is scattered throughout several research areas, a conventional solution would be resorting to distinct discipline-oriented indexes. In a citation index, instead, the disciplinary pertinence has no value because the thematic links emerge spontaneously, beyond conventional boundaries, from the patterns of use of previous literature by authors actively involved on the research front.
4. *Searches by similarity.* Familiar to users of current bibliographic databases and Web search engines through functions like Related Articles or Similar Pages, the search for documents conceptually related to a given, starting document amounts, in a citation-based retrieval system, to the search for documents sharing two or more references or documents co-cited in two or more bibliographies.

Garfield has repeatedly advocated the retrieval potential of citation indexes on the score of their ability to restore cognitive links between documents that elude conventional indexing languages. The main themes of a publication, as expressed in formal subject headings, are undoubtedly significant on a macro (book) or a micro (article) level of analysis, but on a sub-micro or "molecular" level, where the scientist comes to grips with the details of the interactions between lab results and past literature, there are no main themes that can be easily translated into subject headings, but rather specific ideas or events that are connected to other ideas or events.[20] In a citation index, the rarely found ability of profane indexers to recognize and bring to the surface of conventional subject headings such molecular connections is replaced "by the far superior ability of the entire scientific community to do the same thing."[21]

Since the mid-1960s, the effectiveness of bibliographic citations as clues to literature searches alternative or complementary to keywords and subject headings appeared with a certain regularity on the research agenda in the United States. Myer Kessler, at the Massachusetts Institute of Technology (MIT), was among the first to supply empirical evidence to the effect that bibliographic coupling, that is, the indirect link established between papers sharing one or more bibliographic references, entails also a close resemblance between the subject identifiers of the coupled papers. On similar grounds, cutting edge researchers in the field of information retrieval didn't miss the opportunity to investigate the utility of citations in information systems design: Gerard Salton and Michael Lesk experimented, first at Harvard and then at Cornell University, with the potential value added by citations to the SMART retrieval system; M. E. Stevens and G. H. Urban at the National Bureau of Standards, implemented citation enriched indexing in their SADSACT (Self Assigned Descriptors from Self and Cited Titles) automatic system; W. A. Gray and A. J. Harley performed a seminal test on computer-aided indexing of MEDLARS records sustained by citation relatedness; and Cyril Cleverdon, though extremely skeptical about the ability of bibliographic references to identify the main themes of a paper, tested a slightly modified version of Kessler's technique in the evaluative framework of the Cranfield studies and found that recall and precision for bibliographic coupling were comparable to those of conventional subject indexing. All these experiments built on the assumption that the descriptors or index terms assigned to the bibliographic references cited by an author contain useful information on the subject content of the author's own paper, and hence on the index terms that should be attached to it.[22]

In the 1980s, as online access to the *SCI* spread via several hosts, it became increasingly easier to compare quickly and on a larger scale the retrieval effectiveness of citation searches with that of conventional subject or keyword searches. Some empirical studies explicitly oriented toward the comparison of the two approaches, by Miranda L. Pao, Katherine McCain, Carlos Vidal-Arabona, and William Shaw Jr., confirmed, on the whole, Salton's insight that the combined use of citations and keywords increases the performance of an information retrieval system.[23] As a general rule, it turned out that, while citations tend to favor precision at the expense of recall, with a large variability in output quality across different topics, if one performs the same query using alternatively keywords and citations, the two search strategies yield quite different, seemingly complementary results, with a minimal overlap between the two sets. The quality of a citation search, in addition, is strictly dependent on the number, quality, and selection criteria of the seed documents used to trigger the citation cycle. Having this in mind, information scientists have also tried to develop indicators for predicting the retrieval performance of a seed-document (for example, by its position in the source article), to investigate the optimal number of seed-documents, and to design automatic systems for their identification on the basis of users' information needs expressed in natural language.[24]

NOTES

1. The historical roots of citation indexes are outlined in Fred R. Shapiro, "Origins of Bibliometrics, Citation Indexing, and Citation Analysis: The Neglected Legal Literature," *Journal of the American Society for Information Science* 43, no. 5 (1992): 337–39. A detailed history of legal citation indexes, including the antecedents to *Shepard's*, is in Patti Ogden, "'Mastering the Lawless Science of Our Law': A Story of Legal Citation Indexes," *Law Library Journal* 85, no. 1 (1993): 1–47. On the connection between legal indexes and the conservatism inherent in American legal culture, see Stephen R. Heifetz, "Blue in the Face: The Bluebook, the Bar Exam, and the Paradox of Our Legal Culture," *Rutgers Law Review* 51, no. 3 (1999) : 695–711.

2. A general introduction to the theory of indexing and abstracting is the classic by Frederick W. Lancaster, *Indexing and Abstracting in Theory and Practice* (Champaign: University of Illinois, School of Library and Information Science, 1998); in particular, see chapter 15 on automatic indexing and abstracting. Comprehensive monographic studies of information retrieval history are lacking, but a vast amount of secondary literature on the subject has piled up over the past decades, covering both theoretical and technical developments. Two basic reference sources are Trudi Bellardo Hahn and Michael K. Buckland, eds., *Historical Studies in Information Science* (Medford, N.J.: Information Today, 1998), and Mary E. Bowden, Trudi Bellardo Hahn, and Robert V. Williams, eds., *Proceedings of the 1998 Conference on the History and Heritage of Science Information Systems* (Medford, N.J.: Information Today, 1999). On the history of the statistical

approach to information retrieval, see the two outlines set forth, respectively, in Sparck Jones, "Statistics and Retrieval: Past and Future," in *International Conference in Computing: Theory and Applications* (Kolkata: IEEE Computer Society, 2007), 396–405 and Salton, "Developments in Automatic Text Retrieval," *Science* 253, no. 5023 (1991): 974–80.

3. Luhn, "A Statistical Approach to Mechanized Encoding and Searching of Literary Information," *IBM Journal of Research and Development* 1, no. 4 (1957): 310, www.research.ibm.com/journal/rd/014/ibmrd0104D.pdf.

4. Luhn, "The Automatic Creation of Literature Abstracts," *IBM Journal of Research and Development* 2, no. 2 (1958): 160, courses.ischool.berkeley.edu/i256/f06/papers/luhn58.pdf.

5. Salton, "Automatic Text Analysis," *Science* 168, no. 3929 (1970): 339.

6. It is often postulated, for example, that a word randomly distributed according to a Poisson distribution is not as informative about the document in which it appears as a word that exhibits a nonrandom behavior. See Abraham Bookstein and Don R. Swanson, "Probabilistic Models for Automatic Indexing," *Journal of the American Society for Information Science* 25, no. 5 (1974): 312–18. On the vector-space model, see Salton and McGill, *Introduction to Modern Information Retrieval* (New York: McGraw-Hill, 1983).

7. Peter Ingwersen and Kalervo Järvelin, *The Turn: Integration of Information Seeking and Retrieval in Context* (Dordrecht: Springer, 2005), 30. For a comprehensive review of the various theories and definitions developed over the past thirty years on the key notion of relevance, see Tefko Saracevic, "Relevance: A Review of and a Framework for the Thinking on the Notion in Information Science," *Journal of the American Society for Information Science* 26, no. 6 (1975): 321–43; Saracevic, "Relevance: A Review of the Literature and a Framework for the Thinking on the Notion in Information Science, Part II; Nature and Manifestations of Relevance," *Journal of the American Society for Information Science and Technology* 58, no. 13 (2007): 1915–33; and Saracevic, "Relevance: A Review of the Literature and a Framework for the Thinking on the Notion in Information Science, Part III; Behavior and Effects of Relevance," *Journal of the American Society for Information Science and Technology* 58, no. 13 (2007): 2126–44.

8. Kui-Lam Kwok, "A Probabilistic Theory of Indexing and Similarity Measure Based on Cited and Citing Documents," *Journal of the American Society for Information Science* 36, no. 5 (1985): 342–51.

9. The phrase "bits and pieces of experience and insight" is in Garfield, *Citation Indexing: Its Theory and Application in Science, Technology, and Humanities* (Philadelphia: ISI Press, 1983), 6, www.garfield.library.upenn.edu/ci/title.pdf. A detailed reconstruction of the genesis and development of the index is given by Paul Wouters, "The Citation Culture" (Ph.D. thesis, University of Amsterdam,

1999), www.garfield.library.upenn.edu/wouters/wouters.pdf. The history and fortunes of citation indexing have been told, from a variety of perspectives, in a collection of essays issued on the occasion of Garfield's seventy-fifth birthday, by Blaise Cronin and Helen Barsky Atkins, eds., *The Web of Knowledge: A Festschrift in Honor of Eugene Garfield* (Medford, N.J.: Information Today, 2000). More recently, the political, social, and intellectual influences surrounding Garfield's project have been reviewed by Stephen J. Bensman in "Garfield and the Impact Factor," *Annual Review of Information Science and Technology* 41 (2007): 93–155, garfield.library.upenn.edu/bensman/bensmanegif2007.pdf.

10. Bar-Hillel, "The Present State of Research on Mechanical Translation," *American Documentation* 2, no. 4 (1951): 229–37.

11. The outline of the connections between structural linguistics and Garfield's work on automatic indexing is based on Garfield, "Automatic Indexing and the Linguistics Connection," in *Essays of an Information Scientist 1981–1982* (Philadelphia: ISI Press, 1983), 31–38, www.garfield.library.upenn.edu/essays/v5p031y1981-82.pdf, and Garfield, "The Relationship Between Mechanical Indexing, Structural Linguistics and Information Retrieval," *Journal of Information Science* 18, no. 5 (1992): 343–54, www.garfield.library.upenn.edu/papers/jis18(5)p343y1992.html. The latter is a post-publication of a manuscript dating back to the 1960s.

12. Garfield, "Chemico-Linguistics: Computer Translation of Chemical Nomenclature," *Nature* 192, no. 4798 (1961): 192, www.garfield.library.upenn.edu/essays/v6p489y1983.pdf.

13. Garfield, *Citation Indexing: Its Theory*, 7.

14. Garfield, "Citation Indexes for Science: A New Dimension in Documentation Through Association of Ideas," *Science* 122, no. 3159 (1955): 108–11, www.garfield.library.upenn.edu/essays/v6p468y1983.pdf.

15. Eugene Garfield and Alfred Welljams-Dorof, "Citation Data: Their Use as Quantitative Indicators for Science and Technology Evaluation and Policy-Making," in *Essays of an Information Scientist 1992–1993,* 192 (Philadelphia: ISI Press, 1993), www.garfield.library.upenn.edu/essays/v15p188y1992-93.pdf.

16. The idea of a normalized measure of journal impact alternative to total citation counts is developed in Garfield and Sher, "New Factors in the Evaluation of Scientific Literature Through Citation Indexing," *American Documentation* 14, no. 3 (1963): 195–201, www.garfield.library.upenn.edu/essays/v6p492y1983.pdf.

17. Garfield, *Citation Indexing: Its Theory*, 25. On ISI journal selection criteria, see Garfield, "How ISI Selects Journals for Coverage: Quantitative and Qualitative Considerations," in *Essays of an Information Scientist 1990*, 185–93 (Philadelphia: ISI Press, 1991), www.garfield.library.upenn.edu/essays/v13p185y1990.pdf.

18. Garfield and Sher, "Keywords Plus™ Algorithmic Derivative Indexing," *Journal of the American Society for Information Science* 44, no. 5 (1993): 298–99, www.garfield.library.upenn.edu/papers/jasis44(5)p298y1993.html.

19. Garfield, *Citation Indexing: Its Theory*, 41–61. Two examples of complex searches on the online versions of the *SCI* are presented in Anthony E. Cawkell, "Checking Research Progress on Image Retrieval by Shape-Matching Using the Web of Science," *ASLIB Proceedings* 50, no. 2 (1998): 27–31, and Cawkell, "Methods of Information Retrieval Using Web of Science: Pulmonary Hypertension as a Subject Example," *Journal of Information Science* 26, no. 1 (2000): 66–70.

20. This argument, inspired by the Indian information scientist Shiyali Ramamrita Ranganathan, was already present in the 1955 *Science* article, and is reaffirmed in Garfield, *Citation Indexing*: *Its Theory,* 41: "If the book is considered the macro unit of scientific thought, and the journal article regarded as the micro unit, as had been suggested by Ranganathan, the citation index seemed to reach down to the molecular level, which certainly was a useful degree of specificity that was beyond the economic, if not conceptual, capabilities of the subject indexes."

21. Garfield, *Citation Indexing: Its Theory*, 9.

22. Salton processed a set of 200 documents in the field of aerodynamics against a set of user queries adopting two different methods of document representations: a standard content analysis method and a citation-enriched version of the document identifiers. He concluded that, if citations are from documents whose relevance to the query is established in advance, "the augmented document vectors, consisting of standard concepts plus bibliographic citation identifiers appear to provide a considerably better retrieval performance than the standard vectors made up of normal subject indicators only." Salton, "Automatic Indexing Using Bibliographic Citations," *Journal of Documentation* 27, no. 2 (1971): 109. On Kessler's technique and its evaluation by Cleverdon, see Bella Hass Weinberg, "Bibliographic Coupling: A Review," *Information Storage & Retrieval* 10, nos. 5/6 (1974): 189–96. On the other above mentioned experiments, see W. A. Gray and A. J. Harley, "Computer Assisted Indexing," *Information Storage & Retrieval* 7, no. 4 (1971): 167–74.

23. On the comparison between keyword and citation indexing, see Pao and Worthen, "Retrieval Effectiveness by Semantic and Citation Searching," *Journal of the American Society for Information Science* 40, no. 4 (1989): 226–35; McCain, "Descriptor and Citation Retrieval in the Medical Behavioral Sciences Literature: Retrieval Overlaps and Novelty Distribution," *Journal of the American Society for Information Science* 40, no. 2 (1989): 110–14; Vidal-Arbona, "Comparing the Retrieval Effectiveness of Free-Text and Citation Search Strategies in

the Subject of Technology Planning" (Ph.D. dissertation, Case Western Reserve University, 1986); Shaw, "Subject and Citation Indexing, Part 1: The Clustering Structure of Composite Representations in the Cystic-Fibrosis Document Collection," *Journal of the American Society for Information Science* 42, no. 9 (1991): 669–75; and Shaw, "Subject and Citation Indexing, Part 2: The Optimal, Cluster-Based Retrieval Performance of Composite Representations," *Journal of the American Society for Information Science* 42, no. 9 (1991): 675–84.

24. On the determination of the optimal number of seed-documents, see Lanju Lee Yoon, "The Performance of Cited References as an Approach to Information Retrieval," *Journal of the American Society for Information Science* 45, no. 5 (1994): 287–99. The changes in references' value according to their placement in the citing paper are discussed in Henry Voos and Katherine S. Dagaev, "Are All Citations Equal? Or Did We Op. Cit. Your Idem?" *Journal of Academic Librarianship* 1, no. 6 (1976): 19–21. An experiment on automatic identification of seed-documents is reported by Birger Larsen, "References and Citations in Automatic Indexing and Retrieval Systems—Experiments with the Boomerang Effect" (Ph.D. dissertation, Royal School of Library and Information Science, Department of Information Studies, 2004), www.db.dk/dbi/samling/phd/birgerlarsen-phd.htm.

Chapter Three

The Philosophical Foundations of Bibliometrics: Bernal, Merton, Price, Garfield, and Small

After the launching of the *SCI*, which supplied most of the raw data for the construction of citation-based bibliometric indicators, and the publication of *Science Since Babylon* (1961), *Little Science, Big Science* (1963), and *Networks of Scientific Papers* (1965) by Derek Price, scientometrics already had a sound empirical and conceptual toolkit available. Down the path toward disciplinary maturity, it also established a deep kinship with Merton's seminal work in the sociology of science and Thomas Kuhn's *The Structure of Scientific Revolutions* (1962). Price and Garfield, in turn, openly acknowledged the intellectual legacy of the Irish-born scientist John Desmond Bernal in bringing forward the social, economic, and organizational issues integral to the project of a quantitative science of science. Blinded by the sparkling mathematical polish of current scientometric research papers, an external, nonspecialist eye would probably overlook the fact that the conceptual roots of the discipline run deep into the insights of the above authors who, striking out in directions far beyond the bare analysis of the methodological and linguistic aspects of scientific discourse, projected the study of the structure and dynamics of research activity on the background of the concrete social life of scientific communities. Their legacy lives on in every subsequent social theory that takes communication as the nitty-gritty of science both on the informal and the formal sides of its concrete manifestations. The scientometric offspring focused resolutely on the latter and, to the extent that published scientific literature and its internal bibliographic connections allowed the harvesting and analysis of relatively unobtrusive sociometric data that lent themselves to empirically testable generalizations, scientometricians' vocation to infer aspects of the structure and dynamics of communication processes from quantitative patterns circumscribed to the formal expressions, came up as a quite natural extension.[1]

3.1. THE "RED" INFORMATION SCIENCE OF JOHN DESMOND BERNAL[2]

Bernal, professor of physics at the Birkbeck College of London since 1937, is a leading figure in the history of science in every possible sense: as an actor, given his pioneering contributions to X-ray crystallography and molecular biology, and as a director teaching historians of sciences how to study and write about science in its social context. Bernal was a Marxist in philosophy and a communist in politics, two equally important attributes for grasping both his revolutionary approach to information science and his blindness toward the nonsense propagandized by the scientists of the Soviet regime. Being a Marxist, adhering to Engels's dialectical materialism, he believed that science is a social affair, carried out by an international community of networked researchers, and intimately connected to the whole range of human activities. The trust in such a "connected whole" and the impatience with detail, while preventing him from shutting himself up in a laboratory to work at the resolution of a Nobel prize–winning puzzle, influenced his resolution to devote time and effort to the advancement of the material conditions for the achievement of the full cognitive, social, and political potential of science and technology.

In contrast to orthodox Marxists, Bernal didn't undertake to mechanically reduce the existing corpus of scientific theories to dogmas rooted in capitalist ideology. Instead, he considered science and scientific method the chief promoter of social change and the foundation of all valuable human knowledge, whether concerning nature or society. His *The Social Function of Science* (1939), while showing traces of early Soviet investigations on the social aspects of scientific research carried out between the end of the nineteenth and the beginning of the twentieth centuries, triggered a knock-on effect of cross-fertilization between Eastern and Western scientometric traditions. Derek Price's "thermodynamics" of science would be decisively influenced by Bernal's objectification of science products and his advocacy of the "need to apply science to the field of administration";[3] Price's work, in turn, would catalyze Dobrov's and Nalimov's efforts toward the foundation of a sound analytic framework for quantitative science studies; and the Russian commitment to such a project would, again, reverberate in the research agenda of the United States and Europe, leading eventually to a wider scholarly and political legitimization of Western scientometrics.

Bernal's 1939 book, in addition to gaining the reputation of being the bible of the "red science" during the war and postwar periods, aroused the interest of Western information scientists due to its visionary prefiguration of a new, revolutionary information system. Interestingly, the author's case for promoting this radical shift in the structure of scientific communication was backed up by arguments partly similar to those employed a few years before by Samuel Bradford in introducing the law of scattering—i.e., the inability of current bibliographic services to cope with the mass of documentation produced daily by researchers—and partly amenable to a deep understanding of the way science actually works. Bernal, in fact, perfectly knew what practicing scientists tacitly took for granted and later philosophers and sociologists would place at the core of modern critical theories of science: namely, that not all, and hardly even the greater part, of scientific communication is enclosed in published papers. To a larger extent, indeed, "the transference of scientific ideas from one set of scientific workers to another is effected by means of visits, personal contacts, and letters." Thus, even if the obstacles to effective documentation were removed, laboratory life would have priority over written reports, because "there would remain techniques which are impossible to transmit without visual demonstration, and ideas too intangible to be put into writing, yet capable of communication by personal contact."[4] Journal papers, nonetheless, still formed the bulk of the official apparatus of scientific communication, and a new organization was urgently needed to bring their content as close as possible to live research activity.

The revolution had to take place in two nearly simultaneous steps: a destruction offset by a reconstruction of something completely different. In the first stage, publishers' and scientific societies' vested interest in the massive publication of periodical literature would be destabilized; there were too many journals, too many poor-quality papers replicating in different journals, and too much money wasted in mirroring the (badly indexed) collections of existing journals. After all, as many advocates of the open access movement would currently assert, "periodicals exist for science and not science for periodicals."[5] So, for the most part, the existing scientific journals had to be abolished, replaced by the set of all individual papers, or something quite similar to the individual papers detached from journals. Once the journal oligopoly was razed to the ground, the reconstructive effort should be directed toward the building of a central clearinghouse or, alternatively, a network of decentralized clearinghouses in close communication with one another, for the storing, collection, organization, and selective dissemination of scientific information. The scientific paper sent to the central publication office, upon approval by an editorial board of referees, would be microfilmed, and a sort of print-on-demand system set in action thereafter, the master film to be preserved in the central archive and a certain number of copies sent to libraries, for the purpose of storing, and to all the scientists that, in a previously filled-in form, had declared their interest in the subject of the paper. The abstracting system, fueled by the authors themselves, would pursue a similar scheme, with a central set of abstract cards from which as many thematic subsets could be easily obtained as necessary to satisfy the needs of libraries and researchers. The dissemination of the new publication units should be managed by the central distribution service, which Bernal prefigured on the pattern of the Scientific Information Institute proposed in 1940 by Watson Davis for the American information system.[6] The service would ensure the delivery, to each scientist, of the sole useful information with the proper amount of detail: "All relevant information," he pointed out, "should be available to each research worker and in amplitude proportional to its degree of relevance. Further, that not only should the information be available, but also that it should be to a large extent put at the disposal of the research worker without his having to take any special steps to get hold of it."[7]

The scientific archive's boundaries envisaged by Bernal stretched far beyond the narrow domain of individual scientific papers, encompassing four types of records with different structures and functions:

1. ephemeral notices of daily laboratory life, such as accounts of new discoveries, techniques, meetings, and discussions;
2. handbooks and popular works on science, relating scientific progress to common human needs and aspirations;

3. "old-style" journals, serving the limited purpose of giving the latest news from the world of research and discussing the social impact of scientific discoveries; and
4. detailed, comprehensive reports and monographs documenting the advancement of each singular field of science over time as well as the interrelationships of various fields.

A key position in this blueprint was occupied by the reports or reviews, which Bernal considered especially important for keeping up with scientific advancement and helping administrators and technocrats map progress across different disciplines. The culture of literature reviews, which exercised a great influence on Garfield's conceptualization of citation indexes, would ultimately find completion in the plan of a *World Encyclopaedia* pursuing, in the wake of Herbert G. Wells's *World Brain*, the old dream of "a comprehensive and continually revised presentation of the whole of science in its social context."[8]

It is obvious that Bernal's destructive plan, the abolishment of scientific periodicals, was never accomplished. He knew perfectly well that times were not ripe, and perhaps would never be, for such an upset. Hence it is not surprising that, despite a last-minute attempt at rewording the proposal of a central information system so as to render it more palatable to scientific societies, he withdrew the paper exposing the revolutionary plan from consideration by the Royal Society Information Conference of 1948. All the same, the idea of a central service for the selective dissemination of current scientific information was destined to survive, though in a weaker form. A couple of decades later, it was reinvented on a commercial basis by Garfield, who devised an alerting and a reprint distribution service centered on citation indexes to promote the selective dissemination of information and to "systematize a practice common among many authors who send reprints to other scientists they have cited."[9]

In the early stages of the *SCI*, Bernal served on its editorial advisory board. He also wrote a review of Garfield's *Index* in which, though criticizing the exclusion of many high-quality journals from the ISI source list, praised its interdisciplinary design on the ground that all great discoveries in science have deep interdisciplinary roots. Furthermore, he acknowledged the *SCI*'s full potential for the history and sociology of science insofar as "it should enable the poly-dimensional graph on the progress of science to be mapped out for the first time . . . a necessary stage in drawing up or planning any strategy for scientific research as a whole."[10]

3.2. CITATION AS SOCIAL REWARD: ROBERT K. MERTON

As of 1942, while the intellectual independence of scientists was seriously threatened by the nationalistic commitment to war, Merton, a professor of sociology at Columbia University, divulged a small set of norms supposedly placed at the core of the universal *ethos* of science, that is, the complex of prescriptions, prohibitions, and values governing the prevalent (nondeviant) behavior of scientists at all times and everywhere.[11] The norms were summarized under four headings: universalism, communism, disinterestedness, and organized skepticism.

"Universalism" holds that every knowledge claim must be checked against a set of preestablished, impersonal criteria; science, like myth and religion, aims at universal knowledge, but its universality is supposed to be "qualitatively" different from that of mythical tales and religious truths because of the way (purportedly "rational") it is attained. "Communism," antithetical to the secrecy typical of the pseudoscience of wizards and alchemists, involves the view that scientific results are public goods assigned to the community, save for the scientist's right to be individually recognized and properly rewarded for the novelty of a contribution; science is a cumulative and collaborative undertaking, so every privatistic restriction on its final products exposes the community to the risk of slowing down progress. "Disinterestedness" is the absence of economic or personal motivations in the pursuit of knowledge. Properly speaking, the norm is not for the sole use of scientists, since an artist, too, is supposed to operate disinterestedly, but in the case of a scientist there is something more than a psychological attitude: given the public and testable character of science, disinterestedness is embedded in a set of institutional structures that form a neutral, nearly police system of surveillance on the behavior of individual researchers to minimize deviant phenomena, such as fraud, irresponsible claims (quackery), and abuse of expert authority. "Organized skepticism," finally, is the mandate to submit every statement or belief to a methodological doubt so as to check its validity against logical and empirical evidence; logically inconsistent and empirically not verified nor verifiable arguments are banished from the realm of science.

In Merton's opinion, the immense success achieved by modern science in the discovery of natural laws can be explained only if one admits that, thanks to this normative structure, there exists an underlying homogeneity and continuity among the diverse levels of the research activity.

1. Scientific practice, what scientists actually do at the laboratory bench, yields results in the form of new theories or techniques.
2. The results are communicated to the audience of qualified peers in the form of written publications, for example journal papers. Scientists are eager to publish to secure their priority in ideas and discoveries. Unlike private economic interests, indeed, scientific interests are better protected through resource sharing: the more a work of the intellect is made freely available to others, the safer its symbolic property is from larceny and fraud.
3. Published articles are read and assessed by the community of peers, who recognize their value by citing them in their own works. The bibliographic citation, therefore, is the first, elementary building block of the reward system, an "atom of peer recognition."[12]
4. The scientific establishment benefits researchers who publish the most original results enhancing their social (and symbolic) status within the community, for instance through academic chairs and qualifications. Symbolic credit flows spontaneously from documents to authors, who convert it into social prestige and institutional power.

Merton certainly didn't claim that scientists mechanically follow the above scheme in everyday work, nor that the immaculate aspect of journal papers faithfully reflects the operations going on in research practice, with its typical hesitations, deviations, and settlement strategies. Still, he maintained that scientists, on average, operate according to the prescriptions of universalism, communism, disinterestedness, and organized skepticism. It is not even necessary for them to believe in those prescriptions, the success of the enterprise depending simply on their behaving as if they did. The functional equilibrium thus reached by the science system informs the division of labor peculiar to scientific communities, where power and resources are concentrated in the hands of a comparatively small number of individuals. The norms guarantee that, unlike other social systems, stratification and scientific inequalities in science grow by the application of universal criteria, so that the most significant contributors are also the best rewarded.

The system, unfortunately, is not foolproof, its robustness being undermined by the frequent emergence of particularistic strains. For example, when, due to deliberate fraud, errors in judgment, or simply a limited number of available credits, not all those who deserve proper recognition receive it, some get ahead to others' disadvantage. In an influential book published in 1965, Warren Hagstrom put the spotlight on the social influences inducing conformity to or deviance from scientific norms and values. He pointed out that each time, in a specialty with a highly developed formal organization and a strong consensus about the relative importance of research problems, there exists a high number of people potentially able to solve them, the competition for recognition constitutes a major force of deviant behavior.[13]

One important symptom of illness occurs also when, regardless of their levels of performance and the actual value of their contributions, scientists already rewarded for their achievements get a higher chance of being rewarded once again, so that they become part of an elite group enjoying preferential access to scientific resources and facilities. Merton called this success-breeds-success phenomenon by which the rich get richer while the poor get comparatively poorer the "Matthew Effect," after a well-known verse in the Gospel according to Matthew (Matthew 25:29, King James Version).[14] From the standpoint of citation practices, the Matthew Effect entails the propensity of authors to prefer to cite works by awe-inspiring colleagues in the field, neglecting other less visible ones who may be just as (or even more) pertinent. Merton's favorite disciple, Harriet Zuckerman, documented the concrete operation of the Matthew Effect in the social milieu of elite scientists, specifically the recipients of the Nobel Prize, whose post-prize prestige boost didn't seem adequately justified by sustained excellence in research performance.[15] Similarly, having investigated in empirical fashion the occurrence of deviant behaviors among academic scientists, Jonathan Cole and Stephen Cole concluded that "science does to a great extent approximate its ideal of universalism," but in almost all cases where it departs from the ideal the Matthew Effect is at work: "People who have done well at time 1 have a better chance of doing well at time 2, independently of their objective role-performance; the initially successful are given advantage in subsequent competition for rewards."[16]

In bibliometricians' hands, the Matthew Effect lost its negative social connotations and, prior to any involvement with the nuts and bolts of mathematical modeling and performance evaluation, sneaked into the very heart of the theory building (and justifying) process. Mathematically oriented informetricians (as chapter 4 illustrates in greater detail), set out to translate into the language of probability theory its formal resemblance to the mechanism responsible for the highly skewed distributions ubiquitous in information science. On the evaluative side, instead, Manfred Bonitz engaged in the setting up of a skewed "Matthew world" wherein the competition between countries for scientific leadership and their degree of success in promoting scientific talents becomes a measurable property of the global science communication system; here, thanks to one of the many offshoots of Garfield's journal impact factor, the average citation rate for a set of articles issued by a scientific journal acts as a standard, an ideal "expected" citation impact, against which the positive or negative deviation of each country's observed citation impact—the share of "Matthew citations" referred to the same journal—can be gauged.[17]

Merton, who celebrated the potential of citation indexes as a new and long-awaited tool of sociological analysis, played a constant role as mentor and advisor in the perfection of Garfield's creature, though he never personally worked in the field of citation theory and analysis. He also underlined the rudimental character of citation-based scientometric indicators and insisted on possible phenomena causing the loss of significant data in citation analysis, first of all the "obliteration by incorporation," namely the emergence of key documents so important for a research field that they end up embedded into the corpus of currently accepted knowledge, being no longer cited in references. "To the extent that such obliteration does occur," he wrote, "explicit citations may not adequately reflect the lineage of scientific work. As intellectual influence becomes deeper, it becomes less readily visible."[18]

Merton's norms appeared to many later sociologists too generic to capture the complexity of scientific practice. Yet bibliometricians took them tacitly as an intellectual frame of reference for the understanding of citation behaviors, which could be construed, accordingly, as rational and therefore predictable to the extent that they were amenable to a professional ethic imposing, among other things, the recognition of individual credits within the collective undertaking of science.

3.3. CITATION AS CONCEPT SYMBOL: EUGENE GARFIELD AND HENRY SMALL

It is commonplace, above all among detractors of citation analysis, to say that there are plenty of reasons to cite, so many that it would be at least ingenuous to think of citation as an absolute index of intellectual commitment; so diverse that a Mertonian normative theory of citing seems to catch only a small part, if any, of the scientific authors' citation styles. What detractors often forget, however, is that the citer's motivation to summon up a particular document in a reference is just one side of the coin, the other being the cited document itself, its conceptual content living somewhere "outside," independently of its creator and, even more so, of its future or potential citers' reasons to cite. Whatever the individual motives for that content being called into play by one or more citers over time, and whatever the specific informational unit addressed each time, in fact, its symbolic representation in the form of a bibliographic reference is inevitably set into motion by the act of citing, thus becoming an active node of a citation network that unfolds parallel to (and interacts with) the discursive shaping of scientific arguments by individual authors. Once caught in the network, the multiplicity of meanings that can be assigned to the cited document in the citing papers is somewhat constrained by the finite set of meanings conventionally attached to it by the members of a specialized scientific community steeped in a specific intellectual (and linguistic) tradition at a particular stage of its history.

Garfield's 1955 article "Citation Indexes for Science" was a turning point in the way information scientists conceptualized the role of bibliographic citations in the knowledge production process. It pivoted on the idea that citations are the building blocks of a language that reflects, better than conventional subject headings, the deep structure of scientific communication, where past literature (and terminology) is in a constant state of reinterpretation according to transformational rules dictated by disciplinary practices. Bibliographic citations, unlike fixed subject headings, label cited documents from as many viewpoints as exist in the community of the potential citers. Hence, they perform a subject categorization of past literature that Garfield considered, on the whole, more precise and flexible than conventional subject indexing insofar as it uses an unequivocal document pointer to address the scientists' (and not the indexers') units of thought. But for a document, being summoned up by a scientist at the latest stage of a research documentation process—the disclosure of the findings to the community of peers—is far more significant than merely being indexed by a professional indexer, because it means being invoked as an active part of the research process behind the results being communicated. So, systematically reestablishing these connections for hundreds of documents is a clue to the intellectual background of a piece of research and, to the extent that citations are the operation of traceable authors belonging to traceable institutions and collaboration networks, it is also a clue to its social background. Taken collectively, in fact, citations connect cited and citing authors in sociocognitive networks open to historical and evaluative judgment, thereby making citation indexes useful "when one is trying to evaluate the significance of a particular work and its impact on the literature and thinking of the period."[19]

In the 1970s Henry Small, a member of the research team at ISI, further developed Garfield's insights into the ability of citations to mimic the transfer and uptake of ideas typical of more codified forms of language. He advocated the basic cognitive function of bibliographic citations on the ground that, apart from individual reasons to cite, each reference incorporates an idea or concept accounting for the citer's resolution to invoke it in a specific context. The idea may or may not coincide with that of the citer, but, to the extent that it does, as is often the case in scientific papers, the reference itself can be regarded as a simple and relatively stable symbol of that idea, a concept symbol.[20]

Small borrowed the theory of the concept symbol from the British anthropologist Edmund Leach, who, in *Culture and Communication* (1976), applied structuralist analysis to social anthropology, starting from an assumption currently taken for granted: culture, like language itself, is a system of signs to be decoded, and each cultural phenomenon is, more or less consciously, a communicative process. In the wake of Noam Chomsky, Leach held that even the material aspects of human life could be analyzed for the purpose of unveiling their deep grammatical structure:

> All the various non-verbal dimensions of culture, such as styles in clothing, village lay-out, architecture, furniture, food, cooking, music, physical gestures, postural attitudes and so on are organised in patterned sets so as to incorporate coded information in a manner analogous to the sounds and

> words and sentences of a natural language . . . . It is just as meaningful to talk about the grammatical rules which govern the wearing of clothes as it is to talk about the grammatical rules which govern speech utterances.[21]

Simplifying the terminology inherited from the semiotic tradition, Leach drew a fundamental distinction in human expressive actions between "signs" and "symbols." In a "sign," the object or index that works as a sign of a certain entity is contiguous with what is meant (metonymic relation): the crown is a "sign" of sovereignty in the context of European political traditions; the sequence of letters "a-p-p-l-e" is a "sign" for a particular fruit in the context of English-language conventional usage. In a "symbol," instead, there is no contiguity between the object-index and what it stands for, because they belong to different cultural contexts and the relationship between them is arbitrary (metaphoric): the crown used as a trademark of a brewery is a "symbol," not a "sign" of beer; the serpent in the Garden of Eden is a "symbol," not a "sign" of evil. There is, however, a substantial gap between private, occasional "symbols," such as those occurring in dreams or poetic images, which do not convey any public information as long as they are not enhanced by a comment, and standardized "symbols," which convey information in the public domain and tend to communicate quite stable meanings.

Small transposed the above scheme in the bibliographic space, thus further strengthening the kinship of the citation culture with the structuralist tradition of linguistic studies already established by Garfield at an early stage. A bibliographic reference is a "sign" and a "symbol" at the same time. It is a "sign," made up of the usual sequence author–journal–volume–page–year, pointing to a physical object, the cited document, with which it shares some formal features (metonymic relation). It is a "symbol" of the concept or concepts articulated by the cited document (metaphoric relation), in the sense that it invokes the cited document in connection with a specific point of the text, thereby labeling it with the concept it is supposed to represent in that particular context. Concepts are not necessarily abstract entities, but include "experimental findings, methodologies, types of data, metaphysical notions, theoretical statements or equations."[22] Nor are they fixed entities; it might be the case, indeed, that the citing paper does not confer to the cited document the same meaning intended by the latter's author, since "prior literature is necessarily in a constant state of reinterpretation, adapting to changes in knowledge within the field."[23] There exists, in science as well as in literature, a certain degree of autonomy of the intellectual product when it enters the communications circuit, but to Small and the entire bibliometric tradition, the most common case remains that of normative consensus, at least as far as science products are concerned. In some circumstances, moreover, scientists so strongly agree on the cognitive structure of previous literature that they persistently cite the same documents to address the same concept. Thus, while some references are private "symbols" that the author uses to communicate his or her own ideas, "other citations are to documents whose significant content may be shared by a community or group of scientists, and such documents are likely to be frequently cited: in Leach's terminology they are 'standard symbols'."[24] The "standard symbols" are clearly "the product of a dialogue and selection process on the part of many individuals over a period of time," and it can be hypothesized that each scientist "carries with him a repertoire of such collective concepts and their corresponding document-symbols."[25]

If the foregoing premises are accepted and the concept/citation equivalence is assumed to be valid for a comparatively large number of cited documents, then the *SCI* can rightly be considered an index of scientific concepts tantamount to a disciplinary thesaurus. Consequently, it can be trusted as both a retrieval tool for thematic literature searches and a source of sensitive data on the cognitive impact exercised by documents and authors in terms of citation scores. Besides, as explicitly advocated by Garfield,[26] an interdisciplinary citation index could afford an even greater philosophical ambition. The equivalence between citations and concepts, in fact, paved the way for a hypothetical bibliographic unification of the entire body of scientific literature, a shortcut to the project that, since the late 1920s, had engaged the logical empiricists of the Vienna Circle in the construction of a unified language of science. In its initial, rough formulation, later revised after Karl Popper's criticisms, the rationale behind the project was that scientific concepts from various disciplines could be reduced, by means of definitions or reduction sentences, to the primitive, strictly observational terms of the language of physics. Such an assumption evolved, toward the end of the 1930s, in the outline of the *International Encyclopaedia of Unified Science*, whose first volume appeared in Chicago, under the editorial direction of Otto Neurath, in 1938. Garfield was well acquainted with the Vienna Circle's philosophy, although his interest in the construction of an interdisciplinary, unified index was chiefly driven by its immediate practical value as a documentation tool. Bibliographic citations, actually, had a looser structure and far less conceptual stability than the primitive concepts expressed in a physicalistic language, but, as long as their symbolic power was relied on, they seemed to fit a similar purpose quite well, forming, in Garfield's own terms, the building blocks of "a plan for accomplishing what Neurath calls 'an encyclopedic integration of scientific statements,' what I call a 'Unified Index to Science'."[27]

3.4. WEAVING THE NETWORK OF SCIENCE: DEREK JOHN DE SOLLA PRICE

> Why should we not turn the tools of science on science itself? Why not measure and generalize, make hypotheses, and derive conclusions? . . . My approach will be to deal statistically, in a not very mathematical fashion, with general problems of the shape and size of science and the ground rules governing growth and behavior of science-in-the-large . . . The method to be used is similar to that of thermodynamics, in which is discussed the behavior of a gas under various conditions of temperature and pressure.[28]

The declaration of intent set forth by Derek Price in the preface of *Little Science, Big Science* was a clear statement of scientometrics' research program for the generations to come. His subsequent work in the field may be considered the technical fulfillment of Bernal's legacy, as it led, within a few years, to the delineation of a paradigm for science studies, in the very sense that he not only demarcated the boundaries of the new discipline and the necessary conditions for its existence, but also showed concretely how its puzzles could be solved thereafter.

Price, professor at Yale University since 1959, was a British historian of science and technology with an extensive physico-mathematical background. His ideas on the nature and evolution of science rest on two fundamental premises, eventually converging toward a "bibliometric reductionism" that marks the entire tradition of scientometrics. First, science is inherently different from other areas of scholarship by virtue of the objective criteria it adopts in the observation and manipulation of empirical data. That's why "there is in the field of science a cumulative accretion of contributions that resembles a pile of bricks . . . to remain in perpetuity as an intellectual edifice built by skill and artifice, resting on primitive foundations."[29] Second, even though, during the stage of discovery, unsystematic laboratory life as well as social, psychological, and philosophical factors cooperate actively in the shaping of new ideas and techniques, eventually, when the game is up, science amounts to published scientific literature. A scientist, therefore, is not recognized by having received a certain education, or working in a would-be scientific institution, but simply by the fact that, at least once in life, he or she has produced a scientific publication read and approved by the community of peers. Coherently with such premises, the counting, classification, and representation of research publications in the form of temporal series provides a reliable indicator of science evolution, whereas the analysis and comparison of temporal series discloses the quantitative laws governing the growth of knowledge. Historians, then, who up to that point had limited themselves to investigating either the internal, qualitative aspects of research activity, or its social, economic, and political contexts, were now in the position to manipulate historical records, mainly journal papers and the network of citations between them, by the same methods of inquiry usually applied to physical phenomena, in such a way as to disclose the hidden patterns of their behavior.

From the counting and classification of a huge amount of data related to the history of science, including the papers collected in the *Philosophical Transactions of the Royal Society of London* and the references listed in the 1961 edition of Garfield's index, Price came to what he considered the "fundamental law of any analysis of science": whatever numerical indicator of the various sectors and aspects of modern science, from the mid-seventeenth century onward, is taken into consideration—whether the number of scientific journals, articles, abstracts, or the number of universities, scientists, or engineers—its normal growth rate is exponential, that is, it multiplies, in equal periods of time, by a constant factor. Of course, Price knew that the size of science doesn't grow exponentially throughout its escalation and doesn't exhibit the same rhythm in all disciplines. Furthermore, as befits many natural and social phenomena, it is not and cannot reasonably be infinite: at a certain moment, a saturation point is reached when the growth process levels off and the exponential trend turns into a logistical (S-shaped) one. In the saturated state of science, which Price predicted for the second half of the twentieth century, things would look very different from earlier stages because the superabundance of literature and the increasing specialization, coupled with manpower shortages and a diminishing number of talented scientists, would impose politically enlightened choices about objectives and priorities in the deployment of scientific efforts.

Turning from temporal series to the cross-sectional distribution of authors and papers at any given time, Price recognized that the universe of scientific communication is dominated by skewed distributions and followed Galton's elitist philosophy in assuming that there exist only a limited number of people with the talent necessary to become a scientist. He also sought to give a more precise estimation of such skewness through a square-root law of productivity, a sort of modified version of Lotka's Law (discussed in chapter 4). In a simplified fashion, Price's Square-Root Law asserts that about half of all the papers produced by a population of scientists come from a subset of highly productive sources approximately equal to the square root of the total number of authors.[30] This amounted to saying that, more or less, "the number of scientists doubles every ten years, but the number of noteworthy scientists only every twenty years."[31] Hence, if scientific literature grows exponentially, one is forced to conclude that noteworthy scientists are also the most productive. In other words, the advance of science does not depend linearly on the sheer number of researchers recruited at any given period, but on the number, which grows much more slowly than the former, of "good," very productive scientists.

Obviously there is a snag in the above argument, for one cannot simply infer quality (good scientists) from quantity (productive scientists). Price, in fact, admitted that quantity doesn't necessarily amount to quality at the individual level. Publishing in the *Physical Review* is not the same as publishing in the *Annual Broadsheet of the Society of Leather Tanners of Bucharest*. Even more convincingly, "Who dares to balance one paper of Einstein on relativity against even a hundred papers by John Doe, Ph.D., on the elastic constant of the various timbers (one to a paper) of the forest of Lower Basutoland?"[32] All the same, appealing to biographical compilations, such as James McKeen Cattell's *American Men of Science*, and to Wayne Dennis's study on the productivity rates of scientists listed in the National Academy of Sciences *Biographical Memoirs* (1943–1952), he made the point that highly productive scientists are also, on average, those who prevailingly show up as starred entries in dictionaries, being comparatively much more honored and rewarded than their lazier colleagues. So, despite flagrant violations, "on the whole there is, whether we like it or not, a reasonably good correlation between the eminence of a scientist and his productivity of papers."[33] Such correlation is one of the several empirical manifestations of the success-breeds-success principle, already described by Merton as the Matthew Effect, by which some sort of capital, whether material or symbolic, tends to flow and concentrate in the hands of a few who already hold a portion of it. As sections 4.1 and 4.7 show, Price managed also to demonstrate, with the help of Herbert Simon's mathematics, that this structural skewness in the formal communication system of science can be modeled by a suitable probability distribution ("Cumulative Advantage Distribution").

Beyond the number of published papers, a qualitative insight into the concrete operation of the cumulative advantage process came from citation and library usage data. Donald Urquhart's 1956 analysis of the Science Museum interlibrary loan records on one side and the citation patterns hidden in Garfield's newborn *SCI* on the other side, revealed that the distribution of requested articles among journals stored in a library and the distribution of citations among papers exhibit a similar hyperbolic pattern: a few papers and journals attract the bulk of citations and library user requests, opposed to much higher percentage of uncited and unrequested materials. In what would be subsequently recognized as one of the earliest examples of scale-free networks with power-low degree distribution, Price estimated that, in any given year, about 35 percent of all the existing papers are not cited at all, another 49 percent are cited only once, 9 percent twice, 3 percent three times, 2 percent four times, and only 1 percent six times or more, thus leading to conjecture that the number of papers cited n times decreases, for large n, as $n^{2.5}$ or $n^{3.0}$.

On top of all this, the story written between the lines of the 1961 edition of the *SCI* turned out even more instructive than that of a land ruled by true undemocracy. The statistical distribution of citations manifested, indeed, two important underlying regularities, which Price also exemplified, resorting to a citation matrix filled with data from the reference network of a self-contained spurious specialty, the N-Rays research field.

1. Half the references pointed to a loose and somewhat randomly defined portion (about half) of all the papers published in previous years, whereas

the other half referred to a smaller, tightly defined subset of earlier literature. "We may look upon this small part," commented Price, "as a sort of growing tip or epidermal layer, an active research front."[34]

2. The most cited papers were also more recent than the rest, whereas the chance of being cited for any one paper decreased exponentially by a constant factor (about two) in equal periods of time (about 13.5 years); this suggested, as a rough guess, that "about half the bibliographic references in papers represent tight links with rather recent papers, the other half representing a uniform and less tight linkage to all that has been published before."[35]

In Price's view the "immediacy factor," the marked propensity to overcite recent papers at the expense of older ones, in addition to demarcating science from other scholarly activities (i.e., the social sciences and the humanities), accounts for its cumulative and progressive character, and vouches for the existence of an active research front. In 1970, he propounded a simple diagnostic tool to detect the differing levels of immediacy characteristic of the structurally diverse modes of knowledge production occurring in the hard sciences, the soft sciences, technology, and the nonsciences: the "Price's index," i.e., the percentage of references to documents not older than five years at the time of publication of the citing sources.[36]

An early statistical analysis of the references appended to papers in the 1963–1965 volumes of the *Monthly Notices of the Royal Astronomical Society* carried out by Jack Meadows in the mid-1960s showed the concrete operation of the immediacy factor in astronomical literature. The author measured its effect by introducing an "immediacy index," defined as "the total number of citations to literature of the last six years divided by the total number of citations to literature more than twenty years old,"[37] and noticed its variability according to the subject area. Later experiments by Belver Griffith on physics journals and by Stépane Baldi and Lowell Hargens on the reference networks of special relativity and two social science subspecialties (spatial diffusion modeling and role algebra analysis) shed further light on the bibliographic dependence of the research front on past literature. Griffith conjectured that the immediacy effect might have been invented by the science communication system around the 1920s. Baldi and Hargens found, in the social science specialties, a pattern close to the one Price would probably ascribe to the humanities, so they hypothesized he could have underestimated the structural variability of reference networks across scientific domains. They also replicated the N-rays analysis and discovered that, had Price taken into account self-citations, the effect of overcitation to recent papers would have almost faded out.[38] The new technique of co-citation analysis whereby, after Small and Griffith, many authors were seeking either to clarify the specialty structure of science or to track down the contours of such a puzzling entity as Kuhn's paradigms (see chapter 5), revealed one further remarkable property of immediacy, namely its tendency to increase during the periods of intellectual focus usually coupled with scientific revolutions.[39]

As to the research front, Price conjectured it might correspond, by and large, to the work of a few hundred very productive, heavily cited scientists, and the content of a few thousand core journals. Within each field, the research front advancement is driven by an informal communication network of scholars forming a highly interactive, tight group of productive scientists who, beyond national and institutional boundaries, collaborate via face-to-face communication and preprint exchange to define problems, techniques, and solutions. Price labeled these informal clusters comprising "everybody who is really somebody in a field" "invisible colleges," after Robert Boyle's epithet for the pioneers of the Royal Society of London, and claimed that they served essentially a twofold purpose: bypassing the communication difficulties brought about by the explosion of scientific literature and bestowing on each member the share of credit derived from the esteem and recognition of his or her topmost peers.

Price's construct of the invisible college is rather ambiguous. In a footnote to *Science Since Babylon*, in fact, he cast a shadow over the desirability and even the historical necessity of these informal clusters of elite scientists, which he characterized as exclusive "power groups."[40] Moreover, having had the opportunity, in collaboration with Donald DeB. Beaver, to dissect the information flows in a tightly controlled group of people presumably constituting a single cluster in the field of "Oxidative Phosphorylation and Terminal Electron Transport," he realized that the large majority of interconnected authors contribute only minimally to the field, usually by virtue of a floating collaboration with highly productive scientists, and that separate and relatively unconnected groups seem to exist in what would otherwise appear to be a single invisible college.[41] This conclusion, taken very seriously by all subsequent advocates of the research group as the basic unit of any scientometric assessment, handed over to posterity the challenge of determining how concretely the fine structure of an invisible college could be spotted across the vast territory of science. Starting with Diana Crane's *Invisible Colleges* in 1972,[42] the contributions of authors who took up the challenge increased at such a pace that an annotated bibliography explicitly devoted to the subject appeared as early as 1983.[43] Paradoxically, while the initial emphasis was on informal communication patterns and social relations among scientists supposedly belonging to the same clique, subsequent authors dwelled primarily on the formal communication behavior embodied in the network of scientific publications and relied heavily on bibliometric techniques for testing hypotheses referred to the informal substratum.[44] Later sociologists would blame this choice as inadequate for the comprehension of the social processes and information flows that properly govern an invisible college, and would opt instead for methodologies and observation techniques typical of the ethnographic style of investigation. Yet the idea that the communicative structure of science could be made visible with the help of citation analysis inspired, in the years to come, the most fruitful research lines in the area of bibliometric mapping.

Price himself envisioned the drawing of a comprehensive bibliometric map of science as a politically desirable and technically feasible objective. He felt that, if properly manipulated, citation matrixes could favor the automatic identification of "classic" and "superclassic" papers in every research field, whereby the stratified structure of the citation network could be dug out. This, in turn, would be the first, elementary step toward the completion of a detailed topographic map of all research fields, a sort of "war map" eligible for science administration purposes. "With such a topography established—he contended—one could perhaps indicate the overlap and relative importance of journals and, indeed, of countries, authors, or individual papers by the place they occupied within the map."[45] As discussed in chapter 5, a large-scale mapping project of this type was undertaken during the 1970s by Henry Small and Belver Griffith.

NOTES

1. An early remarkable survey that pulled together the threads of previous studies covering both the informal and the formal patterns of scientific com-

munication (including bibliometric sources of evidence) is A. Jack Meadows, *Communication in Science* (London: Butterworths, 1974).

2. Both the title of this section and the discussion on Bernal's information science are indebted to Dave Muddiman, "Red Information Scientist: The Information Career of J.D. Bernal," *Journal of Documentation* 59, no. 4 (2003): 387–409; Muddiman, "Red Information Science: J.D. Bernal and the Nationalization of Scientific Information in Britain from 1930 to 1949," in *Conference on the History and Heritage of Scientific and Technological Information Systems: Proceedings of the 2002 Conference*, ed. W. Boyd Rayward and Mary E. Bowden, 258–66 (Medford, N.J.: Information Today, for the American Society of Information Science and Technology and the Chemical Heritage Foundation, 2004), www.chemheritage.org/pubs/asist2002/21-muddiman.pdf.

3. Bernal, *The Social Function of Science* (Cambridge, Mass.: MIT Press, 1967), 378.

4. Bernal, *The Social Function*, 303.

5. Bernal, *The Social Function*, 300.

6. Watson's proposal is republished as an appendix in Bernal, *The Social Function*, 449–55.

7. Bernal, *The Social Function*, 294.

8. Bernal, *The Social Function*, 306.

9. Garfield, "Citation Indexes—New Paths to Scientific Knowledge," *The Chemical Bulletin* 43, no. 4 (1956): 11, www.garfield.library.upenn.edu/papers/31.html. The awareness service set up by the ISI in 1965 was called ASCA (Automatic Subject Citation Alert). It was different from competing clipping services as it used citations, besides words, authors, and other conventional access points, to build individual interest profiles against which, each week, the newly published literature was matched to build personalized reports. In addition, the ASCA subscriber could receive tear sheets of the articles listed on the report or, alternatively, could order them by using ISI Original Article Tear Sheet (OATS) service. See Garfield and Sher, "ISI's Experiences with ASCA—a Selective Dissemination System," *Journal of Chemical Documentation* 7, no. 3 (1967): 147–53, www.garfield.library.upenn.edu/essays/v6p533y1983.pdf.

10. Bernal, "Review of the Science Citation Index," *Science Progress* 53, no. 211 (1965): 455–59, www.garfield.library.upenn.edu/essays/v5p511y1981-82.pdf.

11. The article, originally published in Merton, "Science and Technology in a Democratic Order," *Journal of Legal and Political Sociology* 1 (1942): 115–26, was subsequently reprinted in Merton, "The Normative Structure of Science," in *The Sociology of Science: Theoretical and Empirical Investigations*, 267–78 (Chicago: The University of Chicago Press, 1973), books.google.it/books?id=zPvcHuUMEMwC.

12.

> A composite cognitive and moral framework calls for the systematic use of references and citations. As with all normative constraints in society, the depth and consequential force of the moral obligation to acknowledge one's sources become most evident when the norm is violated (and the violation is publicly visible). The failure to cite the original text that one has quoted at length or drawn upon becomes socially defined as theft, as intellectual larceny or, as it is better known since at least the seventeenth century, as plagiary. Plagiary involves expropriating the one kind of private property that even the dedicated abolitionist of private productive property, Karl Marx, passionately regarded as inalienable (as witness his preface to the first edition of Capital and his further thunderings on the subject throughout that revolutionary work).

Merton, "The Matthew Effect in Science, 2: Cumulative Advantage and the Symbolism of Intellectual Property," *ISIS* 79, no. 299 (1988): 622, www.garfield.library.upenn.edu/merton/matthewii.pdf.

13. Hagstrom, *The Scientific Community* (New York: Basic Books, 1965).

14. Merton, "The Matthew Effect in Science," *Science* 159, no. 3810 (1968): 56–63, www.garfield.library.upenn.edu/merton/matthew1.pdf.

15. Zuckerman, *Scientific Elite: Nobel Laureates in the United States* (New Brunswick and London: Transaction Publishers, 1996), chapter 8. Note that the first edition of this book dates back to 1977.

16. Cole and Cole, *Social Stratification in Science* (Chicago: University of Chicago Press, 1973), 235.

17. On the definition and properties of the Matthew Effect for countries, see Manfred Bonitz, Eberhard Bruckner, and Andrea Scharnhorst, "Characteristics and Impact of the Matthew Effect for Countries," *Scientometrics* 40, no. 3 (1997): 407–22. For a retrospective résumé of the author's work on the subject, carried out since the mid-1990s mostly in collaboration with Andrea Scharnhorst and Eberhard Bruckner, see Bonitz, "Ten Years Matthew Effect for Countries," *Scientometrics* 64, no. 3 (2005): 375–79, www.viniti.ru/icsti_papers/english/Bonitz.pdf.

18. Merton, "Foreword," in Eugene Garfield, *Citation Indexing: Its Theory and Application in Science, Technology, and Humanities,* vii (Philadelphia: ISI Press, 1983).

19. Garfield, "Citation Indexes for Science: A New Dimension in Documentation Through Association of Ideas," *Science* 122, no. 3159 (1955): 109, www.garfield.library.upenn.edu/essays/v6p468y1983.pdf.

20. Small, "Cited Documents as Concept Symbols," *Social Studies of Science* 8, no. 3 (1978): 327–40.

21. Leach, *Culture and Communication: The Logic By Which Symbols Are Connected: An Introduction to the Use of Structuralist Analysis in Social Anthropology* (Cambridge, Mass.: Cambridge University Press, 1976), 10.

22. Small, "Cited Documents," 329.

23. Small, "On the Shoulders of Robert Merton: Towards a Normative Theory of Citation," *Scientometrics* 60, no. 1 (2004): 75.

24. Small, "Cited Documents," 329.

25. Small, "Cited Documents," 329.

26. See, for example, the retrospective account in Garfield, "Citation Consciousness. Interview with Eugene Garfield, Chairman Emeritus of ISI, Philadelphia, U.S.A," *Password*, no. 6 (2002): 22–25, www.garfield.library.upenn.edu/papers/passwordinterview062002.pdf.

27. Garfield, "A Unified Index to Science," in *Proceedings of the International Conference on Scientific Information* [Washington, D.C., November 16–21, 1958] (Washington, D.C.: NAS/NRC, 1959), 461, www.garfield.library.upenn.edu/essays/v2p674y1974-76.pdf.

28. Price, *Little Science, Big Science . . . and Beyond* (New York: Columbia University Press, 1986), xv–xvi. On the theoretical background and wide-ranging influence exercised by this book throughout the history of scientometrics and, more generally, on science policy studies, see Jonathan Furner, "Little Book, Big Book: Before and after Little Science, Big Science; A Review Article, Part I," *Journal of Librarianship and Information Science* 35, no. 2 (2003): 115–25, and Furner, "Little Book, Big Book: Before and after Little Science, Big Science; A Review Article, Part II," *Journal of Librarianship and Information Science* 35, no. 3 (2003): 189–201.

29. Price, *Science Since Babylon*, enlarged ed. (New Haven, Conn. and London: Yale University Press, 1975), 162. The original edition dates back to 1961.

30. Price, *Little Science*, 47. Soon after the publication of Price's Law, its relationship with Lotka's Law became a matter of intense debate, which culminated in the proof that the former follows from the latter only if a particular (but not unique) assumption is made about the largest number of papers contributed by the most prolific authors. See David Allison et al., "Lotka's Law: A Problem in Its Interpretation and Application," *Social Studies of Science* 6, no. 2 (1976): 269–76. An attempt to assess the empirical validity of the law is Paul Travis Nicholls, "Price's Square Root Law: Empirical Validity and Relation to Lotka's Law," *Information Processing & Management* 24, no. 4 (1988): 469–77. As with other bibliometric regularities, exact and more general expressions of the original form have been worked out, for example, in Wolfgang Glänzel and Andreás Schubert, "Price Distribution: An Exact Formulation of Price's Square Root

Law," *Scientometrics* 7, no. 3–6 (1985): 211–19, and Egghe and Rousseau, "A Characterization of Distributions Which Satisfy Price's Law and Consequences for the Laws of Zipf and Mandelbrot," *Journal of Information Science* 12, no. 4 (1986): 193–97.

31. Price, *Little Science*, 42.

32. Price, *Little Science*, 36.

33. Price, *Little Science*, 37.

34. Price, "Networks of Scientific Papers," *Science* 149, no. 3683 (1965): 512, www.garfield.library.upenn.edu/papers/pricenetworks1965.pdf.

35. Price, "Networks," 514.

36. Price, "Citation Measures of Hard Science, Soft Science, Technology, and Non-Science," in *Communication Among Scientists and Engineers*, ed. Carnot E. Nelson and Donald K. Pollock, 3–22 (Lexington, Mass.: Heath Lexington Books, 1970.

37. Meadows, "The Citation Characteristics of Astronomical Research Literature," *Journal of Documentation* 23, no. 1 (1967): 30.

38. Griffith, "Derek Price's Puzzles: Numerical Metaphors for the Operation of Science," *Science, Technology & Human Values* 13, nos. 3–4 (1988): 351–60; Baldi and Hargens, "Re-Examining Price's Conjectures on the Structure of Reference Networks: Results from the Special Relativity, Spatial Diffusion Modeling and Role Analysis Literatures," *Social Studies of Science* 27, no. 4 (1997): 669–87.

39. The offspring of Price's theory of immediacy is briefly reviewed in Susan E. Cozzens, "Using the Archive: Derek Price's Theory of Differences Among the Sciences," *Scientometrics* 7, nos. 3–6 (1985): 431–41.

40.

> The process of access to and egress from the groups have become difficult to understand, and the apportioning of credit for the work to any one member or his sub-team has already made it more meaningless than before to award such honors as the Nobel Prize. Are these "power groups" dangerously exclusive? Probably not, but in many ways they may turn out to be not wholly pleasant necessities of the scientific life in its new state of saturation.

Price, *Science Since Babylon*, 168.

41. Price and Beaver, "Collaboration in an Invisible College," *American Psychologist* 21, no. 11 (1966): 1011–18.

42. Crane, *Invisible Colleges: Diffusion of Knowledge in Scientific Communication* (Chicago; London: The University of Chicago Press, 1972).

43. Daryl E. Chubin, *Sociology of Sciences: An Annotated Bibliography on Invisible Colleges, 1972–1981* (New York: Garland, 1983).

44. Leah Lievrouw criticized Crane's approach and many subsequent contributions on the subject because of their tendency to describe processes (informal communication dynamics) by means of structures (bibliometric and sociometric patterns derived from published reports). See Lievrouw, "Reconciling Structure and Process in the Study of Scholarly Communication," in *Scholarly Communication and Bibliometrics*, ed. Christine L. Borgman, 59–69 (Newbury Park, Calif.: Sage, 1990). A recent attempt to reconceptualize the structure–process dialectic in light of a new model for the construct of invisible college may be found in Alesia Zuccala, "Modeling the Invisible College," *Journal of the American Society for Information Science and Technology* 57, no. 2 (2006): 152–68, individual.utoronto.ca/azuccala_web/InvisibleCollege.pdf.

45. Price, "Networks," 515.

Chapter Four

The Mathematical Foundations of Bibliometrics

Classifying and counting scientists, books, papers, and citations, as early statistical bibliographers set out to do, remained a fairly extemporary activity as long as data continued to be examined outside a mathematical framework that would let them disclose meaningful patterns in the documentation process. The turning point, or at least the prologue to a turning point, occurred between the 1920s and the 1930s, when three basic bibliometric studies were published: Lotka's work on the distribution of scientific papers among authors; Bradford's contribution on the scattering of papers on a given subject in scientific journals; and Zipf's work on the distribution of words in a text.[1] The mathematical regularities unveiled by these studies and the profusion of technical adjustments, reformulations, and syntheses of the original formulations they triggered, however, should not be regarded as mathematical foundations of bibliometrics in a sense comparable to the way Newton's laws of motion and gravitation are the foundation of classical mechanics. Whereas the latter allow computing and predicting, with the highest degree of accuracy, the motion of every terrestrial and celestial object, whether it be a falling apple or a spacecraft sent to a distant planet, the "laws" of bibliometrics enable the analyst to make sense of the overall structure of several existing datasets but do not support exact predictions about the output of specific communication processes, such as the number of articles an author will actually write before retiring, the number of journals that will publish papers on a given subject, or the number of citations that will accrue to a paper or a scientist's oeuvre over a certain time span. As discussed below, a more realistic dimension of bibliometric laws is to be placed in the world of probabilistic reasoning, where a phrase such as "the given dataset is adequately described by the function *f*' simply means that the observed distribution of empirical values and a given theoretical distribution of exact values seem to match to a plausible degree upon the completion of a suitable statistical test. But even when bibliometric distributions are interpreted in a probabilistic fashion, and an attempt is made to work out a stochastic model with predictive potentials, real predictions in the sense of forecasting probabilities are impossible. The reason is simple: given the necessity to reach a compromise between reality and mathematical simplicity, parameter estimations are based on limited observation windows, while the quantity and quality of the assumptions required for the model to work properly preclude any straightforward test of its validity on datasets produced under different conditions. Yet classic bibliometric distributions, their subsequent axiomatization, and probabilistic generalization play a pivotal role in the foundation of quantitative science studies because they provide a general framework in which the discouraging individuality of documentation processes is reduced to manageable sets of mathematical functions useful to 1) replace inexact empirical formulations with exact mathematical concepts so as to enhance the mutual transparency and comparability of competing models (just a first humble step toward a yet-to-be-developed "grand bibliometric theory"); 2) specify the conditions of applicability of standard statistical tools to the analysis of specific datasets, thereby helping estimate random errors in the measurement of information flows; and 3) connect the mathematical structure of bibliometric processes with that of extra-bibliometric phenomena, such as the patterns emerging in the study of economically and biologically complex systems, so as to help clarify problems having common characteristics and promote the development of common methodologies for their resolution.

4.1. THE MATHEMATICS OF SKEWNESS: A FEW "QUALITATIVE" INSIGHTS

Lotka, Bradford, and Zipf used simple mathematical statements and graphical devices to express the empirical relation between sources and the items they produce in three areas: authors producing papers, journals producing papers on a given subject, and texts producing words with a given frequency. Their common denominator is a striking inequality in the pattern of the information processes under observation: a few authors turn out to be responsible for the largest portion of scientific literature in a given research field; a few scientific journals seem to concentrate the literature required to satisfy their needs; and a relatively small number of recurrent word units govern their (and not only their) linguistic habits. In more abstract, albeit crude terms, what those regularities assert is that, unlike most natural phenomena, as far as the information processes discussed above are concerned, no average productivity value is more likely than any other to have the remaining values neatly distributed around it, but many low-productive sources tend to coexist with few highly productive ones, so the overall source-item frequency distribution is markedly skewed, conforming to a hyperbolic pattern conveniently described by a power law.

Lotka's, Bradford's, and Zipf's statements are usually referred to as "bibliometric laws," somewhat of a misnomer if intended to mimic the universal validity and explanatory power of natural laws in the physical sciences. It is a fairly acceptable label if circumscribed, in Maurice Kendall's terms, to "a pattern of a human aggregate which is observable, reproducible and, as a rule, quantifiable; perhaps only descriptive in character, perhaps explainable in terms of a model, but in any case related to observation."[2] An occasional source of confusion has derived from these "laws" being formalized as either a size-frequency or a rank-frequency function. In the former case, the relation is directly established between the number of sources and the number of items they produce. By way of illustration, given a list of authors with the corresponding number of authored publications, the number $f(n)$ of authors who have produced exactly n publications can be tabulated for $n = 1, 2, \ldots, n$, thus forming a size-frequency distribution of productivity. In the latter case, first sources are ranked in decreasing order of productivity, then the relation is established between the rank and the number of items produced by the source at that rank. If the authors of the previous example are ranked according to the numbers of publications they have written, and if $r = 1, 2, \ldots, n$, are the ranks of, respectively, the first most productive, the second most productive, through the *nth* most productive author, then the number $g(r)$ of publications issued by the author with rank r can be tabulated for increasing values of r, thus forming a rank-frequency distribution. Often cumulative distributions are used in both situations; this means that we are interested either in the number of authors having published at least n

contributions (size-frequency) or in the number of articles published by the *r* most productive authors (rank-frequency). Under specific assumptions, the size- and rank-frequency formulations may be proved to be fully equivalent, both leading to a similar highly skewed pattern in the distribution of items among sources: a few sources are expected to be responsible for the production of a large percentage of the items and to coexist with a much higher number of low-productive sources.

In a typical bibliometric dataset, if productivity data of sources are plotted on a linear scale, with the number of items *n* on the *x* axis and the number of sources producing *n* items on the *y* axis, and if one seeks to work out a function that closely fits the distribution of data points as obtained by sampling, then the basic graph structure yields a hyperbolic or J-reverse or power law function, with an unduly long tail of scattered values falling toward regions of high concentration (see figure 4.1).

If logarithmic scales are used for both *x* and *y* axes, then the basic graph structure for the same dataset is a straight line with negative slope (see figure 4.2).

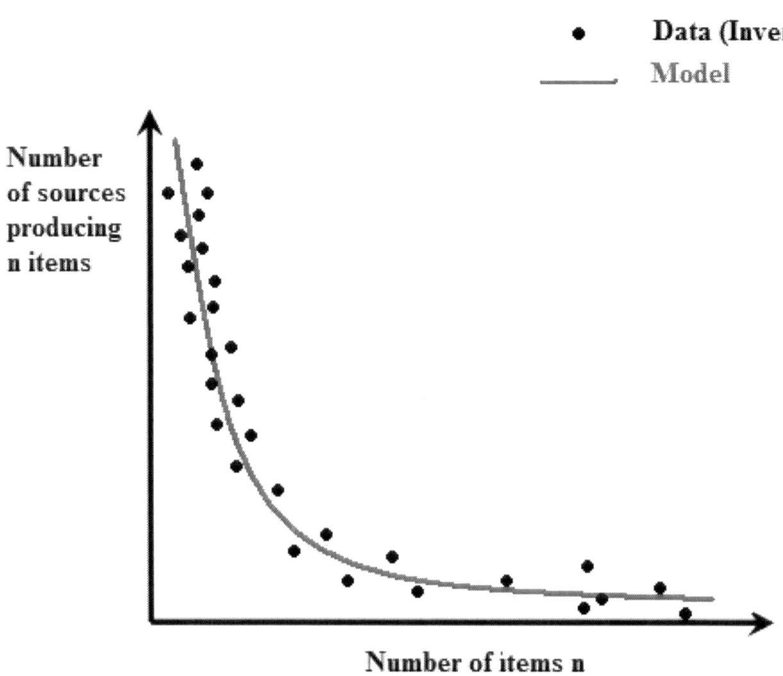

Figure 4.1. Basic hyperbolic size-frequency distribution

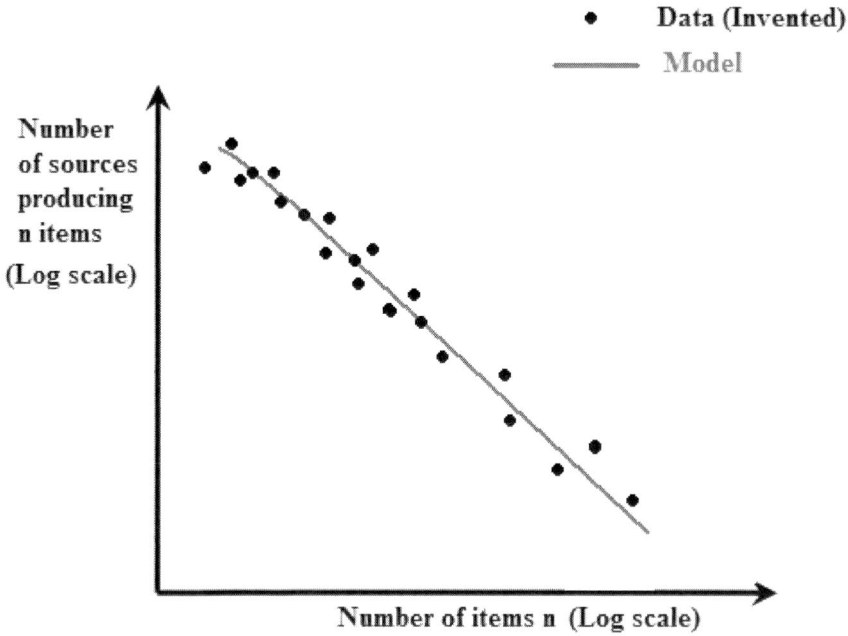

Figure 4.2. Basic hyperbolic size-frequency distribution plotted on a double logarithmic scale

It should be borne in mind that, even though these oversimplified graphical patterns are recurrent, if not ubiquitous, in bibliometric literature, in Bradford's/Lotka's/Zipf's original formulations they were meant to describe not the whole range of the distribution but only the behavior of the most productive sources, which, in statistical jargon, is usually referred to as the "tail" of the frequency distribution. The reader should also be aware that drawing a curve through data points referred to empirical measurements and seeking for its analytical expression is by itself a complex operation, which supersedes actual experience by mathematical function but still belongs to the world of the "more or less" if it is performed by sight or with the help of simple graphical procedures. As such, it is a questionable operation leaving considerable margins of uncertainty about the validity of the final outcome. So, for example, the simple fact that data points arrange themselves in a curve analytically expressed as $y = x$ doesn't imply an underlying linear law. Though useful in conveying the basic form of a dataset, in fact, simple mathematical functions and graphs are not suitable for predictive purposes because they do not account for the fact that real-world information processes are complex systems evolving under the action of causes whose effects can be tackled only within a broader mathematical framework. Therefore, if one wishes to capture the mathematical rationale of empirical formulations, it is necessary to prove that they can be formally derived from a more general theory on the structure of information processes. Such a theory, in turn, can take either a deterministic or a probabilistic form. An important exemplification of a deterministic approach is provided in section 4.7 in connection with Egghe's Lotkaian informetrics. For now, let's concentrate on the approach that, much in tune with the vocation of contemporary physical sciences, considers information production processes as stochastic processes and, accordingly, empirical bibliometric laws as statements about probability distributions or expected values.

A "stochastic process" is a sequence of events governed by probabilistic laws. Operationally, what distinguishes stochastic processes from other representations of real-world phenomena is their behavior being conceptualized by means of random variables, that is, quantities that can take one of a finite or infinite set of values. In the simplest possible cases, a stochastic process amounts either to a sequence of random variables (a "time series") independent of each other or to a sequence in which the dependency of the successive events goes back just one unit in time, so that the future behavior of the process depends only on the present state and is not influenced by its past history (in statistical jargon this is the "Markovian property"). An appropriate description of a stochastic process requires the knowledge of the probability distributions of the random variables at stake. Roughly speaking, a probability distribution is the specification of all the possible numerical values the variable can take over an interval, along with the precise indication of the probability of occurrence associated with each value. The nice thing about being able to fit a theoretical probability distribution to an empirical frequency distribution is that one can summarize the entire range of values by a few parameters, the most important of which are the center or expected value (also called the mean) and the dispersion of the remaining values around the center (the variance).

When interpreted in probabilistic terms, a classic informetric law asserts that the number or density of sources $f(x)$ producing an expected number of items x is proportional to some function of x. Such a shift doesn't remove the aforementioned uncertainty because different probability distributions may account for the structure of the same cloud of data points referred to an empirical frequency distribution of some sort. Many physicists, for example, have followed Derek Price's lead in trying to model the distribution of bibliographic citations disclosed by ISI databases; most have come to the same conclusion that the asymptotic tail of the distribution appears well described by a power law, but alternative, equally acceptable solutions featuring an exponential function have been occasionally suggested.[3]

The shift to probabilistic reasoning, therefore, does not get rid of uncertainty. Still, under specific assumptions, it allows exerting a certain control over uncertainty. Probability distributions are indeed the key to inferential statistics, a complex of techniques intended to draw general conclusions relative to the population under study on the basis of one or more samples derived from it. Specifically, provided the samples are representative of the population in that they are "truly random" samples, a vast array of techniques is available that allows estimating, with a computable degree of accuracy (or margin of error), the main parameters of the population, for instance the mean and the variance of some critical properties, starting from the corresponding sample statistics. Here probability distributions provide the rules against which statistical hypotheses referred to the structure of a specific dataset are tested so as to check whether they disclose significant patterns of the population of origin or rather point to chance effects.

Over the last three centuries, the mathematical description of a huge number of natural and social systems has come to hinge on the nice properties of the normal or Gaussian probability distribution, which was first derived, as a limit to the binomial (or Bernoulli) distribution, in the 1730s by Abraham de Moivre and later perfected by Carl Friedrich Gauss and Pierre Simon Laplace. It is the ubiquitous, well-understood, and easily computable bell-shaped curve, in which an average value (or mean) more probable than others is placed just at the center, the remaining values being arranged on both sides in a symmetrical fashion. In statistical textbooks it is also labeled the "law of error" because it lies beneath the assumption that, each time several measurements of a physical quantity are performed, their arithmetic mean is better qualified to express the true value than any other.

Throughout the first half of the nineteenth century, the belief that every natural and social system could be ultimately reduced to a probability structure governed by the normal distribution ruled the roost. The normal curve became the pillar of inferential statistics, because the very possibility of employing many standard statistical techniques in learning from experience is intrinsically related to quantities that, at least approximately, exhibit a normal distribution. This, in turn, is a surprisingly robust effect of one of the most important theorems in the history of mathematics, the central limit theorem, which ensures that, as the size of a sample increases, whatever the distribution of the (independent) random variables at stake, their sum, provided that their variance is finite, is a random variable with an approximately normal distribution. Two classic statistical techniques taking advantage of the normal paradigm are correlation and regression. Since, in one form or another, both techniques are frequently encountered in bibliometric literature, it is worthwhile to discuss them here.

In statistical parlance, two variables are said to be correlated if some kind of (direct or inverse) linear relationship between them does exist; that is to say, as one variable gets larger, the other gets larger (or smaller) in direct proportion. Graphically, the strength of the relationship between the variables depends on how closely their covariation approximates a straight line when data are plotted on Cartesian axes. Numerically, the varying levels of "closeness to a line" are captured by a correlation coefficient, which can take many forms according to whether the variables are assumed to follow a particular distribution or not. In the case of continuous measures, for instance, a popular choice is the Pearson's r, a dimensionless quantity ranging from -1 (perfect negative correlation) to $+1$ (perfect positive correlation), whose statistical significance can be determined insofar as both variables are random samples from a normal distribution (hardly the case in bibliometric experiments). "Correlation does not imply causation" is nearly a mantra of elementary statistical manuals: if two variables are positively correlated, it might well be the case that neither of them is the cause of the variation in the other variable, and a third, unknown factor is responsible for their joint behavior. All the same, even if not a sufficient condition for causation, correlation is often interpreted as a hint of the existence of a causal relationship, above all if some sort of independent evidence to the same effect is available. Many validation studies of citation analysis in research evaluation, as well as many studies arguing for the beneficial effects of the free online availability of documents on their later citation scores, for example, rely on statistical correlation techniques for demonstrating the positive linear relationship between the first factor, peer ratings and open access, respectively, on citation counts, so as to suggest a possible causal relationship between them.

The limited predictive capability of correlation is partially overcome by another technique very popular in social and behavioral sciences: regression analysis. In the simplest case, it takes the form of linear regression between two variables, where the behavior of one dependent variable is examined in relation to that of one independent variable ("predictor" or "explanatory variable") supposedly "causing" the former under the assumption that such relationship is linear. The analysis is designed to derive an equation (or model) for the line that best fits the data points expressing the relationship between the dependent and independent variable. The equation has the basic form

$$Y = a + bX + e$$

where Y is the dependent variable (the outcome to be explained), X is the independent variable (a factor that "explains" the outcome), b is the regression coefficient to be estimated from empirical data, a is a constant chosen so as to make the overall prediction error as small as possible, and e represents the error that still remains after predicting Y. In real-life situations, there are many distinct explanatory variables whose linear combination is likely to produce a given outcome; hence, it makes much more sense to use multiple instead of simple linear regression. In multiple regression, one seeks to establish whether a set of two or more independent variables explains a proportion of the variance in a dependent variable at a significant level. Such an explanation also permits, under specific assumptions, ranking the relative importance of each predictor. In the case of two independent variables, the multiple regression model becomes

$$Y = a + bX + cZ + e$$

where c is a further regression coefficient to be estimated from data. Many estimation techniques are available for the purpose, the most common being those based on the least squares principle. As in the case of correlation, however, the normal distribution is still behind the scenes. Indeed, when hypotheses based on the parameter estimates have to be tested, the most easily manageable forms of regression analyses, that is, the parametric ones, assume, among other things, that for any given value of the independent variable, the values of the dependent variable are normally distributed about some mean. A classic use of multiple regression in bibliometrics occurs in those studies, which are reviewed in subsection 7.1.3, where an effort is made to estimate the relative importance of different factors, such as author reputation, journal impact, number of references, and so on, in predicting the citation scores of a given set of papers.

A dramatic shift in the probabilistic structure of human knowledge occurred when, in the aftermath of the biometric revolution started in England during the mid-1860s, statisticians realized that normality is nothing but a myth, because a surprising number of biological and social datasets are inherently skewed, with variability playing an even greater role than had been assumed. In a skewed dataset, variance is significantly greater than the mean, and the mean is nowhere near the center of the distribution, so it cannot reasonably be taken as representative of the whole set. In the early phase of the revolution, spurred by the application of statistical tools to Darwin's theory of evolution, the work of such pioneering figures of modern inferential statistics as Francis Galton, Walter Weldon, Karl Pearson, William Gosset (best known by his pen name "Student"), and Udny Yule led to the overthrow of what Stephen Bensman has called the "normal paradigm" and to the introduction of fresh mathematical tools and concepts necessary for coping with the oddities of a skewed universe.[4] Data with heavy "probability tails" were increasingly collected in fields as diverse as economics, linguistics, sociology, information science, telecommunications, and physics of condensed matter. Thus the awareness also gained ground outside biometrics that normal and non-normal probability distributions conceal basic and irreducible differences in the structure of the events, the former being adequate to represent phenomena derived from the aggregation of many random, independent patterns of behavior, and the latter being more consonant with processes whose randomness is constrained, as it were, by the appearance of extraordinary, out of the norm values, which maximize variability to the point of suggesting the existence of some substantial causal processes at work. What Lotka, Bradford, and Zipf did, in this connection, is to definitely make room, also in the information science arena, for distributions completely devoid of normality or, if one prefers, whose "normality" consists exactly of the coexistence of many small but highly probable values with a core of a few large but improbable ones. As a result, skewed patterns issued an unprecedented challenge to social and information scientists because they compromised the descriptive and inferential utility of basic statistics, such as arithmetic mean and variance, and jeopardized the reliability of standard statistical tests of significance, or at least of those working well under the assumption of normality.

To get a broad idea of what structural skewness means in practical situations, let's examine the average citation rates for two journals computed over the same period of time—what in chapter 6 will be explained as the impact factor—and, supposing the numerical difference between those averages to be great, let's wonder whether such difference conveys a statistically significant picture of individual articles' quality (in terms of citation impact). In elementary probabilistic terms the answer would be definitely no; indeed, given the well-known skewness in the citation distribution of individual articles (i.e., there will be few heavily cited papers and many papers scoring zero or nearly zero in both journals), the probability that a randomly selected paper from the first journal has at least as many citations as a randomly selected paper from the second journal will be higher than intuitively expected on the basis of (journal) average scores.

Now let's switch to another staple of classic inferential reasoning—the concept of random sample—and think of the curve in figure 4.1 as a frequency distribution of papers (items) produced by a hypothetical cohort of scientific authors, say at the aggregate level of world universities (sources): the head of the distribution, where many small similar values typically occur, includes the less-productive universities, which form the bulk of the entire population; the long tail contains the few most productive universities whose productivity rates, unlike those in the head, are scattered in a rather "fuzzy" fashion, with rare ties and many gaps between successive frequency values. In this scenario, a statistical analysis aimed at drawing general conclusions on the basis of sample productivity data would encounter serious difficulties from the very beginning because, although the construction of randomly sampled sets is a precondition for many conventional statistical procedures to work properly, any sampling procedure which, in the name of randomness, failed to include the few highly productive sources, such as Harvard, Cambridge, Stanford, or MIT, would inevitably result in incorrect generalizations.

Turning to a real-life, historical case, social scientists' disappointment in the mathematical intractability of hyperbolic datasets is well exemplified, in the area of economic studies, by the several attempts to find a rationale for the distribution of wealth originally captured in a formula by the Italian economist Vilfredo Pareto and later popularized as the 80/20 rule of thumb (about 80 percent of the income goes to only 20 percent of the population). Pareto's Law has long puzzled economists because of both its failure to adequately model the middle range of incomes and its apparent lack of justification in the existing theoretical framework. For Benoît Mandelbrot, this inadequacy was tantamount to the admission that classic statistical thinking had very little to offer the social sciences, to the point that, as early as the late 1950s, he suggested a possible way out of the impasse, drawing economists' attention to the explanatory potential of stable non-Gaussian probability distributions, a solution that required the counterintuitive notion of infinite variance and, in some cases, an infinite mean to be metabolized.[5] In the mainstream, however, faith in the central limit theorem and the distributional robustness of many standard tests enabled social scientists as well as bibliometricians to take up the challenge on the firm ground of classic probability theory.

The "bipolar" nature of several datasets encountered in social sciences seems even to suggest that, although the head of the empirical frequency distribution can be processed by standard statistical methods, the tail should undergo a different treatment, possibly within a framework more resilient to the occurrence of unusual and rare events (in statistical jargon, "extreme values"). Extremes are treated as "outliers" in many standard statistical procedures. Accordingly, they are usually ignored and easily removed by ad hoc statistical software functions so as to prevent the Gaussian scaffolding from falling apart. Nevertheless, as the uniqueness of many socially relevant phenomena came to the foreground, statisticians started developing analytical tools to make outliers theoretically relevant and partly predictable, in the hope of helping to estimate the risk of such unusual events as stock market crashes, wars, and natural disasters. One of these tools, actually a special case of extreme value statistics, is the statistics of exceedances, implemented by John Huber for modeling the upper tail of the continuous (theoretical) distributions of the rate of publication and career duration in true cohorts of scientific authors and inventors.[6]

Taken in their entirety, skewed datasets are usually made more tractable by a series of techniques that either point to enhanced symmetry in the whole distribution or bypass the assumptions of normality outright. At the most basic level, skewed data may undergo a simple transformation, such as the popular logarithmic transformation, which stabilizes the variance of the sample, allowing approximately standard normal variables to be managed. Far from being neutral, though, this transformation can significantly affect the outcome of descriptive statistics when applied in information science for classification purposes.[7]

At a deeper layer, skewness is tamed by the adoption of nonparametric statistical models and techniques. Such devices pursue a higher degree of robustness by keeping the number of underlying assumptions as weak as possible, above all by avoiding the pretense that data are drawn from the normal or any other probability distribution. In the computation of statistical correlation, for example, when the scale of measurement of the two variables is an ordinal, as is the case with ranked units for which no clear numerical interpretation is possible, then a rank correlation coefficient is often employed, such as the Spearman's rank correlation or Kendall's τ (tau). Likewise, if non-"metric" variables are at stake and there is reason to believe that the dependent variable y cannot result from a simple linear combination of the predictors, then more flexible regression models are conveniently adopted, drawing upon the family of the generalized linear models (such as logistic regression for binary outcomes, Poisson and binomial regression for modeling counts) or the family of the nonlinear regression models. Finally, if the dependence of the outcome variable on the predictors cannot be traced in advance to any conceivable function, whether linear or nonlinear, then nonparametric regression models become a suitable choice.

At the highest level of mathematical sophistication, skewness becomes a theoretically predictable effect within the framework of full-fledged stochastic models of information production processes. A stochastic model is a mathematical device situated on a higher plane of abstraction than the probability distribution, in the sense that, in very rough terms, it features a probabilistic mechanism from which the joint probability distribution of the outcomes of a particular event at any point in time can be estimated. In many such models, the explanatory mechanism is a mixture of simple counting processes, such as the Poisson process, but a popular alternative consists of the stochastic "birth-death process," such as the one embedded in the urn scheme used by Derek Price to formalize his cumulative advantage principle. Both the Poisson and stochastic birth-death processes are special cases of continuous-time Markov processes, and both succeed, under specific constraints, in explaining (mathematically) the large inequalities of output observable in information systems, for example, scientific bibliographies: a few authors publishing a lot and being proportionally much more highly cited than the rest.

Poisson Models

A stochastic process is a counting process if it represents the total number of events that occurred up to a certain time *t*. For example, the number of citations accruing to a paper at or prior to time *t* is a counting process in which an event corresponds to the paper being cited. Hence, in stochastic counting processes the mechanism by which a source produces items consists of a sequence of events occurring at random in time. If the probability of occurrence is constant (probabilistic homogeneity), and the number of events occurring in a time interval is independent of the number of events occurring in any other disjointed period of time ("memoryless" or "Markov property"), then the process is a Poisson process and the number of events between time t_1 and time t_2 follows the Poisson distribution. But if the above restrictions are released, as would seem more appropriate for real-life situations characterized by marked differences in productivity rates over time, then the overall process is better construed as a "mixture" of the individual counting processes, and it seems quite natural to model it by "mixing" the parameters related to individual sources' productivity rates. In so doing, more complex distributions than the simple Poisson are called for, which arise, under particular assumptions, as mixtures of Poisson distributions. The most popular of these in bibliometrics are the negative binomial distribution, the Waring distribution, and the generalized inverse-Gaussian-Poisson distribution (GIGPD). Mixtures of Poissons, it is worth noting, did not originate in bibliometrics. They had already been investigated during the 1920s, in connection with accident data. Prior to their investiture with general informetric powers, announced in Brookes's seminal papers on the mathematical explanation of Bradford's Law and the Law of Anomalous Numbers, they slipped into the sphere of information studies through the early stochastic models of library circulation patterns put forth by Philip Morse in the 1970s and, later on, by Burrell and Cane, who in 1982 proposed a simple stochastic model based on a gamma mixture of Poisson processes to obtain a reasonable fit for the circulation data of books in a variety of university libraries. Burrell and Cane also used the model to test the 80/20 rule for library collections, according to which 80 percent of the circulations are accounted for by about 20 percent of the holdings. Their conclusion that, unless one is examining single-year circulation data, that rule doesn't seem to be valid for longer time periods, shed more light on the dynamic nature of information processes and prefigured an increasing concern about their time dependence.[8]

Success-Breeds-Success

A natural generalization of the Poisson process consists of letting the chance for an event to occur at a given instant of time be dependent on the number of events that have already occurred. This is called "stochastic birth" and is perfectly exemplified by the reproduction of living organisms (hence the name), in which, under proper conditions of sustainability, the probability of a birth at a given time is proportional to the population size at that time. A particular type of stochastic birth process also underpins Derek Price's conceptualization of the success-breeds-success principle on the pattern of George Pólya's and Florian Eggenberger's urn model.[9] Its operation can be easily visualized through the same example put forth by Price. Imagine an urn containing two balls, one red ball (signifying a success) and one black ball (signifying a failure). At regular intervals, the players draw a ball randomly. Of course, the initial probability for a red-success ball to be drawn is exactly the same as for the black-failure ball and, if the composition of the urn remains fixed—i.e., the ball is replaced after each drawing—the odds do not vary for the different players. But if, from the first round on, the rules are changed in such a way that the player drawing a red-success ball, after replacing it, adds one more red-success ball to the urn and gets to continue in the game, while the player drawing a black-failure ball is banned from the game, then obviously the odds get progressively better for the first player with each succeeding round. After the first round, for example, in an ideal situation where the experiment is repeated an infinite number of times, half the players would withdraw, having drawn a black-failure ball; the other half would continue putting a new red-success ball in the urn and drawing again. Because their urn now contains two red-success balls and one black-failure ball, the probability of drawing a red-success ball increases for them from 1/2 to 2/3. If the player is an author, and the prize of admission to the game, as well as the prize awarded each time a red-success ball is drawn, is a scholarly publication, then a simple mathematical formula can be worked out that is capable of approximating quite well the skewed frequency distribution of productivity formally described by Lotka in 1926 and later observed in countless disciplinary bibliographies. Price showed how such a formula could be derived from the Simon/Yule probability distribution, which he christened the "Cumulative Advantage Distribution" (CAD).

Both kinds of stochastic arguments—counting as well as birth processes—have been successfully applied to model inequality in science, and what makes particular probability distributions, such as the negative binomial, so appealing to bibliometricians is exactly the possibility that they can be derived either way. One cannot say, of course, which process is "really" at work behind the observed pattern when data string out in a negative binomial fashion. One cannot say, for example, whether a paper is being more and more cited because of the intrinsic quality of its content or due to the social standing and increased visibility of its author. Yet, despite the neutrality of stochastic models in relation to possible causal explanations, it has been tempting to speculate on the sociological meaning of the mathematical structures revealed by means of those arguments. As noted elsewhere, for instance, the process leading some sources to produce disproportionate numbers of items at the expense of steadily low-productive sources is formally analogous to the Matthew Effect investigated by Merton, the Coles, and Zuckerman: greater increments of recognition and greater bibliographic visibility, along with easier access to research facilities (funds, costly instrumentation, stimulating colleagues, and so on), accrue to scientists already enjoying a good reputation in a sort of spiraling escalation of rewards, bringing about an ever-increasing dispersion of productivity. In a somewhat opposite way, if a Poisson distribution is found to provide a reasonable fit to the rate of production of scientific papers across authors' careers in a given area, then a plausible interpretation of the discrepancy between random fluctuations in individual publication records and sharp inequality in the overall distribution of productivity could be the admission of a structural heterogeneity: some authors are "naturally" endowed with higher (mental, scientific, creative) abilities than others; hence, inequality is largely determined by the unfolding of this "sacred spark" throughout individual careers' duration; in other words, it's a matter of destiny.

Of course, destiny and cumulative advantage are not mutually exclusive, and extant evidence in support of either hypothesis is hardly conclusive. During the 1980s, for example, Paul Allison and colleagues found evidence for increasing inequality in publication counts of a cohort of scientists as it grows older, but somewhat surprisingly, the same cumulative advantage didn't show up in citation counts.[10] Later on, having tested a new method for measuring scientific productivity throughout individual authors' career duration across samples drawn from a wide range of fields of human creative performance (physicists, mathematical logicians, inventors, and composers), Huber concluded that—contrary to any cumulative advantage postulate—authors tend to publish at a constant rate, with random fluctuations that fit the Poisson distribution, while both career duration and rate of production are approximately exponentially distributed within each group. On account of these findings, creativity and talent, as captured by crude bibliometric measures of productivity, revealed a random time pattern scarcely affected by individual and environmental factors, but this, as Huber himself admitted, "is a minor mystery, compared to the mystery of how creativity is first manifest."[11]

4.2. LOTKA'S LAW

Outside the field of bibliometrics, Alfred Lotka is celebrated for having conceived, simultaneously with the Italian scientist Vito Volterra, though independently, a mathematical model of animal population dynamics explaining the fluctuations in the predator–prey ratio. Turning to science studies, he sought to determine to what degree scientists of different caliber, involved in the struggle for life on the forefront of scientific communication, contributed to the advancement of knowledge. Lotka addressed the task simply, classifying, by number of authored publications, the scientists indexed in the *Chemical Abstracts* (1907–1916) and in Auerbach's *Geschichtstafeln der Physik* (covering the whole history up to 1900), the count being limited to senior authors in the ambiguous case of papers written in collaboration by several scientists. Having plotted "the frequencies of persons having made 1, 2, 3, . . . contributions, against these numbers 1, 2, 3, . . . of contributions" on a double logarithmic scale, he realized that the data points representing the two variables in the sample appeared "rather closely scattered about an essentially straight line having a slope of approximately two to one."[12] Then he utilized the least-square method to calculate the slope of the best-fitting line and concluded with a statement corresponding to the original formulation of Lotka's Law: the number of authors producing n contributions is approximately equal to $1/n^2$ of the number of authors that produce only one contribution. In simplest terms, this amounts to saying that, if sixty authors out of one hundred in a subject area produce only one paper, then fifteen out of one hundred will produce two papers, seven out of one hundred three papers, and so on. The gift of scientific productivity, thus, is not equally divided among colleagues; there are a few very productive scientists outnumbered by the far greater number of researchers with low productivity rates. Subsequent empirical tests showed that the exponent at the denominator—the "Lotka's exponent"—is not necessarily 2 but ranges approximately from the original Lotkaian value of 2 to 4, so if one replaces its actual value by the generic symbol a, the more general form for Lotka's Law is

$$p(n) = \frac{C}{n^a}$$

where p is the number of authors that produce n contributions, and C is a constant characteristic of the research field. Incidentally, it should be noted that the scientist's mind does not inspect the world in watertight compartments: the Lotka-Volterra model for the predator–prey ratio mentioned above doesn't belong to a universe other than that governed by the inverse power law at the level of scientific productivity. It has been demonstrated, indeed, that biological, demographic, and economic systems of the Lotka-Volterra type entail power law relationships at the individual level, for example in the relative individual wealth distribution and in market returns.[13] Lotka himself, though, knew that power laws lurk beneath many natural and social phenomena without telling the whole story about them; these types of frequency distributions, he admitted, "have a wide range of applicability to a variety of phenomena, and the mere form of such a distribution throws little or no light on the underlying physical relations."[14]

Lotka's work went unnoticed for many years. Then, beginning in the 1970s, as the increased availability of journal cumulative indexes and standardized catalog entries facilitated publication counts, it met with a huge number of applications, refinements, criticisms, and alleged confirmations. Contributions on the subject piled up at such a pace that, as early as 1978, a bibliography of 437 works on Lotka's Law and related statistical regularities appeared in the first issue of the journal *Scientometrics*.[15] Ever since, the convergence of productivity patterns with Lotka's formula has been advocated in as diverse areas as library and information science (including bibliometrics), computer science, semiconductor and microcomputer research, medicine, biochemistry, theoretical population genetics, entomology, econometrics, patent literature, library card catalog entries, and web hyperlink distributions.

As usual when something works without knowing exactly why it works, many authors took for granted the alleged evidence flowing from previous studies without paying attention to the details of the data-collection methods and the appropriateness of the statistical techniques surrounding its production. Yet, also thanks to the keen collection of Lotka-compliant bibliographies done by the Czech information scientist Jan Vlachý throughout the 1970s, it didn't take long to realize that the original inverse square law inadequately represented the actual distribution of productivity in most of the reported datasets, above all in the lowest and highest ranges of productivity, that is, in the heads and tails of the empirical curves. Many previously published tests turned out to be methodologically flawed and, even worse, a similar drawback was traced back to the source itself by William Potter in a 1981 review paper, where the application of a goodness-of-fit test to Lotka's original data showed that the inverse square formula fitted only a portion of them.[16] That's why subsequent authors felt it necessary to establish a sort of protocol or standard (and replicable) testing procedure that would permit meaningful statistical comparisons between different tests of Lotka's Law by explicitly defining the minimal requirements for each step involved in any validation exercise: 1) specification of the model; 2) data collection (i.e., measurement and tabulation of the variables); 3) estimation of the unknown parameters in the model equation (the constant C and the slope a in the original Lotka's formula); and 4) testing of the conformity of the observed data to the theoretical distribution by means of a goodness-of-fit test.[17]

Unfortunately, none of the foregoing steps is constrained in one single direction, but different choices are consistent with the formal correctness of the entire validation procedure; as discussed in section 4.7, different Lotka-type models can be put to test beyond the original one and, as documented in section 7.3, different ways to allocate credit to authors are equally plausible (counting senior authors only, all coauthors, or fractional counting), which bear strongly upon the resulting distribution. Likewise, a number of alternative criteria are available both for estimating the unknown parameters (the method of moments, the linear least squares, the maximum likelihood, etc.) and for goodness-of-fit testing (chi-square, Kolmogorov-Smirnov test, etc.). Even more important, the fourth and decisive step of the procedure, namely testing whether the sampled population is distributed in a specific way by means of a statistical hypothesis test, pivots on the usual assumption that data constitute a random sample drawn from a well-defined larger population, which is not actually the case for most bibliometric studies of author productivity, where both randomness and a clear specification of the population of sources are quite impossible to achieve. Consequently, some have opted for an alternative modeling approach that starts with empirical data, not hypotheses, then draws simple generalizations from peculiar aspects of the data, finally introducing and putting to test simple mechanisms and more elaborate theories to explain those generalizations.[18]

Notwithstanding the foregoing drawbacks, the observation that scientists' productivity patterns tend to follow "some kind" of Lotka's Law in a wide range of contexts, no matter the peculiar social conditions and the time span under consideration, led to the supposition that its underlying mechanism could be derived from more fundamental mathematical principles. The encounter with other critical manifestations of the inverse power law eventually confirmed this supposition. In the aftermath of the encounter, nonetheless, even the assumption that the accuracy of Lotka's

Law in fitting empirical datasets can be established by means of classic statistical methods would be called into question.

4.3. BRADFORD'S LAW OF SCATTERING

Samuel Bradford, chief librarian at the London Science Museum, shared with Paul Otlet and Henry La Fontaine the ideal of a collective catalog of all human knowledge enabling scientists to achieve universal bibliographic control over technical and scientific documentation. A major obstacle to the project, in his opinion, was the inability of contemporary abstracting and indexing services to manage the "documentary chaos" by enclosing the available literature in a homogeneous grid of classified subject headings. Fragmentation and lack of standards resulted in poor-quality information services and duplication of efforts: "Examination of these abstracts—he complained—showed that they related to only about two hundred and eighty thousand different papers, each paper being abstracted on an average about 2.7 times. This means that less then half the useful scientific papers published are abstracted in the abstracting periodicals and more than half the useful discoveries and inventions are recorded, only to lie useless and unnoticed on the library shelves."[19]

The dissatisfaction with poor-quality indexing prompted Bradford to the task of determining its possible causes, one suitable candidate being the way articles on a given subject are scattered among scientific journals. The analysis and classification of entries in two geophysics bibliographies, the *Current Bibliography of Applied Geophysics* (1928–1931) and the *Quarterly Bibliography of Lubrication* (1931–1933), led him to the following observation: if journals carrying articles relevant to a given subject are ranked in decreasing order of productivity, and the number of papers contributed by each of them is computed, the result will be a core or nucleus of a few journals accounting for most of the articles on that subject, followed by other groups containing the same number of articles as the nucleus, but spread over an ever-increasing number of journals. In the applied geophysics bibliography, for example, Bradford found that

1. the first group of 9 titles contributed altogether 429 articles;
2. the next group of 59 journals contributed altogether 499 articles; and
3. the last group of 258 titles contributed altogether 404 references.

Each group, then, produced fairly similar proportions of articles, though by no means exactly the same number. The number of journals required to obtain such proportions could be roughly expressed as

1. 9 titles in the core of most productive sources;
2. 9×5 titles in the second group (equals 45 titles, a fairly acceptable approximation to 59); and
3. $9 \times 5 \times 5 = 9 \times 5^2$ titles in the third group (equals 225 titles, a fairly acceptable approximation to 258).

Thus, the ratios of title groups contributing roughly the same number of articles in the nucleus and succeeding zones could be written as

$$9 : 9 \times 5 : 9 \times 5^2$$

In the general case—i.e., for numbers of titles in the core different from 9 and for multipliers different from 5—if we divide the entire expression by 9 and replace 5 with a variable m without necessarily restricting the number of zones to three, the result is

$$1 : m : m^2 : \ldots$$

Here m is the "Bradford multiplier" and depends on the specific collection of journals; though admitting it is "by no means constant," Bradford hypothesized it could approximate a constant, at least for the larger groups.

The above sequence of ratios amounts to saying that the bulk of articles on a given topic is concentrated in a small set of core journals and then scattered across other journals to such a degree that, if the set of relevant articles is subdivided in groups or zones containing the same number of items as the core, an exponentially increasing number of journals will be required to fill the succeeding zones. This simple theoretical statement is the "verbal" version of Bradford's Law. It specifies a geometric progression in the cumulative number of source journals necessary for a subject bibliography to be "complete," and its graphical output is easy to obtain in Bradford's style: journals are ranked 1, 2, 3, . . . , n in decreasing order of productivity and arranged on the x axis on the common logarithmic scale; then the cumulative totals $R(n)$ of articles carried by the first n journals are marked along the *y* axis. So, for example, $R(1)$ is the number of articles contributed by the first-ranked (most productive) journal, $R(2)$ is the sum of the number of articles contributed by the first-ranked journal plus the articles contributed by the second most productive title, and so on; finally, $R(n)$ is plotted against the logarithm of n. When the instructions are followed to the letter, Bradford's verbal formulation predicts a gently rising (convex) curve, by no means perfectly linear, which nonetheless approximates a straight line over a large part of the range. "We can thus say," commented the author, "that the aggregate of references in a given subject, apart from those produced by the first group of large producers, is proportional to the logarithm of the number of sources concerned, when these are arranged in order of productivity."[20]

Bradford also supplied a graphical version of the observed regularities which, in addition to explicitly delimiting the abnormal productivity of the nuclear zone, helped him to work out the algebra of the above "straight-line" logarithmic law. Its pattern is similar to the one displayed in Zone 1 and Zone 2 in figure 4.3: an initially rising or convex curve,

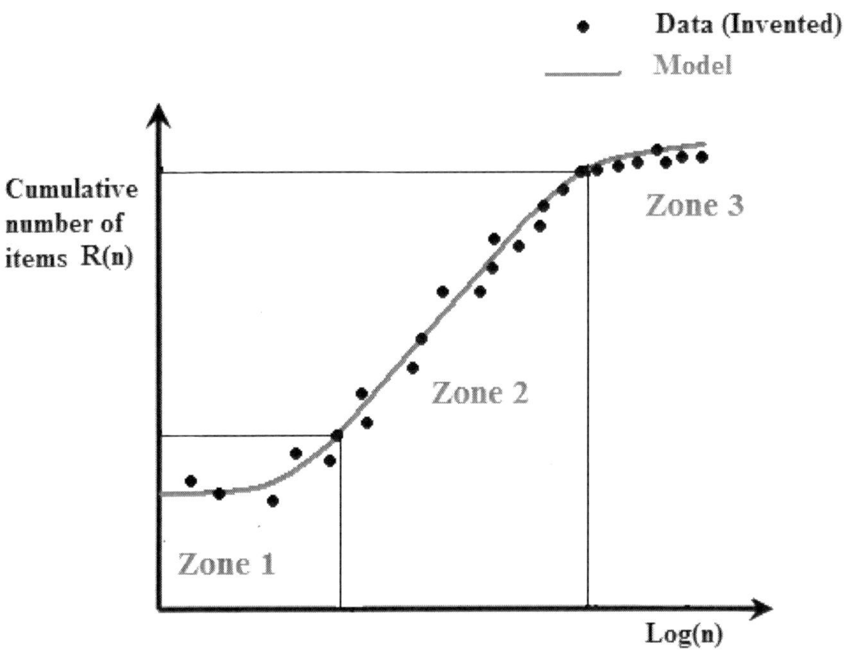

Figure 4.3. Bradford-type distribution: sources are ranked in decreasing order of productivity and the logarithm of the rank is marked on the x axis

representing the nuclear zone of exceedingly productive journals, turns rather abruptly, at a certain critical point, into a straight line running smoothly toward the zones of decreasing productivity.[21]

Bradford did not release a clear analytical expression of the graphical law of scattering, because he implicitly assumed its equivalence with the verbal statement, but Brian Vickery demonstrated, as early as 1948, that such equivalence doesn't hold, and the algebraic expression corresponding to the graphical proof predicts only the upper straight-line portion of the observed curve. Consequently, the verbal and graphical formulations followed, in the decades to come, two seemingly different trajectories.[22]

4.3.1. Early Theoretical Developments

In 1960, having applied Bradford's Law to a bibliography on operational research, Kendall found a linear pattern even more pronounced than in Bradford's own graph.[23] A few years later, building on Vickery's work, Ferdinand Leimkuhler issued a simple function for Bradford's distribution, which was named after him:

$$R(n) = a\log(1 + bn)$$

Here again $R(n)$ is the cumulative number of articles issued by the n most productive journals, while a, b are parameters related to the subject and completeness of the journal collection. Leimkuhler treated previous equations derived from Bradford's original statements as linear approximations of the above formula and, echoing Kendall, came to the conclusion that "Bradford's Law and Zipf's Law are essentially just two different ways of looking at the same thing."[24] With hindsight, it can be said that the search for a mathematical rationale of Bradford's distribution initiated by early information scientists placed them on a road toward mathematical abstraction that later bibliometricians would also travel in pursuit of a general mathematical framework for Bradford's Law and related statistical patterns.

Starting in the late 1960s, several mathematical formulations, models, and syntheses of previous statements related to Bradford's Law have been put forth, but very little agreement exists about which model is the best. The degree of fitting achieved by a single formula, in fact, depends largely on the selection of the initial sample and is severely hampered, in most datasets, by the occurrence of the "Groos droop," a downward deviation at the upper end of the Bradford curve (see figure 4.3, Zone 3) named after the author who first reported it in 1967. A statistical test of the existing models for Bradford's Law conducted by Liwen Qiu demonstrated that even the best of them could not fit those datasets where the droop showed up.[25]

A deep impact on this line of investigation was exercised by Bertram Brookes's obsession with the social and mathematical implications of Bradford's Law. Whereas Leimkuhler had focused on the verbal formulation of the law, Brookes explicitly addressed the issues raised by its "hybrid," graphical version and subsumed all the previous variants under a formula composed of two parts accounting, respectively, for the raising curve in the nuclear zone and the linear range:[26]

$$R(n) = an^b$$

$$R(n) = k\log(\frac{n}{s})$$

Here $R(n)$ is still the cumulative number of articles contributed by the first n journals, k is a constant dependent on the document collection, s is the intercept on the x (log rank) axis, a is the number of relevant papers published in the most productive journal, and b is an additional parameter that approximates a constant only for bibliographies restricted to relatively short time spans.

Brookes dismissed as "one true law syndrome" the previous attempts to pursue a global unification of bibliometric distributions through allegedly universal functions. A universal formula, in fact, appeared hardly achievable in the existing conceptual framework, both because of the impossibility of deriving Bradford's Law from assumptions of classic probability theory and because of the way data originate: the law is expressed in rank-frequency instead of conventional size-frequency terms; and ranking is more primitive and more congenial to social studies than counting, as it incorporates qualitative information alien to the physical statistics' classification system, so treating ranks as cardinals would cause crucial information on items' individuality to get lost. Frequency-rank distributions of the Bradford type, by contrast, "can be analyzed and interpreted without dependence on any statistical theory or underlying probabilistic assumptions; they can be 'theory free' in use."[27]

In line with these premises, Brookes set out to show the equivalence between an "empirical Log Law" derived from Bradford's Law (without the nucleus) and a simple hyperbolic formula dubbed "inverse square law," under which were subsumed also Lotka's and Zipf's formulations, the "anomalous" Law of Number, and Laplace's Law of Succession.[28] The generality of the Log Law and the inborn superiority of the logarithmic representation of information units were sanctioned, in his opinion, by the simple observation that, each time a homogeneous set of sources is involved in some sort of (discrete) homogeneous process of item production, the resulting frequency-rank distribution is logarithmic. Such pervasiveness led him to hypothesize that, beneath the proliferation of hyperbolic patterns in biological and socioeconomic phenomena, there stands a single, all-embracing success-breeds-success mechanism, whose mathematical explanation is well accomplished by means of a "mixed Poisson" stochastic model. Bradford's Law turned by this way into "a particular example of an empirical law of social behavior which pervades all social activities,"[29] and its unrelatedness to conventional mathematical ideas became the ideal justification for a yet-to-be developed "calculus of individuality" capable of competing with the standard calculus of classes in the foundation of a new statistical theory for the social sciences.[30]

4.3.2. Garfield's Law of Concentration

Bradford's Law transposes into an elementary statement a rather intuitive fact that is well known to library acquisitions departments: given a specialized field of study, a small number of journals produces approximately 90 percent of the literature essential to its disciplinary survival, while if the purpose is to achieve 100 percent coverage, a much higher number of sources has to be added to the initial set. To define in exact mathematical terms the dispersion of items over sources in view of facilitating comparison between different datasets, bibliometricians resort to measures of concentration akin to those employed by economists for the unequal distribution of wealth and income within societies. The most popular are the Gini Index, named after the Italian economist who introduced it in the early 1900s, and its bibliometric counterpart, the Pratt's Index, issued in

1977 by Allan Pratt. Both are independent of the unit of measurement and apply to a Lorenz curve, that is, the curve obtained plotting the amount of resources held by equal segments of the population (or, in bibliometric terminology, the number of items produced by equal portions of sources). In the ideal situation, in which each segment produces the same number, the Lorenz curve is represented by a 45-degree line. In real situations, on the other hand, any deviation from equal distribution causes the curve to dip below the 45-degree straight line. The Gini coefficient captures the difference between the ideal cumulative distribution for the line of equality and the actual curve; hence, its value is set between 0 (no concentration, total equality) and 1 (no equality; the winner takes it all).[31]

In spite of the increasingly refined measures of concentration, Bradford's original formulation has attracted many criticisms because of its limited practical value for collection management and database design purposes. For one thing, the law is articulated in terms of groups of journals, which seem to be an artifact of the law itself and do not relate in any way to the main variables at stake when journal content is handled in concrete situations, such as the research field and the specific properties of the journal collection.[32] At a deeper level, much of the inadequacy has been ascribed by some critics to an unproblematic operationalization of the concept of subject made by Bradford and his supporters. Before estimating the scattering of literature on a given subject in scientific journals, in fact, one has to determine plausible criteria for assigning subject descriptors to individual documents. Yet, as showed in section 2.1, the solutions to this fundamental problem of information science, whether derived from human judgment, automatic full-text indexing, or citation links, do not necessarily overlap to the point of ensuring a univocal identification of core journals in the first zone of a Bradford curve. And things do not seem to get clearer from the perspective of online retrieval systems if, turning again to the massive literature on Bradford's Law, one contrasts a 1981 article by Howard White, wherein the opportunity is welcomed to "Bradfordize" the search output from online databases by simply resequencing the hits by journal titles in decreasing order of number of hits contained in each journal, with the lessons learned from experiments focused on online searching behavior like the Getty Online Searching Project, carried out in 1989–1990 at the Getty Center for the History of Art and the Humanities in Santa Monica, California.[33] The Getty experiment testified to the uselessness of the concept of "core journals" in the design of information services in the humanities for the simple reason that, being already familiar with a high number of individual publications in their field, the monitored scholars were much more inclined to judge positively the usefulness of the hardly predictable, unfamiliar hits produced by the low concentration sources in the outer zone (the tail) of the Bradford curve. A Bradford analysis is thus exposed to the risk, recently underlined by Jeppe Nicolaisen and Birger Hjørland, of passing off as a rule that journals publishing the largest number of contributions on a topic are the only ones worth reading by virtue of an undue transubstantiation of their quantitative prominence into qualitative eligibility.[34]

Measuring inequality in a Bradford curve, moreover, is not sufficient to answer the fundamental questions, Where precisely should one set the pole to demarcate the core of truly fundamental sources? How can the size of Bradford's nucleus actually be computed for specific collections? How, in addition, can one estimate the size of the nonobserved class of journals, which, having not yet carried relevant articles on a subject at a given point in time, might do so at any moment in the future? Bensman has hypothesized that the inherent vagueness of Bradford's sets might be properly conceptualized by means of Lotfi Zadeh's fuzzy set theory, while Egghe and Rousseau have operatively made use of fuzzy sets and measures of concentration to work out exact core membership criteria for Bradford's zones.[35] The fuzziness of subject sets, which ultimately confirmed Bradford's faith in the unity of science as reflected by the lack of clear boundaries between disciplines and journals, explains why, despite the initial enthusiasm of many librarians for the managerial potential of Bradford's Law in serials selection, its practical implementations are scarcely reported, with just one noteworthy exception: ISI databases.

Toward the end of the 1960s, Eugene Garfield reaffirmed the validity of Bradford's Law and, based on evidence drawn from the *SCI* (1969) and the *Current Abstracts of Chemistry and Index Chemicus* (1969), extended its range of applicability from single disciplines to science as a whole. His empirical investigation of journal citation scores revealed that science can be construed as a super-set of disciplinary areas intersecting and overlapping at various points but still preserving a relatively small, multidisciplinary core of documentary sources. "So large," he contended, "is the overlap between disciplines, in fact, that the core literature for all scientific disciplines involves a group of no more than 1000 journals, and may involve as few as 500."[36] The observation that, in any field of science, articles are concentrated within the same group of multidisciplinary journals was subsumed by its author under a new general law, Garfield's Law of Concentration, though he himself recognized its status as being more that of an axiom than of a law in the strict sense of the word. Metaphorically speaking, where Bradford could have depicted the literature on a subject like a comet, with the nucleus containing the few core journals, and the tail comprising all the other journals that sporadically publish important articles on the same subject, Garfield's Law turned the tail of a discipline into a mixture of the documentary nuclei of the other disciplines. Further analyses also demonstrated that not only did the core journals publish comparatively more articles than peripheral ones, they also tended to preserve a high degree of stability in citation performance over the years, along with the ability to attract the most highly cited papers in every field. So whereas, from a global classificatory perspective, the Law of Concentration exasperated the fuzziness of journal sets caused by the intrinsic interdisciplinarity of scientific research, on the information retrieval (and evaluation) side it also legitimized the idea that, contrary to the myth of the scientist submerged by a flood of information spread throughout an ever-increasing bulk of literature, a limited number of journals produce most of the relevant scientific results in any field. This conclusion dramatically reduced the dimensions of the (virtually infinite) documentary universe in such a way that any attempt to achieve complete coverage would be both financially unadvisable and conceptually wrong. Consequently, multidisciplinary indexes like the *CC* and the *SCI*, whose source selection criteria were openly inspired by Bradford's premises, could rightly claim a higher effectiveness in the retrieval of really relevant literature than any other subject-oriented index.

As early as the 1970s, the community of information scientists realized the full convergence of Garfield's Law with a similar regularity encountered in the area of library services by the British librarian Donald Urquhart. Analyzing the interlibrary loans of journals from the Science Museum Library in 1956, Urquhart found that less than 10 percent accounted for approximately 80 percent of the requested items, and suggested the Poisson distribution as a plausible stochastic model for circulation data. Because the number of interlibrary loans of each journal had a strong positive correlation with the holdings of the same titles at major British libraries, he generalized the result in a law, stating that the number of interlibrary loans of a journal is a reliable measure of its total use. Subsequent experiments basically confirmed Urquhart's findings, which also appeared perfectly in tune with analogous patterns of concentration in the local usage of library materials revealed by a series of studies conducted in American libraries since the late 1960s. Thus, what Garfield held true for the journal communication system, Urquhart had previously shown to be valid also for the library services arena: a relatively small set of journals satisfies most of the supra- and intralibrary demand and consists of the same core titles most widely held by academic libraries. Library usage phenomena, as Bensman would suggest in a 1985 paper, are no less driven by a cumulative advantage process than publications and citations, and there is good reason to believe that their skewed patterns are ruled by the same laws operative in the social stratification of scholarship.[37]

In some very fundamental ways, this structural inequality posited by Bradford-like regularities at the core of the science communication system is the source of all good and evil alike in the history of scientometrics. What is good is its ability to render technically and financially affordable the tracking of excellence in the jungle of scientific information; were Bradford's Law completely wrong, the *SCI* wouldn't exist and be so useful in descriptive and evaluative bibliometrics as well. What many judge evil is its proneness to perpetuate and theoretically justify the existing order of things—marked by an unfair allocation of social, economic, and symbolic capital—by telling a story that Garfield's citation indexes would have brought to a logical completion, the story of a stratified scientific literature produced by a stratified community wherein scientific excellence is confined to a few minds that live and work in a few institutions and deliver the fruit of their creativity to a few, high-quality, "must-buy" international journals owned by a decreasing number of multinational corporations ruling the roost in the global information market.

As an ideal bridge between the apparent abstractedness of Bradford's Law and the concreteness of interests revolving around the lively world of evaluative bibliometrics, two further examples are in order. One is the "evaluating-the-evaluators" perspective at the journal level inaugurated during the 1980s by Braun's "gatekeeping indicators," which address the national structure and selection criteria of journal editorial boards entrusted with the task of securing the quality of published research by a permanent activity of control and screening at the content layer; any

Bradford-type pattern in the journal publication process is ultimately the distilled product of such critical gatekeeping routine.[38] The other is the long-lasting debate over the exclusion of the Third World (or developing countries) from the new scientific information order subject to the monopoly of Western countries; in a Bradford-compliant universe, there is only one valid world science, and those who possess it are expected to hand over its secrets to those who don't. Questionable as it may be, this assumption is at the roots of both a deeper understanding of the structural limitations of citation databases/indicators in research evaluation on a worldwide scale and a stronger participation by Western and non-Western authors in the stream of bibliometric research which, starting from Price's, Moravcsik's, and Garfield's pioneering concern for the subject, ends up in the project of mapping the ever-growing presence and impact of Third World research in international scientific journals.[39]

4.4. ZIPF'S LAW AND MANDELBROT'S REINTERPRETATION

Zipf's Law is apparently outlying in comparison with Lotka's, Bradford's, and Garfield's laws because it has nothing to do with the traditional, external actors of the documentation scenario (authors, papers, journals, citations). Rather, it pertains to the inner and somewhat impenetrable act of generating and shaping the linguistic expression of a text or speech, whatever its scope. Word frequency distributions, in addition, have long attracted statisticians' attention not only because, unlike other skewed distributions recurrent in nature and society, they feature large numbers of rare, low-probability events (words), but also for the readily available empirical environment they provide when the effect of the non-randomness typical of human discourse on probabilistic models imbued with assumptions of randomness is to be ascertained. It's no accident, then, that written text, specifically the alternation of vowels and consonants in Pushkin's poem *Eugene Oneghin*, served as the main testing ground for the groundbreaking extension of probability calculus to the outcomes of trials dependent on each other, performed by the Russian mathematician Andrey Markov at the beginning of the twentieth century. And such extension was all the more remarkable because, some decades later, Markov processes would govern individual studies in bibliometrics almost the same way classic bibliometric laws would take the lead in group studies.

George Zipf, professor of philology at Harvard University, considered language not simply an elaborate system of signaling and communication, but a complex tool of behavior, whose structure cannot be isolated from the personal, social, and cultural background of the speakers. He held that, to grasp its hidden structure, one should quantitatively investigate its patterns of use in everyday speech, which are "by no means essentially incommensurable with the patterns of style, of metrics, even of music."[40] Such premises allowed him "to investigate speech as a natural phenomenon . . . in the manner of the exact sciences, by the direct application of statistical principles to the objective speech-phenomena."[41] It is of little importance that Zipf was not a skilled mathematician and supported his word counts with arguments appealing to some mystical force acting in nature. The history of science, as is well known, teems with examples of fruitful research programs and valid scientific results born of (or grafted onto) a controversial core of metaphysical beliefs.

Zipf's linguistic law is just one of the several empirical regularities disclosed by the author's visionary human ecology. Formerly stated by the French stenographer Jean Baptiste Estoup and the American physicist Edward Uhler Condon, the law applies to words in their material, full-inflected form as it appears in a text (word-tokens), each word-token being a particular occurrence of the corresponding lexical unit or dictionary entry (word-type). For example, in the sentence "child is child is child is child," there are two types ("child" and "is") and seven tokens (all the distinct occurrences of "child" and "is"), so a Zipfian analysis would count all the seven items, assigning rank 1 to the word-type "child" occurring four times and rank 2 to the word-type "is" occurring three times. Zipf's Law states that in a relatively long text, if one ranks the word-types in a decreasing order of frequency, assigning rank 1 to the first most frequent word, rank 2 to the second most frequent word, and so on, and then multiplies each rank by the number of occurrences of the corresponding word-tokens, the product is roughly equal to a constant, at least for values placed in the middle to low ranks. In symbols,

$$r \cdot f = C$$

where r is the rank of the word-type, f is the number of times it occurs, and C is a constant related to sample size. The same equation is most commonly given in the logarithmic form

$$\log r + \log f = \log C$$

which, plotted on a double logarithmic chart, yields the characteristic graph of a straight line with a slope approximately equal to -1 (see figure 4.2).

At its simplest, the law amounts to saying that the vast majority of text words appear only a few times, and a limited number are extremely frequent. The distribution of words in a text is skewed, closely resembling that of scientific papers between authors (Lotka) and articles between journals (Bradford). Indeed, if Zip's Law is written in the more general form

$$r^B \cdot f = C$$

where B is a constant with value less than or equal to 1, it is clear that the number f of occurrences of a word on rank r is a decreasing power function of r.

Zipf knew that his law fitted some empirical datasets poorly. The word frequency distribution in "nonstandard" texts produced by children and psychiatric patients, for example, revealed a distinct deviation from the straight line. Nonetheless, he tirelessly set out to corroborate the ubiquity of the inverse rank-frequency relationship by analyzing frequencies counts of words in a huge number of different natural language corpora, including four Latin plays by Plautus, Joyce's *Ulysses*, and a text in colloquial Chinese. From a word index of Joyce's *Ulysses*, for instance, he discovered that the tenth most frequent word in the novel was mentioned 2,653 times, the one hundredth most frequent word occurred 265 times, and the five thousandth most frequent word occurred 5 times. If one multiplies $10 \times 2{,}653$, then 100×265, and finally $5 \times 5{,}000$, the result is approximately constant, or at least variable within a small range, thus confirming the law's validity even in the case of a highly experimental and linguistically subversive literary text. Zipf ascribed the skewness of word distribution to the human intrinsic tendency toward an equilibrium in the ordering of the language's elementary components, a tendency ultimately amenable to the basic explanatory factor that he posited at the origins of all personal and collective actions: the principle of economy or "principle of the least effort." An intuitive consequence of this principle is, for example, the preference for usual, short, and easy to pronounce words over rare, long, and difficult to pronounce ones.

Zipf's model of language did not catch linguists' attention until recent times, when the importance of a holistic, quantitative, and discourse-based approach to language structures has been reappraised. From an information science perspective, his theories and sound research methodology are less noteworthy than the extra-linguistic fortune of the empirical formula they underpinned. Since the 1950s, indeed, Zipf's Law has undergone various refinements and attempts at rigorous mathematical foundation from perspectives as diverse as Mandelbrot's uptake of Shannon's information theory, Bruce Hill's adaptation of Bose-Einstein statistics, Herbert Simon's and Derek Price's derivations of hyperbolic distributions from a stochastic birth process, and Gustav Herdan's project to translate into mathematical functions De Saussure's dichotomy *langue-parole*.[42] In hindsight, the Simon/Price's stochastic pathway and Mandelbrot's deterministic approach exercised the strongest impact on subsequent informetric modeling. Mandelbrot introduced a more general formulation of Zipf's Law:

$$(r + m)^B f = C$$

where f is still the frequency of a word, r its rank, m, B, C are constants dependent on the corpus. The additional parameters m and B could be adjusted to better fit empirical data, especially for the high-frequency words. It has been suggested, for instance, that m could account for different uses of a given language, while B could reflect the difference in frequency patterns between natural and artificial languages (such as DIALOG and Elhill).[43]

Mandelbrot's formula resulted, as a first approximation, from interpreting Zipf's principle of the least effort in terms of minimization of communication costs in accordance with Shannon's theory of communication; that is, language aims at transmitting the largest variety of signals with the least delay, and the cost of a word can be thought of as the number of letters (plus additional space) composing it. Zipf's Law, specifically, was a necessary consequence of the probabilistic structure of human discourse, conceived as a highly complex system of phonemic units (and pauses) regulated by stochastic rules. During the 1950s, Mandelbrot had also specified how these rules could be figured out, advocating the urgency of studying scaling distributions of the Pareto and Zipf-type within the framework of stable, non-Gaussian distributions. In trying to derive Zipf's Law from general premises, he simply connected its hyperbolic pattern to the scaling property of the lexicographic tree, each branch of the tree being approximately a reduced-scale version of the whole. Such capability of preserving a given pattern or degree of complexity at different magnitudes would become the hallmark of the fractal objects, whose mathematical theory and graphical advertising have been attached to Mandelbrot's name since the mid-1970s.

Fractals are one of the most ingenious devices enabling scientists to cope with the mathematical structure of complex systems. In daily life, it is quite common to refer to elementary geometrical shapes, such as lines, squares, or cubes, taking them to be, respectively, one-, two-, or three-dimensional. In a sense, this apparently trivial characterization points to three different degrees of complexity associated with those shapes. But if the measure of an irregular shape of fractal type, let's say the highly involved curve of a coastline, is attempted, then different results will be expected depending on the resolution of the measuring instrument, because each portion of the coastline is, statistically, a reduced-scale image of the whole. Stated alternatively, the fractal-coastline is a self-similar object calling for a characterization in terms of fractional dimensions incompatible with the conventional picture of uni-/bi-/tri-dimensional shapes encountered in everyday life. This peculiar resilience to complexity ensured that, after Mandelbrot's seminal work, fractals progressively gained credit as a conceptual and mathematical tool for modeling many chaotic phenomena occurring in natural, social, and artificial environments. Bibliometrics did not evade this trend, although one basic enigma steadily resists it: notwithstanding the variety of models adaptable to Zipf-like distributions, nobody could explain convincingly why they exist and are so widespread. We can't explain, for example, why Zipf's Law seems to be valid even in randomly generated texts, such as a text typewritten by a monkey hitting keys at random.

4.5. THE GROWTH OF SCIENTIFIC LITERATURE

The growth rate of science is not merely a matter of speculative interest. With the obvious proviso that scientific growth and scientific literature growth are far from being coincident, the rate at which the size of the literature in a specific subject area increases is by itself a variable affecting the outcome of several other variables deeply embedded in flesh and blood, including individual scientists' career opportunities in that area (the more rapidly a field expands, the more likely young aspiring researchers will hold a position in it), citation performance (the faster the growth rate, the higher the portion of recent citable papers), and national productivity (a logistic curve limiting the growth of science is in principle unacceptable if, as in the Soviet Union, science is raised to the rank of key driver of the national economy). This explains why, for want of more viable alternatives, Derek Price's exponential and logistic models of scientific growth are still widely discussed and have undergone the same treatment reserved to classic bibliometric laws: they have been tested against empirical (bibliographic) evidence, and their goodness-of-fit has been compared with that of competing or complementary models across many specialties. Predictably, it turned out that literature growth follows different, sometimes intermingled patterns within different research fields, so that linear, exponential, logistic, power, or mixed models appear necessary, from time to time, to obtain the best fit for empirical data series.[44] Most important, it became clear that any measure of the size of science is strongly affected by several obstacles of technical and methodological nature. One of them is the difficulty of determining the exact point in time at which a research field is supposed to grow out of previous literature: a historical problem mostly lacking an immediate solution. Equally critical points concern the geographic and linguistic biases of journal literature databases and the existence of many fields in which journal papers are not the primary means of communication nor the best starting point for estimating knowledge production rates.

In the early 1960s, an original solution for dealing quantitatively with literature dynamics came from William Goffman, who, working solo or in collaboration mainly with Vaun Newill and Kenneth Warren, developed a set of conceptual and mathematical tools inspired by an epidemic model of diffusion. This model compares the transmission of ideas within a population to the transmission of an infectious disease, the written scientific article being the intermediary vector for the transference of infectious materials (ideas) between "infectives" (authors, researchers) and "susceptibles" (those who can become infectives upon contact with infectious material). Though admitting the feasibility and even desirability of a (more realistic) stochastic variant based on finite state Markov processes, the authors issued a deterministic version of the model featuring a set of differential equations applied to the analysis of growth, stability, and decay of communication processes in a given research field.[45] Among the manifold subsequent applications and extensions of Goffman's theory, the mathematical description of the time-evolution of six (natural sciences) research fields in terms of authors and publications recently carried out by Luís Bettencourt and colleagues conveys further evidence of the self-similar structure of science implicit in the ubiquity of power laws. In each field, independently of whether the growth pattern is logistic or not, productivity per author remains stable also when the field grows in size by orders of magnitude, thereby "suggesting that recruitment (of susceptibles via author pool growth) is the fundamental driver of scientific development."[46]

Here, as elsewhere in the history of bibliometrics, a major hindrance to the generalization of results is the difficulty of crossing the boundaries of pure quantity to enrich the analysis with quality criteria that might help distinguish the sheer growth of literature from the more desirable growth of (good, useful) knowledge. Price was aware of this gap when, building on "rather tenuous hypotheses and measurements,"[47] he formulated his square root law. He also wished to exert a better quality control on the measurement process by suggesting that, in testing the exponential growth hypothesis, it is more advisable to count "the heads of whichever papers are listed by one of the great abstract journals or bibliographies" rather than simply to count the journals, which recurrently publish papers "immersed in nonscientific material."[48] Other authors applied preestablished qualitative criteria to the counting units themselves, as in the case of Kenneth May, who developed a predefined classification scheme for distinguishing, in a bibliography of mathematical papers on determinants, the units bearing new ideas and results from duplications, trivia, and papers with applicative, historical, or educational character. He could thus compute the time trends for each class of materials separately. Later on,

Jean Tague and colleagues adapted a similar approach to a bibliography on the obsolescence of library materials.[49]

Along this line, unavoidably, the observation of a literature growth trend in a research area raises the question of whether and how it relates to changes in the knowledge content. It is just a short step further, then, to go beyond the sheer problem of determining the size increase of science and meet some true epistemological dilemmas about what causes scientific change and what separates good science from garbage. Crossing the quantity/quality threshold one finds, first and foremost, Lakatos's emphasis on the occurrence, throughout the history of science, of progressive and degenerative problem shifts; Popper's claim that scientific change is not a cumulative process, nor simply a sequence of increasingly refined deductive systems, but rather a progression from problem to problem driven by error elimination; and Kuhn's idea that knowledge advancement is the result of scientific revolutions bringing on a drastic reorganization of the existing conceptual framework.

Zooming out from individual disciplines to the global dynamics of science, in the late 1990s, van Raan repeated, on a larger scale, Derek Price's analysis of the references in an annual edition of the *SCI*.[50] He examined the age distribution of nearly fifteen million references from ISI-covered papers published in 1998. Except for a couple of remarkable differences in the interwar and post–World War II periods, the results substantially confirmed Price's conclusions: the evolution of science-in-the-large conforms to a super-exponential distribution, which can be formally approximated by the same model equation describing unlimited tumor growth (Gompertz equation). For shorter time intervals, instead, a normal exponential equation fits the data quite well. What differentiates van Raan's model from other proposals is that, taking advantage of the mathematics of complex systems, he uses the fractal structure to explain the evolution and differentiation of science-in-the-large. Science is portrayed, accordingly, as a complex, self-organizing, biologically growing system, comprising many self-similar subsystems or research fields whose size distribution conforms to a power law. Each research field originates from an important breakthrough in scientific research and each, in turn, gives rise to other breakthroughs, hence to new research fields, with a probability proportional to its size. The philosophical implications of Van Raan's and Price's models sound like an anti-Kuhn manifesto: science doesn't evolve by dramatic, socially and psychologically driven paradigm switches marking the transition from revolutionary to normal (cumulative) scientific activity; rather, new ideas and research fields, which may well be deemed revolutionary in their own right although on different scales of magnitude, develop almost linearly from antecedent ones, so that "there is no 'normal' science alternated with well-defined periods of 'revolutionary' science in which new paradigms start to dictate the rules. Science is *always* revolutionary, but by the typical statistics of complex systems, there are mostly smaller and only rarely big breakthroughs."[51]

Quantitative analyses cannot answer fundamental questions on scientific change better than philosophical or sociological ones. Nevertheless, as testified by Price's and van Raan's examples, they offer philosophers and sociologists just one more perspective, whose aim is not to supersede qualitative investigations, but to "stretch the method to the full and examine critically any benefits which might thereby accrue."[52]

4.6. THE OBSOLESCENCE OF SCIENTIFIC LITERATURE

Scientific literature, like living beings, grows old and loses, partially or totally, its initial power of attractiveness. Authors give up using documents when they no longer meet their requirements or, on the other hand, when they become so essential that their content is silently incorporated in everyday practice. Intuitively, moreover, due to its self-corrective character and a wider agreement of researchers on empirical and methodological issues, scientific literature ages more quickly than literature in the social sciences and the humanities. Bibliometricians call "obsolescence" or "aging" the process by which the chance of a document being currently used declines with its age. This, in turn, is taken as an indicator that, on average, the validity or utility of the information contained therein declines as well. An exact determination of the rate of obsolescence in different fields is a highly desirable task not only for better differentiating the various types of scholarly activity and their respective archival needs, but also for placing in a more meaningful context any indicator of scientific value based on quantitative criteria. The process of aging, in other words, is critical both for librarians or database managers striving to improve management criteria of ever-growing (print or electronic) collections and for information scientists seeking hidden, field-dependent patterns of decreasing information flows over time.[53]

Although obsolescence of scientific literature can be evaluated in various ways, for example by examining the local circulation records in a library, Price linked it closely to the time distribution of citations: if a document is cited soon after its publication and then quickly forgotten, as typically happens to research front papers, its obsolescence is high, whereas if it continues to be cited in the years to come, as is usual in social sciences and the humanities, its obsolescence is low. At the 1958 International Conference on Scientific Information, an active interest already existed in "half-life," a concept shaped by analogy with radioactive decay and made popular by Richard Burton and Richard Kleber in 1960. It is defined as the time during which half the total use of a given literature has been made. At its simplest, if usage is estimated by citations, it is computed for a set of source documents published in a given year by subtracting that year from the median publication year of the papers citing the documents. A typical "radioactive" pattern of aging exhibits an exponential decay of usage hits; that is, the expected number of citations accrued to the source items decreases year to year by the same aging factor.[54] Later bibliometricians welcomed the invitation to play the game through citations and assumed the exponential model of decay as the launchpad for more complex formulations using either (or both) of two perspectives: looking forward, from a point in the present time, to the age distribution of documents citing a fixed set of source documents (diachronous obsolescence) or, alternatively, looking backward to the age distribution of references cited in a given set of sources (synchronous obsolescence).

Brookes canonized Burton and Kebler's model, further specifying the conditions under which a negative exponential distribution (or a geometric distribution, if one works with discrete quantities) can be applied to describe and predict aging phenomena. Yet he also argued that neat mathematical solutions could be relied on only to the extent that they fitted citation data referred to narrow scientific literatures over relatively short time spans. In real library contexts, otherwise, observations are not expected to conform with theoretical distributions, because any practical measure of obsolescence has to cope with the singularities emerging from samples of actual literature usages by local user groups.[55] In the aftermath of Brookes's systematization, a lively debate took place on the subject of literature obsolescence, resulting in a polarization of perspectives alongside two contrasting positions that would still be recognizable in later contributions.

On the conservative front, many authors continued to adopt citation statistics to investigate the aging trajectory of various research areas, to further corroborate exponential decay in literature usage, or to work out best-fitting mathematical functions. In the early 1980s, the Japanese bibliometrician Hideshiro Nakamoto conducted a large-scale study on cited references that appeared in publications between 1961 and 1984 and were indexed by the *SCI*. He confirmed the exponential decrease in citation by age and claimed the equivalence of synchronous and diachronous approaches on the ground that they generated similar curves.[56] More recently, Vincent Larivière and colleagues used ISI citation data to examine, at different time periods, the age distribution of cited references in natural sciences and engineering journals over more than one hundred years (1900–2004). Interestingly enough, besides the slowing down of the exponential growth during the last three decades, they also found a constant increase in the median age of cited literature since the mid-1960s. That is to say, despite the increasing availability of recent, ready-to-access, and ready-to-cite documentation in electronic format, scientists in those areas have been relying on an increasingly old set of source materials. In the authors' view, this is the sign that current researchers have constantly been adding puzzle-solutions to firmly grounded and durable paradigms built on past scientific achievements.[57] Striking out in the consolidated direction of advanced mathematical modeling, others have tried to express the actual citation history of individual documents as the realization of a stochastic process, which works under special assumptions, leading to a gradual decrease in their use. Though alternative solutions have also been proposed, stochastic models seem preferable to competing mathematical treatments because, at least in principle, they allow for the translation of the diverse factors influencing aging into parameters that can be estimated from empirical data with a specified margin of error.[58]

On the front of detractors, some authors sharply criticize the concept of obsolescence, emphasize its methodological shortcomings, and question even the possibility of measuring it in any meaningful way by means of citations. For one thing, it is contended that linking obsolescence to the progress of science, hence to the loss of validity of older documents, and taking usage hits as direct indicators of validity is simply incorrect. Usage and validity are not necessarily related; documents no longer used might contain valid information that is simply incorporated in later documents or pertains to an unfashionable subject area. Likewise, information no longer valid doesn't necessarily cease to be useful, because it might be embedded in widely accessed documents (it is the food for historians, indeed). In the second place, as argued, among others, by Maurice Line and Alexander Sandison, obsolescence rates measured by crude citation counts are only apparent, because they don't take into account the confusing role of literature growth: more citations being made to recent than to older literature might simply reflect the availability of many more recent citable documents instead of an actual decline in use of the older ones. A corrective factor is required if citation rates are to be adjusted for changes in the size of the citing population (or of the citable population in the case of synchronous obsolescence).[59] In addition, citation practices vary widely across disciplines. A study by Helmut Abt in 1981, for example, showed that citations to astronomical papers reach their peak five years after publication and then decrease at a linear rate, still preserving a half-maximum rate twenty years after publication; biochemistry articles, on the other hand, have been found to age according to a negative exponential function, with a median citation age of three years, but with dramatic variations across content typologies.[60] The latter result also suggests that even the subject area is a variable when aging patterns are to be determined, but it would not take too much to extend a similar confounding role to several other variables as well:

1. *Publication types:* A research article ages differently from a letter.
2. *Level of abstraction:* A methodological contribution, even Galileo Galilei's, might still be cited for many years after the disappearance into oblivion of many empirical case studies.
3. *Author reputation and social status:* Few would dare to omit citing even the most trifling paper published thirty years before by the senior author in a field, above all if he or she is likely to sit on future tenure committees.

4. *Journal centrality:* Prominent journals attract high-quality and enduring papers, but the opposite holds true as well, namely that papers published in prominent journals age slower just because of the annexed Matthew Effect.
5. *Degree of interest or fashionableness of the subject:* An unfashionable subject attracts few citations regardless of the validity of contributions.

The true challenge for a bibliometric model of aging, then, would be to account for the complex interaction of the above factors in the aging process. From this point of view, literature aging is a construct as puzzling and difficult to manage as those of impact, influence, and reception, given the impossibility of constraining it in the close-mesh net of a ready-made mathematical framework.

4.7. THE UNIFICATION(S) OF BIBLIOMETRIC LAWS

Frequently in the history of science the quest for certainty and sound theoretical foundations has been fulfilled by the reduction of a complex matter to a small set of basic principles from which specific statements are then derived through logic or mathematical reasoning. This happened also when, persistently from the 1970s onward, bibliometricians sought to demonstrate that, under certain conditions, the laws of Lotka, Bradford, and Zipf are mathematically equivalent, that an exact relation between the corresponding parameters can be given, and that, in the last analysis, they can be "explained" (in a mathematical sense) on the basis of more fundamental principles, whether in a stochastic or a deterministic fashion.[61] More often than not, such explanations rely on real numbers and real analysis techniques, notably integration and differentiation, which supply a mathematical approximation to the discrete reality of informetric data (authors' ranks are natural numbers, and so are productivity and citation scores), but alternative approaches have also been tested, such as Chen and Leimkuhler's demonstration that, if an index is assigned to each entry of a ranked set of bibliometric data, the equivalence of Lotka's, Bradford's, and Zipf's laws can be derived by means of natural numbers and without resorting to goodness-of-fit methods.[62]

Bibliometricians, as discussed in section 4.1, were not alone, nor were they the first to face the mathematical harshness of a skewed universe. Before touching them, the effects of the British biometric revolution on early twentieth-century statistical thinking turned up dramatically in linguistic and socioeconomic areas of inquiry, where critical patterns concerning the skewed distribution of word frequencies in natural languages and of wealth among populations appeared regularly on the research agenda. It was almost natural, then, that the way biologists, linguists, and economists had handled similar situations provided bibliometricians with useful indications of how to set up a mathematical toolkit adequate to the information science arena. This holds true, for example, with reference to Herbert Simon's uptake in econometrics of G. Udny Yule's statistical work, to Joseph Irwin's implementation of the Waring distribution in the analysis of long-tailed biological datasets, to Gustav Herdan and Herbert Sichel's reappraisal of mixed Poisson processes in modeling word frequency distributions, and above all to economists' concern with the oddities of Pareto's distribution.

The occurrence of a hyperbolic pattern "is so frequent," contended Herbert Simon as early as 1955, "and the phenomena in which it appears so diverse, that one is led to the conjecture that if these phenomena have any property in common it can only be a similarity in the structure of the underlying probability mechanism."[63] Simon issued a model based on a stochastic birth process similar, in many respects, to the one developed a few years earlier by the British mathematician David Champernowne for income distribution. It is basically a success-breeds-success model built on two assumptions that he exemplified in terms of word frequencies. Given a text that is being written and that has reached a definite length, 1) the probability that the next word being written is a word that has already occurred x times is proportional to the total number of occurrences of all words that have appeared exactly x times, and 2) there is a constant probability that the next word will be a new word at all. Simon demonstrated that, if managed in a coherent probabilistic framework, the two assumptions lead to the Yule distribution. Of course, they were not the sole assumptions from which the final equation could be derived, but they were sufficiently plausible to enable a mathematically consistent treatment of "nonrandomness" in biological and social phenomena. Despite Mandelbrot's criticisms, Simon's paper spurred a great deal of work on stochastic modeling in information science. It also drew the attention of Kendall who, in the aforementioned 1960 study on the bibliography of operational research, underlined the structural equivalence between Bradford's and Zipf's laws and praised the Simon/Yule approach as "a completely different way of looking at this topic which subsumes the Bradford-type distribution under a general class."[64] Robert Fairthorne followed the line in a 1969 classic review article expressly invoking the profound kinship between the Lotka/Bradford/Zipf formulations and those devised by Pareto and Yule in the economic and biological domains, respectively.[65] Shortly thereafter, an article published in *Nature* by the Indian physicist S. Naranan featured a general, power law version of (the verbal) Bradford's Law. Inspired by Fermi's theory of the acceleration of cosmic rays, the author linked its emergence to the growth dynamics of scientific literature on the basis of three assumptions, none of which were necessarily "true": the number of journals producing articles on a subject grows exponentially; the number of articles within each journal augments exponentially as well; and such growth rate of articles is the same for all journals.[66] Quite similar assumptions would be used, many years later, by Egghe to describe information production processes as self-similar fractals in the context of Lotkaian informetrics.

In the late 1970s, the Russian mathematician A. Yablonsky embraced Mandelbrot's cause as far as the intractability of skewed distributions in the conventional probabilistic framework was concerned. Alternatively, he used non-Gaussian statistics to derive Zipf's Law as one of the possible approximations of stationary (time-invariant) scientometric distributions, endeavoring to demonstrate that, for the asymptotic behavior of a distribution to be Zipfian, the distribution must necessarily be non-Gaussian.[67] Sergej Haitun further specified and extended Yablonsky's work, arguing for the non-Gaussian nature of scientific activity in general.[68] His argument on the inadmissibility of Gaussian statistics in the mathematical explanation of Zipfian phenomena was enthusiastically welcomed by Brookes as the decisive call for a new paradigm in quantitative social studies. At about the same time, by contrast, Sichel refuted Haitun's radical contention, showing that a variety of long-tailed datasets encountered in the social sciences can be fitted, still retaining the finiteness of variance as in conventional probabilistic thinking, by a generalized inverse-Gaussian-Poisson distribution (GIGPD).[69] Not only did the GIGPD fit many of the existing datasets, including those refractory to the Simon/Yule and the Waring distributions, but almost all of the previously issued theoretical functions could be derived as its special or limiting cases. One major obstacle in implementing the distribution is that the zero-class, that is, the number of nonactive sources that can potentially generate items at any time, is not actually observed. Such is the case for the number of journals that can carry relevant articles on a given subject, the number of unproductive scientists who can produce research papers, and the number of words that a poet could have used in his or her poems but didn't. Sichel had hinted at one possible solution on the basis of a recursive procedure. Following his path, Burrell and Fenton confirmed the possibility of obtaining good statistical estimations of the GIGPD parameters, thanks to numerical methods of approximation made easier by enhanced computing facilities.[70] Thus, thanks to Sichel, skewed bibliometric distributions recovered a seat in classic probability theory just when they seemed on the verge of losing it and, despite the daunting mathematical form of the GIGPD, the impact of his way to unification on information science was deep, as testified by Brookes's hesitation in publishing a long-planned book on Bradford's Law due to fundamental issues raised by the Haitun/Sichel debate.

The idea that the place of informetric laws in statistics should be much the same as that of the Gaussian distribution in classic probability theory is also pivotal in Abraham Bookstein's informetric work. Output measurement in social sciences is normally steeped in ambiguity, uncertainty, and arbitrariness in the definition of the variables at stake, but the uncanny stability with which, no matter the specific content and the way data are conceptualized, Lotka-type patterns resurface under a vast range of circumstances and modifications, points to one fundamental virtue of classic bibliometric laws: "resilience" to ambiguity and imperfect measurement. In the mid-1970s, inspired by the unifying role of symmetry principles in particle physics, Bookstein issued a general bibliometric model that complied with the above resiliency and encompassed the empirical formulations of Lotka, Bradford, and Zipf as special cases. Starting from Lotka's Law under the specific constraints imposed by symmetry (or invariance) considerations, he argued that, for that law to emerge in such a wide variety of social conditions and independently of the observational time window, the only realistic function expressing the expected number $f(n)$ of authors producing n papers in a disciplinary sector over a definite period of time is obtained when $f(n)$ is proportional to an inverse power of n. An entire set of probability distributions can be shown to generate a similar function and account for the random component inherent in its empirical manifestations, the most general being the class of compound Poisson distributions.[71]

In 1976, Derek Price showed that the same model equation issued by Simon could be derived from a modification of the Pólya Urn scheme. The urn model, already sketched out in section 4.1, was meant to accommodate the skewed empirical distributions of words in text corpora, of authors' and journals' productivity rates, and of citations to scientific papers. Four years later, Ravichandra Rao elaborated on Price's solution to demonstrate that, under the same conditions posited by Bookstein, a negative binomial distribution is the most suitable model for Lotka-type patterns of scientific productivity. Shortly thereafter, Tague corroborated Rao's conclusion through the generalization of Price's urn scheme in a multiple-urn model that leads asymptotically to a binomial distribution containing the Simon/Yule distribution as a special case.[72]

Insistently since the 1980s, a further dimension added to the complexity of mathematical modeling in the bibliometric workshop: time. In a sense, time would be a very desirable component of every truly useful model of a natural or social system because the correct implementation of a time parameter opens the possibility for predicting the future states of the system. As noted elsewhere, though, the stochastic nature of informetric processes and the multiplicity of interconnected causes influencing their dynamic behavior prevents any simplistic description of such evolution. That is why, except for Bradford's and Kendall's early concern with the predictive problem of estimating the new journals appearing during a certain period as well as the "invisible" class of extant journals, which could produce relevant articles on a given subject at any moment, the time scale has gone almost completely overlooked, at least during the first decades of bibliometric history. The classic laws of Lotka, Bradford, and Zipf, for example, are static and purely descriptive, and so is Price's CAD, which depends only on the size of the population and the mean number of successes per item. Interestingly, despite a critical note by Paul Kantor on how a time-dependent reformulation could improve upon the initial model, Price held on to the belief that it is the first pulse that makes the difference. In the case of citations, for example, he felt that soon after publication, in as long as it takes for a paper to reach the members of the targeted visible and invisible colleges, its destiny is marked: "The paper is weighed by peer and in its incunable period produces a first pulse of citations which in most cases probably determines all future citation history."[73]

Since the late 1970s, the prevalent trust in the long-term stability of informetric distributions has met with the growing awareness that time conditions the probability structure of information flows. This concern, already manifested in a 1976 paper on Lotka's Law by Jan Vlachý and in a 1978 report by Gilda Braga on the peculiar form assumed by Bradford's bibliogaph on nearly seventy years of Brazilian literature on schistosomiasis, came to the fore in connection with at least two developments. One is the discovery that Bradford's distribution takes different shapes when journal productivity rates are computed over extended or distinct time intervals.[74] The other is the quite similar fate that befell the success-breeds-success principle when put through the sieve of individual scientists' age-related patterns of productivity: evidence exists that if productivity data are collected for cohorts of authors with the same duration of scientific activity in a given discipline and not, as in many best-fitting exercises relative to Lotka's Law, for cross-sectional groups of scientists with careers lasting different lengths of time, then the skewness of the resulting distributions decreases as the duration of participation increases. In other terms, the longer the career of the sampled authors in a scientific area, the smaller the differences between their productivity (and the less the probability that a cumulative advantage process leading to a Lotka-type inequality will occur).[75]

In response to the challenges issued by the intrusion of temporal instability into the structure of information processes, in some authors bibliometric distributions took the shape of dynamic objects, their description being complemented by a formal representation of the possible changes occurring in sources' productivity rates at different time periods. The stochastic birth processes and the mixtures of counting processes advocated in many modeling exercises were naturally predisposed to perform such a representation in either parametric or empirical Bayes models, but in no case do they permit exact predictions, because their featured time dimension is nowhere near as "objective" as that encountered in the mathematical description of physical and natural systems. So, for instance, one of the most sophisticated informetric models ever conceived, Sichel's GIGPD, explicitly incorporates time as a parameter even if, in the last resort, its concrete exemplifications lack any predictive capacity. All the same, the trend is noticeable, for it testifies to the desire of leading bibliometric theoreticians to render theory more adherent to the actual dynamics of complex systems. In this connection, at least three contributions are worth mentioning.

In a 1984 paper, Schubert and Glänzel extended Simon's stochastic birth process to predict the correlation between the frequency distribution of sources' productivity, modeled by means of the Waring distribution, and population growth. They applied the model to a sample of scientific authors' productivity data and derived a coefficient supposedly measuring the cumulative advantage effect.[76] Likewise, building on the premise that individual sources' productivity rates (e.g., the rate at which individual authors produce scientific papers) change over time according to a negative-binomial function, Burrell employed a form of the generalized Waring distribution to predict the resulting production process for the overall population under study.[77] In the same vein, he has recently argued that, because of the dynamic changes occurring in the concentration of sources' productivity over time, the traditional measures of inequality within a single dataset, the Gini Index above all, need to be complemented by comparative measures capable of expressing the similarity of concentration between two datasets obeying different bibliometric distributions.[78] In a parallel stream of research, among the many results of the long-term collaboration between Egghe and Rousseau is the proposal of a "generalized success-breeds-success model," which allows for a description of the parameters determining the evolution of an information production process more flexible than in Simon/Price's original scheme. Here the probability of a new item being produced by a new source is not assumed to be constant (a truly reasonable extension if one thinks of journals that slow down publishing new articles on a given subject as it becomes less fashionable), nor is it necessary to hypothesize that, in the case of a new item being produced by an already active source, the probability of this item originating from a source having already *n* items be proportional to the total number of sources with *n* items. The admissibility of different values for the latter probability enables the authors to explain a wide variety of frequency distributions, including a time-dependent version of Lotka's Law, and also "situations where success does not necessarily breed success, or where failures produces new failures."[79]

In Egghe's *Power Laws in the Information Production Process* (2005), the generalized success-breeds-success principle is packed in the theoretical framework of Lotkaian informetrics.[80] Here Lotka's Law is the point of departure for a thoroughly developed mathematical theory of information production processes (IPPs), wherein an attempt is made to prove that even apparently well-established principles, such as the success-breeds-success idea incorporated in early stochastic models by Simon and Price, can be reduced to more elementary statements. In the late 1980s, having written his Ph.D. thesis under Brookes' supervision, Egghe found a way out of his mentor's difficulties with the weaknesses of Gaussian statistics in social studies through a completely new approach of a deterministic (instead of probabilistic) nature. Its rationale is not to accept or take for granted a probabilistic interpretation of classic laws functional to the derivation of a model equation fitting as many datasets as possible, but to show that the existing regularities in informetrics can be formally derived from as few and simple assumptions as possible. To obtain such a generality, one needs a toolset of well-established mathematical functions and a small set of axioms, that is, propositions taken as true starting points of the deductive apparatus, though not necessarily "true" in the absolute sense. Axioms are the building blocks of mathematical models, whose chief task is to make sense of a complex, otherwise elusive reality, and whose value is determined as much by the ability to predict empirical patterns as by the internal consistency and the number and quality of applications to similar patterns of events emerging in cognate areas of inquiry. The most fundamental step toward the axiomatization of bibliometric theory consists, according to Egghe, in assuming the validity of a duality principle similar to the duality existing in geometry between straight lines and points. Sources and items, in other words, have to be thought of as interchangeable; that is, every function applicable to sources producing items in this order becomes also automatically extendible to the reverse order of items "producing" the sources from which they actually spring. In the case of authors (sources) producing papers (items), for example, it is assumed that articles can be formally treated as sources of the authors (items) that produced them. Similarly, papers can be considered sources of the citations they receive (items), but also items derived from the citing sources' references.[81] Once the dual perspective is implemented, the scaffold of Lotkaian informetrics requires two further axioms:

1. All information production processes can be reduced to a size-frequency function f that gives, for every $n = 1, 2, \ldots,$ the number $f(n)$ of sources with n items; so, if sources are journals and items are the articles they publish, then $f(n)$ is the number of journals publishing n articles.
2. The size-frequency function f is a (decreasing) power function, a generalization of Lotka's Law:

$$f(n) = \frac{C}{n^a}$$

A power law function has many advantages over competing models. First of all, assuming that one is not working with discrete quantities, such as a set of countable articles, but with continuous variables, for example a hypothetical continuous density of journals producing articles, it can be demonstrated that a power type size-frequency function comprises all other informetric functions, including Zipf-type rank-frequency relations and exponential functions. Second, the form of a power law doesn't change with different values of n, thus allowing the comparison of results at different scales (self-similarity or scale-free property). Such a scale-free property is especially important in connecting informetric laws to the mainstream of contemporary mathematical research on fractal objects and dynamic complex systems. In a fractal object, echoes of larger shapes appear within smaller parts of the shape so that, zooming in on any part of a fractal, one finds the same amount of detail as before, as in snowflakes or coastlines inspected at different heights, even though the self-similarity can be approximate and, in many instances, only statistically defined. A power law curve also produces similar shapes at whatever magnitude in such a way that, given an IPP governed by a power law, the same size-frequency distribution is found at low- or high-productivity values. Thus, if the emergence of order in a complex system depends on the correlations existing between different levels of scale, the complexity of an IPP can be properly expressed by means of fractal geometry, and it can be shown that, if a is Lotka's exponent, $a = 1$ is the fractal dimension, that is, the degree of complexity, of the self-similar fractal associated with the Lotkaian IPP.

Stochastic as well as deterministic mathematical models exhibit one obvious limitation in the eyes of laypeople: they don't tell anything about possible physical or social causes behind the observed inequalities in informetric distributions. Descriptive, predictive, and explicative power, after all, are not complementary nor necessarily coextensive: for centuries astronomers have described and predicted the apparent positions and paths of heavenly bodies by means of geometric combinations of uniform circular motions, without ever worrying about the physical structure of the universe or even maintaining a false belief in geocentricism, but limiting themselves to "save phenomena." Some authors, however, have tried to graft the formal treatment of bibliometric laws onto a more general theory of information structures and dynamics rooted in physical analogies or philosophical premises. These attempts include the following:

1. Zipf's already recalled "principle of the least effort," and its adaptation to Shannon's information theory in Mandelbrot's early writings.
2. The model set forth in the late 1970s by the Rumanian engineer Aurel Avramescu, who pursued a foundation of bibliometric laws by an information diffusion theory modeled on Fourier's treatment of heat conduction, with information transfer processes assimilated to the transfer of thermal energy through matter; the transfer is driven by a "potential gradient" assimilated to the interest for published papers measured by citation rates, and the time dimension plays a pivotal role in the mathematical description of its effects.[82]
3. The rather speculative approach of the Bulgarian information scientist Ludmila Ivancheva, who revives Brookes's and Haitun's obsession with the non-Gaussian nature of information processes in trying to infer the mathematical structure of bibliometric distributions from the axioms of the "General Theory of Sciences" proposed in the 1990s by Georgi Stankov.[83]
4. The unified scientometric model recently developed, on the philosophical groundwork of Michel Callon and Bruno Latour's Actor-Network Theory, by Rafael Bailón-Moreno, whose goal to achieve a better fit to empirical data and a stronger deductive power in the derivation of previous formulations of Lotka-type functions is pursued by the adoption of a "fundamental equation" capable of generating, from time to time, power, exponential, or hybrid power-exponential expressions.[84]

On the whole, it can't be denied that the foregoing authors managed to translate the success-breeds-success principle into a class of functions capable of explaining the mechanism underpinning data production across a wide range of bibliographic and extra-bibliographic contexts. But, in a sense, it has been a Cadmean victory for the simple reason that, just as the same empirical dataset can always be approximated by different theoretical distributions, the same formulas can likewise be derived from quite distinct mathematical hypotheses. So there are clearly different and "non-unifiable" ways to pursue the unification. And if, on the one hand, the ubiquity and propensity of bibliometric laws to follow from different sets of assumptions testifies to their robustness or, as Bookstein puts it, "resilience to ambiguity," on the other hand, it is forbidden to infer the validity of a given set of assumptions by its sheer ability to describe the empirical manifestation of one of those formulations.

NOTES

1. Lotka, "Statistics—The Frequency Distribution of Scientific Productivity," *Journal of the Washington Academy of Sciences* 16, no. 12 (1926): 317–25; Bradford, "Sources of Information on Specific Subjects," *Engineering* 137, no. 3550 (1934): 85–86; Zipf, *The Psycho-Biology of Language: An Introduction to Dynamic Philology* (London: Routledge, 1936). Bradford's paper started to receive attention by the community of information scientists only after the publication of Bradford, *Documentation* (London: Crosby Lockwood, 1948).

2. Kendall, "Natural Law in the Social Sciences," *Journal of the Royal Statistical Society. Series A (General)* 124, no. 1 (1961): 3.

3. A power law model for the tail of the citation distribution is proposed, for example, in Sidney Redner, "How Popular Is Your Paper? An Empirical Study of the Citation Distribution," *European Physical Journal B* 4, no. 2 (1998): 131–34, arxiv.org/abs/cond-mat/9804163; alternatively, the use of a stretched exponential law in fitting the whole range of the citation distribution is documented in Jean Laherrere and Didier Sornette, "Stretched Exponential Distributions in Nature and Economy: Fat Tails with Characteristic Scales," *European Physical Journal B* 2, no. 4 (1998): 525–39, arxiv.org/abs/cond-mat/9801293.

4. Bensman, "Probability Distributions in Library and Information Science: A Historical and Practitioner Viewpoint," *Journal of the American Society for*

Information Science 51, no. 9 (2000): 816–33, www.garfield.library.upenn.edu/bensman/jasistprobart.pdf.

5. Mandelbrot, "The Pareto-Levy Law and the Distribution of Income," *International Economic Review* 1, no. 2 (1960): 79–106; Mandelbrot, "New Methods in Statistical Economics," *Journal of Political Economics* 71, no. 5 (1963): 421–40.

6. Huber, "Inventive Productivity and the Statistics of Exceedances," *Scientometrics* 45, no. 1 (1999): 33–53.

7. The negative effect of log normalization on the statistical classification of journal-journal citation patterns is discussed in Leydesdorff and Bensman, "Classification and Powerlaws: The Logarithmic Transformation," *Journal of the American Society for Information Science and Technology* 58, no. 11 (2006): 1470–86, users.fmg.uva.nl/lleydesdorff/log05/.

8. Burrell and Cane, "The Analysis of Library Data," *Journal of the Royal Statistical Society: Series A (General)* 145, no. 4 (1982): 439–71; Burrell, "The 80/20 Rule: Library Lore or Statistical Law?" *Journal of Documentation* 41, no. 1 (1985): 24–3. Brookes's work on Bradford's distribution is reviewed in subsection 4.3.1.

9. The model is developed in Price, "A General Theory of Bibliometric and Other Cumulative Advantage Processes," *Journal of the American Society for Information Science* 27, no. 2 (1976): 293–296, www.garfield.library.upenn.edu/price/pricetheory1976.pdf.

10. Allison, Krauze, and Long, "Cumulative Advantage and Inequality in Science," *American Sociological Review* 47, no. 5 (1982): 615–25.

11. Huber, "Invention and Inventivity Is a Random, Poisson Process: A Potential Guide to Analysis of General Creativity," *Creativity Research Journal* 11, no. 3 (1998): 240. Empirical findings supporting the author's thesis for a wide range of "performers" can be found in Huber, "Cumulative Advantage and Success-Breeds-Success: The Value of Time Pattern Analysis," *Journal of the American Society for Information Science* 49, no. 5 (1998): 471–76, topology.eecs.umich.edu/archive/huber_99.pdf, and Huber, "A New Method for Analyzing Scientific Productivity," *Journal of the American Society for Information Science and Technology* 52, no. 13 (2001): 1089–99.

12. Lotka, "Statistics," 317.

13. Sorin Solomon and Peter Richmond, "Stability of Pareto-Zipf Law in Non-Stationary Economies" (paper presented at the 5th Workshop on Economics and Heterogeneous Interacting Agents, Marseille, June 15–17, 2001), arxiv.org/abs/cond-mat/0012479v2.

14. Lotka, "Statistics," 323.

15. Jan Vlachý, "Frequency Distributions of Scientific Performance: A Bibliography of Lotka's Law and Related Phenomena," *Scientometrics* 1, no.1 (1978): 109–30.

16. Potter, "Lotka's Law Revisited," *Library Trends* 30, no. 1 (1981): 21–39, hdl.handle.net/2142/7189.

17. An ideal continuation of Potter's work on the factors to be taken into account when attempting to test the applicability of Lotka's Law may be found in Miranda Lee Pao, "An Empirical Examination of Lotka's Law," *Journal of the American Society for Information Science* 37, no. 1 (1986): 26–33, and Paul Travis Nicholls, "Bibliometric Modelling Processes and the Empirical Validity of Lotka's Law," *Journal of the American Society for Information Science* 40, no. 6 (1989): 379–85. Both studies substantially confirm the validity of the law, having tested it, respectively, against forty-eight and seventy datasets of empirical author productivity distributions.

18. Such is Herbert Simon's approach, which is applied to the modeling of Lotka's Law in Ye-Sho Chen, "Analysis of Lotka's Law: The Simon-Yule Approach," *Information Processing & Management* 25, no. 5 (1989): 527–44.

19. Bradford, *Documentation*, 108.

20. Bradford, "Sources of Information," 85.

21. The graphical formulation corresponds originally to two figures, labeled respectively "Fig. 1" and "Fig. 2" by Bradford both in the 1934 paper and in the book *Documentation*. Whereas the former is the actual curve connecting the data points from the two bibliographies, the latter is an idealized representation in which the point of transition from curve to line is clearly determined so as to define the upper bound of the core of the most productive journals.

22. Vickery, "Bradford's Law of Scattering," *Journal of Documentation* 4, no. 3 (1948): 198–203. The "bipolar" nature of the law was further discussed by Elizabeth A. Wilkinson, "The Ambiguity of Bradford's Law," *Journal of Documentation* 28, no. 2 (1972): 122–30. Here it is suggested that the verbal formulation expressed Bradford's theory, while the graphical formulation expressed his observations.

23. Kendall, "The Bibliography of Operational Research," *Operational Research Quarterly* 11, nos. 1–2 (1960): 31–36.

24. Leimkuhler, "The Bradford Distribution," *Journal of Documentation* 23, no. 3 (1967): 207. The equivalence between Leimkuhler's and Vickery's formulations is proved in John J. Hubert, "Bibliometric Models for Journal Productivity," *Social Indicators Research* 4, no. 4 (1977): 441–73. The search for an exact formulation of Bradford's Law started by Vickery and Leimkuhler was further pursued by many other authors. An early classification of bibliometric

laws and their extension from an end-user-oriented perspective can be found in Adedoyin Soyibo and W. Olabode Aiyepeku, "On the Categorization, Exactness and Probable Utility of Bibliometric Laws and Their Extension," *Journal of Information Science* 14, no. 4 (1988): 243–51. For a comprehensive review of the mathematical evolution of Bradford's Law, see Vesna Oluić-Vuković, "Bradford's Distribution: From the Classical Bibliometric 'Law' to the More General Stochastic Models," *Journal of the American Society for Information Science* 48, no. 9 (1998): 833–42.

25. Qiu, "An Empirical Examination of the Existing Models for Bradford's Law," *Information Processing & Management* 26, no. 5 (1990): 655–72. The "Groos droop" is reported in Ole V. Groos, "Bradford's Law and the Keenan-Atherton Data," *American Documentation* 18, no. 1 (1967): 46. Various hypotheses have been proposed on the causes of the S-shape imposed on Bradford's distribution by the droop, including the incompleteness of the dataset (Brookes) and the merging of pure Bradfordian patterns typical of interdisciplinary fields (Egghe and Rousseau). Conclusive evidence in favor of a single hypothesis doesn't exist. However, the incompleteness argument has been refuted, on the basis of a computer-simulated experiment, in Qiu and Tague, "Complete or Incomplete Data Sets: The Groos Droop Investigated," *Scientometrics* 19, no. 3 (1990): 223–37.

26. Brookes, "The Complete Bradford-Zipf 'Bibliograph'," *Journal of Documentation* 25, no. 1 (1969): 58–60. See also Brookes, "Bradford's Law and the Bibliography of Science," *Nature* 224, no. 5223 (1969): 953–56.

27. Brookes and Griffith, "Frequency-Rank Distributions," *Journal of the American Society for Information Science* 29, no. 1 (1978): 13.

28. The equivalence is discussed in Brookes, "Ranking Techniques and the Empirical Log Law," *Information Processing & Management* 20, nos. 1–2 (1984): 37–46.

29. Brookes, "Theory of the Bradford Law," *Journal of Documentation* 33, no. 3 (1977): 180.

30. On the implications of Bradford's Law for the social sciences, see Brookes, "The Bradford Law: A New Calculus for the Social Sciences?" *Journal of the American Society for Information Science* 30, no. 4 (1979): 233–34; Brookes, "The Haitun Dichotomy and the Relevance of Bradford's Law," *Journal of Information Science* 8, no. 1 (1984): 19–24; Brookes, "Towards Informetrics: Haitun, Laplace, Zipf, Bradford and the Alvey Programme," *Journal of Documentation* 40, no. 2 (1984): 120–43.

31. Concentration formulae for the classic bibliometric distributions of Zipf, Mandelbrot, and Lotka, and the geometric distribution have been derived by Egghe, who sought also to 1) establish the exact relation between Pratt's Index

and the 80/20 rule; 2) escape the circularity inherent in the use of a concentration index for defining concentration itself by fixing some basic principles any concentration measure must comply with; and 3) specify under which conditions concentration indexes independent of the size of the population can be developed. See Egghe, "Pratt's Measure for Some Bibliometric Distributions and Its Relation with the 80/20 Rule," *Journal of the American Society for Information Science* 38, no. 4 (1987): 288–97; Egghe and Rousseau, "Transfer Principles and a Classification of Concentration Measures," *Journal of the American Society for Information Science* 42, no. 7 (1991): 479–89; Bonckaert and Egghe, "Rational Normalizaton of Concentration Measures," *Journal of the American Society for Information Science* 42, no. 10 (1991): 715–22.

32. The gap between empirical and theoretical considerations of the phenomenon described by Bradford's Law has been pointed out in Carl M. Drott, "Bradford's Law: Theory, Empiricism and the Gaps Between," *Library Trends* 30, no. 1 (1981): 42–52, hdl.handle.net/2142/7189.

33. White, "'Bradfordizing' Search Output: How It Would Help Online Users," *Online Review* 5, no. 1 (1981): 47–54; Marcia J. Bates, "Document Familiarity, Relevance, and Bradford's Law: The Getty Online Searching Project Report," *Information Processing & Management* 32, no. 6 (1996): 697–707.

34. Nicolaisen and Hjørland, "Practical Potentials of Bradford's Law: A Critical Examination of the Received View," *Journal of Documentation* 63, no. 3 (2007): 359–77, dlist.sir.arizona.edu/2123/.

35. Bensman, "Bradford's Law and Fuzzy Sets: Statistical Implications for Library Analyses," *IFLA Journal* 27, no. 4 (2001): 238–46, www.garfield.library.upenn.edu/bensman/iflabradfordslawfuzzysets.pdf; Egghe and Rousseau, "A Proposal to Define a Core of a Scientific Subject: A Definition Using Concentration and Fuzzy Sets," *Scientometrics* 54, no. 1 (2002): 51–62, hdl.handle.net/1942/762. See also Burrell, "Defining a Core: Theoretical Observations on the Egghe-Rousseau Proposal," *Scientometrics* 57, no. 1 (2003): 75–92.

36. Garfield, *Citation Indexing: Its Theory*, 23. See also two previous contributions: Garfield, "The Mystery of the Transposed Journal Lists Wherein Bradford's Law of Scattering Is Generalized According to Garfield's Law of Concentration," in *Essays of an Information Scientist 1962–1973*, 222–23 (Philadelphia: ISI Press, 1977), www.garfield.library.upenn.edu/essays/V1p222y1962–73.pdf; and Garfield, "Bradford's Law and Related Statistical Patterns," in *Essays of an Information Scientist 1979–1980*, 476–83 (Philadelphia: ISI Press, 1981), www.garfield.library.upenn.edu/essays/v4p476y1979–80.pdf.

37. Bensman, "Journal Collection Management as a Cumulative Advantage Process," *College & Research Libraries* 46, no. 1 (1985): 13–29, www.garfield.library.upenn.edu/bensman/crl85cum.pdf. In a 1976 study commissioned by

Maurice Line, then director of the British Library Lending Division, Pauline A. Scales resorted to the same dataset upon which Garfield had built his theory of concentration and examined the correlation between citation patterns and the results of a local use survey of journal issues. At a first glance, the low score obtained for this correlation pointed at the uselessness of citation-based ranked lists of journals for acquisition policies, but in the sharp dispute that followed there emerged the flawed use of nonparametric statistical techniques lurking behind that score. See Bensman, "Urquhart's and Garfield's Laws: The British Controversy over Their Validity," *Journal of the American Society for Information Science and Technology* 52, no. 9 (2001): 714–24, www.garfield.library.upenn.edu/bensman/jasisturqgar.pdf. Bensman has devoted a series of papers to the key role played by Urquhart in the development of a probabilistic framework for the analysis of scientific and technical journal use. Much of this material is freely available on Garfield's website at www.garfield.library.upenn.edu/bensman/bensman.html.

38. For an outline of the methodology, see Braun, "Keeping the Gates of Science Journals: Gatekeeping Indicators of National Performance in the Sciences," in *Handbook of Quantitative Science and Technology Research: The Use of Publication and Patent Statistics in Studies of S&T Systems*, ed. Henk F. Moed, Wolfgang Glänzel and Ulrich Schmoch, 95–111 (Dordrecht: Kluwer Academic, 2004).

39. For an appraisal of the potential and limitations of citation analysis in measuring Third World science, see S. Arunachalam and K. Manorama, "Are Citation-Based Quantitative Techniques Adequate for Measuring Science on the Periphery?" *Scientometrics* 15, nos. 5–6 (1989): 393–408. An early experiment of Third World science citation analysis is reported in Garfield, "Mapping Science in the Third World," *Science and Pubic Policy* 10, no. 3 (1983): 112–27, www.garfield.library.upenn.edu/papers/mapscithirdworldp112y1983.pdf. The article contains the text of the Magnus Pyke Lecture delivered in December 1982 at the U.S. Embassy in London. A recent approach to the analysis of African scientific literature from the standpoint of a leading scientometric school can be found in Tijssen, "Africa's Contribution to the Worldwide Research Literature: New Analytical Perspectives, Trends, and Performance Indicators," *Scientometrics* 71, no. 2 (2007): 303–27, www.springerlink.com/content/e40126m448gv0127/?p=7f6f962118034e1d8e6449a2632c6eb9&pi=9.

40. Zipf, *The Psycho-Biology of Language: An Introduction to Dynamic Philology* (London: Routledge, 1936), 6.

41. Zipf, *The Psycho-Biology*, v.

42. For a global perspective on these models, see Ronald E. Wyllys, "Empirical and Theoretical Bases of Zipf's Law," *Library Trends* 30, no. 1 (1981):

53–64, hdl.handle.net/2142/7189, and Jane Fedorowicz, "The Theoretical Foundation of Zipf's Law and Its Application to the Bibliographic Database Environment," *Journal of the American Society for Information Science* 33, no. 5 (1982): 285–93.

43. Mandelbrot, "An Information Theory of the Statistical Structure of Language," in *Communication Theory*, ed. Willis Jackson, 486–502 (London: Butterworths Scientific Publications, 1953); Mandelbrot, "Structure Formelle Des Textes Et Communication," *Word* 10 (1954): 1–27. On the adaptation of Mandelbrot's version to artificial languages, see Charles T. Meadow, Jiabin Wang, and Manal Stamboulie, "An Analysis of Zipf-Mandelbrot Language Measures and Their Application to Artificial Languages," *Journal of Information Science* 19, no. 4 (1993): 247–58.

44. In the wake of Price's breakthrough, an early investigation of the differing growth rates in various research fields, their possible causes, and effects on the lives of scientists is Henry W. Menard, *Science: Growth and Change* (Cambridge, Mass.: Harvard University Press, 1971). A review of growth models can be found in Albert N. Tabah, "Literature Dynamics: Studies on Growth, Diffusion, and Epidemics," *Annual Review of Information Science and Technology* 34 (1999): 249–76. Specifically, the offspring of Price's exponential and logistic models is reviewed by Antonio Fernandez-Cano, Manuel Torralbo, and Monica Vallejo, "Reconsidering Price's Model of Scientific Growth: An Overview," *Scientometrics* 61, no. 3 (2004): 301–21. Critical views on the quantitative measures of science growth can be found in, among others, Manfred Kochen, "Stability in the Growth of Knowledge," *American Documentation* 20, no. 3 (1969): 186–97; Michael J. Moravcsik, "Measures of Scientific Growth," *Research Policy* 2, no. 3 (1973): 266–75; G. Nigel Gilbert and Steve Woolgar, "The Quantitative Study of Science: An Examination of the Literature," *Science Studies* 4, no. 3 (1974): 279–94. A model for classifying existing models of scientific growth is propounded in Egghe and Ravichandra Rao, "Classification of Growth Models Based in Growth Rates and Its Implications," *Scientometrics*, 25, no. 1 (1992): 5–46. It is applied to the literature of theoretical population genetics from 1907 to 1980 by B. M. Gupta and C. R. Karisiddappa, "Modelling the Growth of Literature in the Area of Theoretical Population Genetics," *Scientometrics*, 49, no. 2 (2000): 321–55.

45. Goffman and Newill, "Generalization of Epidemic Theory: An Application to the Transmission of Ideas," *Nature* 204, no. 4955 (1964): 225–28.

46. Bettencourt et al., "Population Modeling of the Emergence and Development of Scientific Fields," *Scientometrics* 75, no. 3 (2008): 495–518, web.mit.edu/dikaiser/www/BKKCW.PopModels.pdf.

47. Price, *Science Since Babylon*, 189.

48. Price, *Science Since Babylon*, 170.

49. May, "Growth and Quality of the Mathematical Literature," *ISIS* 59, no. 4 (1968): 363–71; Tague, Beheshti, and Rees-Potter, "The Law of Exponential Growth: Evidence, Implications and Forecasts," *Library Trends* 30, no. 1 (1981): 125–49, hdl.handle.net/2142/7189.

50. van Raan, "On Growth, Ageing, and Fractal Differentiation of Science," *Scientometrics* 47, no. 2 (2000): 347–62, www.cwts.nl/TvR/documents/AvR-Growth-Scientometrics.pdf.

51. van Raan, "On Growth," 360.

52. Price, *Science Since Babylon*, 163.

53. Two useful reviews of obsolescence studies from a library service perspective are D. Kaye Gapen and Sigrid P. Milner, "Obsolescence," *Library Trends* 30, no. 1 (1981): 107–24, hdl.handle.net/2142/7189, and Line, "Changes in the Use of Literature with Time—Obsolescence Revisited," *Library Trends* 41, no. 4 (1993): 665–83, hdl.handle.net/2142/7869.

54. Burton and Kebler, "The Half-Life of Some Scientific and Technical Literatures," *American Documentation* 11, no. 1 (1960): 19. The origins of the term and concept of literature "half-life" have been frequently attributed to this article, but the groundlessness of such attribution and a careful analysis of earlier uses are documented in Endre Száva-Kováts, "Unfounded Attribution of the "Half-Life" Index-Number of Literature Obsolescence to Burton and Kebler: A Literature Science Study," *Journal of the American Society for Information Science* 53, no. 13 (2002).

55. Brookes, "The Growth, Utility and Obsolescence of Scientific Periodical Literature," *Journal of Documentation* 26, no. 4 (1970): 283–94. Brookes's method was applied to computing the obsolescence of scientific literature indexed by the *SCI* up to 1979 in Griffith et al., "The Aging of Scientific Literature: A Citation Analysis," *Journal of Documentation* 35, no. 3 (1979): 179–96. The dependence on time of the aging factor, which Brookes assumed as constant even though recognizing the limits of such an assumption, is discussed in Egghe and Ravichandra Rao, "Citation Age Data and Obsolescence Function: Fits and Explanations," *Information Processing & Management* 28, no. 2 (1992): 201–17.

56. Nakamoto, "Synchronous and Diachronous Citation Distributions," in *Informetrics 87/88: Select Proceedings of the First International Conference on Bibliometrics and Theoretical Aspects of Information Retrieval*, ed. Egghe and Rousseau, 157–63 (Amsterdam: Elsevier, 1988), doclib.uhasselt.be/dspace/handle/1942/837. A best-fitting exercise on Nakamoto's data was performed by Thus Pollmann, "Forgetting and the Ageing of Scientific Publications," *Scientometrics* 47, no. 1 (2000): 43–54.

57. Larivière, Archambault, and Gingras, "Long-Term Variations in the Aging of Scientific Literature: From Exponential Growth to Steady-State Science (1900–2004)," *Journal of the American Society for Information Science and Technology* 59, no. 2 (2008): 288–96, www.ost.uqam.ca/Portals/0/docs/articles/2008/JASIST_Aging.pdf.

58. Two remarkable stochastic models of literature aging based, respectively, on a mixture of Poisson processes and an inhomogeneous stochastic birth process, were developed during the 1980s and 1990s. The former can be found in Burrell, "A Note on Ageing in a Library Circulation Model," *Journal of Documentation* 41, no. 2 (1985): 100–15; for the latter see Glänzel and Schoepflin, "A Stochastic Model for the Ageing of Scientific Literature," *Scientometrics* 30, no. 1 (1994): 49–64. More recently, a test on the applicability of different models (log-normal, Weibull, log-logistic) has been performed, with tools borrowed from the analysis of survival data, in Jesús Basulto Santos and Francisco Javier Ortega Irizo, "Modelling Citation Data with Right Censoring," *Scientometrics* 62, no. 3 (2005): 329–42. For a review of stochastic modeling on literature obsolescence up to the mid-1980s, see James P. Coughlin and Robert H. Baran, "Stochastic Models of Information Obsolescence," *Mathematical and Computer Modelling* 11 (1988): 760–65. An alternative solution, which keeps aloof from Poisson-derived formulations while opting for a more manageable sum of exponential functions, can be found in George A. Barnett, Edward L. Fink, and Mary Beth Debus, "A Mathematical Model of Academic Citation Age," *Communication Research* 16, no. 4 (1989): 510–31; recently, the same model has been adapted to the Internet environment in Barnett and Fink, "Impact of the Internet and Scholar Age Distribution on Academic Citation Age," *Journal of the American Society for Information Science and Technology* 59, no. 4 (2008): 526–34.

59. The problem of controlling the effect of growth on obsolescence was raised in Line, "'Half Life' of Periodical Literature: Apparent and Real Obsolescence," *Journal of Documentation* 26, no. 1 (1970): 46–52, and Line and Sandison, "'Obsolescence' and Changes in the Use of Literature with Time," *Journal of Documentation* 30, no. 3 (1974): 283–350. It was subsequently addressed, on more firm mathematical grounds, in Egghe, "On the Influence of Growth on Obsolescence," *Scientometrics* 27, no. 2 (1993), and Egghe and Rousseau, "Aging, Obsolescence, Impact, Growth and Utilization: Definitions and Relations," *Journal of the American Society for Information Science* 51, no. 11 (2000): 1004–17, hdl.handle.net/1942/786.

60. Abt, "Long-Term Citation Histories of Astronomical Papers," *Publications of the Astronomical Society of the Pacific* 93, no. 552 (1981): 207–10, adsabs.harvard.edu/abs/1981PASP...93..207A; Robert T. Bottle and Y. T. Gong, "A Bibliometric Study on the Ageing and Content Typology Relationship of

the Biochemical Literature," *Journal of Information Science* 13, no. 1 (1987): 59–63.

61. For an accessible introduction to the subject, see Bookstein, "Explanation of the Bibliometric Laws," *Collection Management* 3, no. 2/3 (1979): 151–62, and Rousseau and Rousseau, "Informetric Distributions: A Tutorial Review," *Canadian Journal of Information and Library Science* 18, no. 2 (1993): 51–63. An outline of informetric models can be found in Ravichandra Rao, "Classical Laws of Informetrics: An Overview" (paper presented at the DRTC Workshop on Informetrics and Scientometrics, Bangalore, March 16–19, 1998), https://drtc.isibang.ac.in/bitstream/1849/143/2/AB.pdf.

62. Chen and Leimkuhler, "A Relationship Between Lotka's Law, Bradford's Law, and Zipf's Law," *Journal of the American Society for Information Science* 37, no. 5 (1999): 307–14.

63. Simon, "On a Class of Skew Distribution Functions," *Biometrika* 42, nos. 3/4 (1955): 425. For Mandelbrot's criticisms of Simon's model, see Mandelbrot, "A Note on a Class of Skew Distribution Functions: Analysis and Critique of a Paper by H. A. Simon," *Information and Control* 2, no. 1 (1959): 90–99; Mandelbrot, "Final Note on a Class of Skew Distribution Functions: Analysis and Critique of a Model Due to H. A. Simon," *Information and Control* 4, nos. 2–3 (1959): 198–216.

64. Kendall, "The Bibliography of Operational Research," 34.

65. Fairthorne, "Empirical Hyperbolic Distributions (Bradford-Zipf-Mandelbrot) for Bibliometric Description and Prediction," *Journal of Documentation* 25, no. 4 (1969): 319–43.

66. Sundaresan Naranan, "Bradford's Law of Bibliography of Science: An Interpretation," *Nature* 227, no. 5258 (1970): 631–32.

67. Yablonsky, "On Fundamental Regularities of the Distribution of Scientific Productivity," *Scientometrics* 2, no. 1 (1980): 3–34; Yablonsky, "Stable Non-Gaussian Distributions in Scientometrics," *Scientometrics* 7, no. 3–6 (1985): 459–70.

68. Haitun, "Stationary Scientometric Distributions, Part I: Different Approximations," *Scientometrics* 4, no. 1 (1982): 5–25; Haitun, "Stationary Scientometric Distributions, Part II: Non-Gaussian Nature of Scientific Activities," *Scientometrics* 4, no. 2 (1982): 89–104; Haitun, "Stationary Scientometric Distributions, Part III: The Role of the Zipf Distribution," *Scientometrics* 4, no. 3 (1982): 181–94.

69. Sichel, "A Bibliometric Distribution Which Really Works," *Journal of the American Society for Information Science* 36, no. 5 (1985): 314–21.

70. Sichel, "Anatomy of the Generalized Inverse Gaussian-Poisson Distribution with Special Applications to Bibliometric Studies," *Information Processing & Management* 28, no. 1 (1992): 5–17; Burrell and Fenton, "Yes, the GIGP Re-

ally Does Work—and Is Workable!" *Journal of the American Society for Information Science* 44, no. 2 (1993): 61–69.

71. The original model is developed in Bookstein, "Patterns of Scientific Productivity and Social Change: A Discussion of Lotka's Law and Bibliometric Symmetry," *Journal of the American Society for Information Science* 28, no. 4 (1977): 206–10. A further specification appeared in a series of subsequent articles: Bookstein, "Informetric Distributions, Part I: Unified Overview," *Journal of the American Society for Information Science* 41, no. 5 (1990): 368–75, topology.eecs.umich.edu/archive/bookstein_I.pdf; Bookstein, "Informetric Distributions, Part II: Resilience to Ambiguity," *Journal of the American Society for Information Science* 41, no. 5 (1990): 376–86, topology.eecs.umich.edu/archive/bookstein_II; Bookstein, "Informetric Distributions, Part III: Ambiguity and Randomness," *Journal of the American Society for Information Science* 48, no. 1 (1997): 2–10; Bookstein, "Implications of Ambiguity for Scientometric Measurement," *Journal of the American Society for Information Science and Technology* 52, no. 1 (2001): 74–79. The originality of Bookstein's model has been questioned, on the score of its indebtedness to previous formulations, in Hubert, "General Bibliometric Models," *Library Trends* 30, no. 1 (1981): 71–72, hdl.handle.net/2142/7189, and objections to the mathematics and the general validity of his theory are raised in Burrell, "Ambiguity and Scientometric Measurement: A Dissenting View," *Journal of the American Society for Information Science and Technology* 52, no. 12 (2001): 1075–80.

72. For Rao's and Tague's models see, respectively, Ravichandra Rao, "The Distribution of Scientific Productivity and Social Change," *Journal of the American Society for Information Science* 31, no. 2 (1980): 111–22, and Tague, "The Success-Breeds-Success Phenomena and Bibliometric Processes," *Journal of the American Society for Information Science* 32, no. 4 (1981): 280–86.

73. Price, "A General Theory," 303. On the above-mentioned dispute, see Kantor, "A Note on Cumulative Advantage Distributions," *Journal of the American Society for Information Science* 29, no. 4 (1978): 202–4, and the reply in Price, "Cumulative Advantage Urn Games Explained: A Reply to Kantor," *Journal of the American Society for Information Science* 29, no. 4 (1978): 204–6.

74. After Vlachý, "Time Factor in Lotka's Law," *Probleme de Informare si Documentare* 10, no. 2 (1976): 44–87, and Braga, "Some Aspects of the Bradford Distribution," *Proceedings of the American Society for Information Science* 15 (1978): 51–54, the effect of time and historical changes at the disciplinary level on Bradford's distribution has been investigated in several bibliographic datasets, including

1. a sample of references from Croatian authors in the humanities in Oluić-Vuković, "Impact of Productivity Increase on the Distribution Patterns of Journals," *Scientometrics* 17, nos. 1–2 (1989): 97–109;
2. sample references from Croatian chemists and physicists in Oluić-Vuković, "Journal Productivity Distribution: Quantitative Study of Dynamic Behavior," *Journal of the American Society for Information Science* 43, no. 6 (1992): 412–21;
3. productivity patterns of journals in mathematical logic (1891–1990) and mathematics (1801–1990) in Roland Wagner-Döbler, "Time Dependencies of Bradford Distributions: Structures of Journal Output in 20th-Century Logic and 19th-Century Mathematics," *Scientometrics* 39, no. 3 (1997): 231–52;
4. sixteen bibliographic datasets covering both technical and nontechnical topics in S. R. Coleman, "Disciplinary Variables That Affect the Shape of Bradford's Bibliograph," *Scientometrics* 29, no. 1 (1994): 59–81;
5. Hjerppe's bibliography of bibliometrics in Bluma C. Peritz, "A Bradford Distribution for Bibliometrics," *Scientometrics* 18, nos. 5–6 (1990): 323–29;
6. computer-simulated bibliographic data in Ye-Sho Chen, P. Pete Chong, and Morgan Y. Tong, "Dynamic Behavior of Bradford's Law," *Journal of the American Society for Information Science* 46, no. 5 (1995): 370–83; and
7. three bibliographies dealing, respectively, with the Lanczos and conjugate gradient algorithms, information theory applications to information science, and earthquakes, in S. K. Sen and S. K. Chatterjee, "Bibliographic Scattering and Time: An Empirical Study Through Temporal Partitioning of Bibliographies," *Scientometrics* 41, no. 1–2 (1998): 135–54.

75. This is, for example, the result of an experiment on authors' productivity patterns in mathematical logic over nearly two centuries, reported in Wagner-Döbler, "Where Has the Cumulative Advantage Gone? Some Observations About the Frequency Distribution of Scientific Productivity, of Duration of Scientific Participation, and of Speed of Publication," *Scientometrics* 32, no. 2 (1995): 123–32. Before him, a similar drawback of Lotka's Law had been discussed by Allison, "Inequality and Scientific Productivity," *Social Studies of Science* 10, no. 2 (1980): 163–79.

76. Schubert and Glänzel, "A Dynamic Look at a Class of Skew Distributions: A Model with Scientometric Applications," *Scientometrics* 6, no. 3 (1984): 149–67.

77. Burrell, "Modelling the Bradford Phenomenon," *Journal of Documentation* 44, no. 1 (1988): 1–18. For an exemplification of both parametric and Bayes modeling options, see Burrell, "On the Growth of Bibliographies with Time: An

Exercise in Bibliometric Prediction," *Journal of Documentation* 45, no. 4 (1989): 302–17.

78. The "relative concentration coefficient" and the "co-concentration coefficient" are the two comparative measures introduced for this purpose in Burrell, "Measuring Similarity of Concentration Between Different Informetric Distributions: Two New Approaches," *Journal of the American Society for Information Science and Technology* 56, no. 7 (2005): 704–14.

79. Egghe and Rousseau, "Generalized Success-Breeds-Success Principle Leading to Time-Dependent Informetric Distributions," *Journal of the American Society for Information Science* 46, no. 6 (1995): 427. For a deeper stochastic analysis of the principle, see Egghe and Rousseau, "Stochastic Processes Determined by a General Success-Breeds-Success Principle," *Mathematical and Computer Modelling* 23, no. 4 (1996): 93–104.

80. Egghe, *Power Laws in the Information Production Process: Lotkaian Informetrics* (Amsterdam: Elsevier Academic Press, 2005).

81. Egghe, "The Duality of Informetric Systems with Applications to the Empirical Laws," *Journal of Information Science* 16, no. 1 (1990): 17–27.

82. Avramescu, "Actuality and Obsolescence of Scientific Literature," *Journal of the American Society for Information Science* 30, no. 5 (1979): 296–303; Avramescu, "Theoretical Foundation of Bradford's Law," *International Forum on Information and Documentation* 5, no. 1 (1980): 15–22.

83. Ivancheva, "The Non-Gaussian Nature of Bibliometric and Scientometric Distributions: A New Approach to Interpretation," *Journal of the American Society for Information Science and Technology* 52, no. 13 (2001): 1100–1105.

84. Bailón-Moreno et al., "The Unified Scientometric Model: Fractality and Transfractality," *Scientometrics* 63, no. 2 (2005): 231–57. This model is the companion to the "quantitative model of translation" for the structure and dynamics of scientific networks issued in the late 1990s by Rosario Ruiz Baños's team. See Ruiz- Baños et al., "Structure and Dynamics of Scientific Networks, Part I: Fundamentals of the Quantitative Model of Translation," *Scientometrics* 44, no. 2 (1999): 217–34; Ruiz-Baños et al., "Structure and Dynamics of Scientific Networks, Part II: The New Zipf's Law, the Clusters of Co-Citations and the Model of the Descriptor Presence," *Scientometrics* 44, no. 2 (1999): 235–65.

Chapter Five

Maps and Paradigms: Bibliographic Citations at the Service of the History and Sociology of Science

It is perhaps because scientists behave as if they are discovering something independent of their mental and physical constructs, something already out there, that a common way to portray science makes use of metaphors inspired by modular objects and surfaces with clear-cut boundaries. So, for instance, science is often depicted geographically as an aggregation of research areas or fields, or, to stress the artificial character of any classification, as a mosaic, or a jigsaw puzzle,[1] comprising a certain number of individual units (documents) growing together into subject-related repositories (journals) that, in turn, evolve into the documentary source of well-established, eventually institutionalized specialties and disciplines. From the perspective of science studies, it's quite reasonable to assume that the jigsaw puzzle's big picture is unknown to scientists, who are used to seeing and handling only a few pieces at a time, and that a suitable method of analysis can help to carry the game through. What makes the task challenging, if not unfeasible, is that it is a moving, shifting jigsaw puzzle in which the picture outline, as well as the form of the individual pieces, are not fixed, but in a constant flux. The overall structure changes, accordingly, on the side of both the subject and the object. It changes because of the internal transformations of the communication patterns under investigation, and it inevitably varies in response to the theoretical perspectives and specific methods applied to the study of those communication patterns.

To make the jigsaw puzzle fall into place, that is, to build a reliable model of knowledge domains, is an old dream of scientometricians, sociologists, and science policymakers. Appealing to the aforementioned geographic analogy, they usually qualify this task as the mapping of an intellectual landscape. A map of science is a spatial representation of the relationships among disciplines, fields, specialties, and individual papers (or authors) as reflected in some formal, strictly quantifiable properties of scientific literature at a given time. A map is a desirable object for various reasons. First, it supplies a description of the intellectual structure of a research area independent of subjective judgments and relevance criteria, so that managers and science administrators can rely on it for tracing and evaluating the relative position and strength of the actors on the stage (research groups and individual scientists, universities, nations). Second, it's a potential aid in searching for and retrieving relevant information out of large data collections through user-friendly interfaces that take advantage of the human mind's familiarity with the spatial organization of objects and concepts. Finally, it provides an empirical basis for testing the significance of such abstract constructs as "discipline," "specialty," "paradigm," and "scientific community."

It's quite a commonplace, in light of current historiography, to contend that maps do not faithfully reproduce a landscape, nor can they be trusted as truly objective representations of an outer reality. Indeed, even the most apparently unobtrusive kind of cartography, the design and production of geographical maps, has been deconstructed by historians, who have managed to show that a map is a model, a strategic simplification of reality that, by choosing what to represent, what to discard, and how to portray a segment of the physical territory, betrays its social context of production and its latent identity of technology of power.[2] Bibliometric maps are no exception insofar as the very possibility to construct them and the value attached to their use are strictly dependent on a series of nontrivial theoretical premises concerning the type and quality of the information conveyed by the cartographic units.

To build a two- or three-dimensional map of a research field, bibliometricians generally exploit the co-occurrence of textual or bibliographic data: the more two documents are cited together (co-cited) or, alternatively, the more two elements, such as bibliographic references, title words, keywords, and author names, appear together in a given set of documents, the more likely it is that their repeated co-link has something to say about the sociocognitive structure of the subject area to which the papers belong. Consequently, their position is represented by nearby points on the surface of the map. If, for instance, the cartographic units are bibliographic citations and one accepts Garfield and Small's postulate that citations are words of a highly specialized symbolic language governed by Mertonian rules, then the reconstruction of the citation and co-citation network among scientific documents in a specialized area over a given period of time may be thought to reflect its sociocognitive structure and evolution. Similarly, if the cartographic units are text words and one assumes that the way authors combine words in their articles reflects their world-building strategies instead of an abstract set of methodological rules, then word co-occurrences become an indicator of the different poles of interests that concur to shape the structure and the dynamics of a research field.

5.1. POWER OF TEXTS, POWER OF AUTHORS: CO-WORD ANALYSIS FOR MAPPING THE DYNAMICS OF SCIENCE AND TECHNOLOGY

Scientometrics is not completely citation-addicted, nor is it necessarily the by-product of a normative theory of science. Scientists produce textual stuff to outlive the struggle for life on the research front. Text itself or its surrogates (titles, abstract, keywords), therefore, can be analyzed and measured, instead of citations, to detect constant associations of scientific concepts for delineating subject areas, growing subfields, or disciplinary patterns. In its most codified expression, moreover, text is packaged into journal papers, which, in addition to knowledge claims, also contain relevant information on various aspects of the organization of research, such as authors, institutions, and journals, thereby supplying science studies with a promising tool of investigation at the crossroads of different analytic dimensions. The best-known technique for scientometric text mining is co-word analysis, developed during the 1980s within the orbit of the Centre de Sociologie de l'Innovation of the École Nationale Supérieure des Mines of Paris.[3]

In simplest terms, a co-word analysis involves three steps, discussed below.

Co-word Extraction

Given a text corpus, typically a set of scientific documents, the frequency of co-occurrence of all the meaningful pairs of words (or noun phrases) is calculated. Two words, W_i and W_j are said to co-occur in the corpus if there is at least one document containing both W_i and W_j. The strength of the link between W_i and W_j is given by the number C_{ij} of documents in which the couple (W_i, W_j) appears. Often keywords assigned manually by professional indexers are analyzed instead of title, abstract or full-text words. This solution, however, typical of early co-word experiments, has been criticized on the score of the "indexer effect" caused by the artificiality of the indexing lexicon, its delay with respect to the concrete dynamics of scientific language, and the subjectivity of the indexer's own preclassification of the document content. Similar shortcomings also hamper the effectiveness of many information retrieval systems, but it is worth remembering that the co-word perspective is not comparable to that of information retrieval. Here words are not meant to sum up ideas or information; they are not taken for their meaning, nor are the syntax and logical connections between them a matter of some concern. Rather, words count as indicators of links between documents, they are "macro-terms" defining and imposing, through the alchemy of their combinations, particular sets of problems (and solutions). Co-word analysis, then, rests on the assumption that scientific authors pick their terms out of a somewhat finite and codified repertoire, that they use different terms whenever they postulate nontrivial relationships between concepts, and that the recognition of the same relationships by different authors tells something about the cognitive structure of a research area.[4]

Co-word Classification

Once all the pairs of words have been counted, the C_{ij} values are arranged in a co-occurrence symmetrical matrix, which undergoes a wide variety of algebraic manipulations for the extraction of quantitative information on the patterns of word pair frequencies. In the most general case, one seeks to determine both the hierarchy of research problems and the minor but potentially growing areas of the scientific domain represented in the document collection; hence, it is necessary to define classification criteria of inclusion, proximity, and equivalence of words based on the relative strength of the co-occurrence links. This is done by calculating the inclusion index and the proximity index. The inclusion index is defined as

$$I_{ij} = \frac{C_{ij}}{\min(C_i, C_j)}$$

where C_{ij} is the number of documents in which the word pair (W_i, W_j) appears; C_i and C_j are, respectively, the occurrence frequencies of the words W_i and W_j in the set of documents; and $min(C_i, C_j)$ is the minimum of C_i and C_j. If, for instance, C_j is the minimum, then the inclusion index I_{ij} measures the probability of finding W_i in an article, given that W_j is already present. In the extreme case of $I_{ij} = 1$, the occurrence of W_j implies that W_i is also present in each article. In terms of co-word analysis, this perfect correspondence is interpreted as a full inclusion of the first word in the second. Of course, this is not a typical case; different levels of inclusion usually exist between the word pairs, and the core of the mapping procedure consists of calculating and classifying them. At the peripheral level, the situation in which some words, though having a relatively low occurrence frequency, preserve significant relationships with other words is captured by the proximity index, defined as

$$P_{ij} = \frac{C_{ij}}{C_i C_j} \cdot N$$

where C_i, C_j, C_{ij}, have the same meaning as above, while N is the number of documents in the set. This index allows one to classify word pair frequencies that point to the emergence of minor but potentially expanding topics.

Clustering and Visual Displaying

The word pairs are clustered into groups (or clusters) according to the value of the above indexes. Provided a conventional threshold value limiting the number of words in each cluster is set in advance, the resulting clusters can be arranged in network maps displaying the core topics (inclusion maps) and the links between secondary topics (proximity maps). The words located at the top level of inclusion maps are the "central poles" of interest in the mapped region, while the words clustered under the central poles, including other words at lower levels, are "mediator words." Proximity maps are built in a similar way, using proximity indexes instead of inclusion indexes. The comparison of network maps for different time periods is supposed to highlight the dynamic aspect of scientific development within the subject field. In the second generation co-word analysis, the strategic position of research topics with respect to local and global contexts is defined and mapped in strategic diagrams with the help of two additional indexes of density (strength of the internal links of a cluster) and centrality (strength of the links between clusters belonging to different subject areas within a global research network). A map with central poles, mediator, and peripheral words is quite similar to a Russian nesting doll; it displays networks of themes of various sizes and degrees of connectedness at different levels. In some fields, a clear hierarchy of core themes can emerge, probably hinting at the supremacy of a Kuhnian paradigm, but it might also be the case that no consolidated hierarchy succeeds in providing a clear structure to the observed frequencies.

Co-word analysis supporters claim that co-words are more comprehensive and call for fewer theoretical assumptions than do citations and co-citations. Unlike citations, whose patterns of use are often ambiguous and whose availability depends on the source selection criteria of a proprietary database revolving around journal papers, text is ubiquitous and conceptually richer than the bibliographic section of journal papers, thus providing a clue to the live formulation and negotiation of research problems. The co-word processing machine, however, is no less driven by theoretical premises than citation analysis. In a sense, it can be construed as the technical implementation of a sociological theory, the actor-network theory (also known as sociology of translation), elaborated by a team including Bruno Latour, Michel Callon, and John Law.[5]

In the actor-network theory, as in much sociological research, the focus is on the mechanics of power and organization regulating social interactions. But unlike conventional schematizations of power relationships, interaction is here at the very beginning of the story. Nothing lies outside the network of relations; the actors, conversely, take their shape and strengthen their position in the network by virtue of the relationships they maintain with other actors and the ability to occupy strategic positions. Prior assumptions about the role of the actors are not needed, save that some kinds of interactions are more successful than others in stabilizing and reproducing themselves: the tycoons as well as the wretched, the rich as well as the poor (in a scientific arena: the eminent as well as the mediocre scientist; the revolutionary as well as the ordinary text) are consequently analyzed in exactly the same terms. Not only human actors are put on the same level, but nonhuman entities, such as machines, animals, and matter in general, also receive a similar treatment. Both are "actants" and cooperate actively in shaping relations that are simultaneously material (involving people and objects) and semiotic (involving signs that refer to concepts). Almost all our interactions with other people, indeed, are mediated by several kinds of nonhuman entities, which play an important role in the sociocognitive game. Let's take an extreme example.

This book I'm writing addresses a potential reader through a text. The text is written by means of a computer and, before undergoing the printing, publishing, and distribution processes—all involving different objects and people embedded in their own peculiar networks of relationships—will be revised and validated by an editorial staff relying on the judgment of experts. Many of the experts are probably the same people cited in the reference section of the book, and this entails a potential conflict of interests: Am I softening the criticisms to capture their benevolence? All these networks of people, objects, and interests cooperate to model the social life of the book and, to a certain extent, the destiny of its author, whose hope is that they will eventually contribute to overcome the readers' reluctance to read the text.

In this framework, knowledge is no more than a particular (and provisional) kind of order resulting from the interactions of a large number of different strategies followed by individual actors, each being the carrier of a particular mixture of personal interests and points of view. Power, accordingly, emerges when the ability of an actor to form strategic alliances with other actants aligns them to his or her interests. The complex of methods by which an actor identifies other actors or elements, places them in relation to one another, and recruits them is called "translation." Each actor builds a "translated" world comprising actants that are linked together and made dependent on him or her. Other actors, of course, behave the same way and build parallel networks comprising their own. The tensions and conflicts generated by this permanent process of translation/counter-translation shape the identity and strategic position of the actors, define the problems that are worthwhile addressing, and establish hierarchies between collective and individual entities. An actor-network, then, is an interrelated set of entities, both human and material, that have been successfully "translated" by an actor who borrows their force and acts on their behalf (or with their support) while being itself part of the same network.

The apparently abstract definitions conveyed by the sociology of translation are perfectly exemplified by the way scientific communication works in practice. Science has nothing special in comparison with other social activities and institutions. A scientific author writing an article on a specialized topic, for example, enrolls various actants to make the arguments set forth in the article unassailable, thereby strengthening the fortress of personal interests. The careful choice of a particular set of words defining the main problems at stake and the proposed solutions, the empirical dataset produced at the laboratory bench, and the bibliographic references appealing to the authority of colleagues and predecessors are just a few, perhaps the most powerful, of the recruited allies. It follows that one cannot distinguish on a priori grounds between science (searching for timeless truth) and politics (the reign of history and power) or, in equivalent terms, between internal and external factors of scientific change. Science is by no means a pure intellectual activity carried out by a disinterested group of norms-driven individuals, set aside from daily and profane activities; rather, it is a locus of strategic action where, inextricably, new knowledge claims and new social relations are constantly under production and modification. The best thing to do, then, would be to dog scientists' footsteps in the laboratory and trace the complex of strategies by which scientific and technological facts are constructed and made part of a successful toolkit in the actor-network dynamics.

A basic complement to the ethnological study of the laboratory site is the analysis of scientific texts. "Literary inscriptions"—as Latour and Woolgar prefer to call them—indeed play a crucial role in scientific communication.[6] They come to laboratory life from the outset, at first in the form of rough drafts, working notes, computer printouts, and similar items; then, at a later stage, with the polite dress of journal articles, conference papers, and books. Acting at a distance, literary inscriptions set in motion the actor-scientist's world-building strategies, whereby sociocognitive linkages going far beyond the laboratory space are created and propagated. Scientific texts establish equivalences among concepts, thematic areas, and problems, but this operation is not neutral, nor is it confined solely to the cognitive space of ideas. In fact, along with other rhetorical devices, such as personal acknowledgments and bibliographic references, problem defining is by itself a strategy to enroll an army of potential allies as long as it entails the creation of obligatory passage points. The equivalence between two problems, for example, forces those who wish to solve one of them to accept a proposed solution for the other. This peculiar form of translation is dubbed "problematisation"; the definition of problems and their relationships, consequently, entails the creation of a "problematic network."

In scientific communication, problematic networks are constantly being constructed, deconstructed, and rebuilt in the never-ending effort of the actors to shape literary texts according to their interests. Yet scientists are too committed to their private concerns and particular points of view to reach a global perspective on the knowledge structure of their field. The structure, after all, is a global effect of local network interactions, but the scientists' view and that of sociologists examining their behavior alike is necessarily biased by the limits of their local horizon. "Objective" procedures, consequently, are needed to mine the problematic networks out of the surface of textual rhetoric. Co-word analysis is the method designated by the supporters of actor-network theory to perform such a difficult task. The rationale behind the choice is that, when considered from the point of view of problematic network construction, literary texts can ultimately be reduced to strategic combinations of words that act as "translation operators," obliging the reader to take part in the author's game of world-building. Given that, at the level of a subject area, knowledge production is the combined effect of a large number of individual actors' strategies, each assembling words from a distinct perspective, co-word analysis is a convenient method to reconcile the heterogeneity of scientific world-building procedures with the task of identifying common patterns of word associations among the actors, of characterizing their strategies, and of tracking the convergence or intersection of their interests.

The validity of co-word analysis has been questioned from different perspectives, the most intriguing being that of Loet Leydesdorff. Science is a complex, multidimensional, and multilayered activity in which communication processes are in a state of constant transformation. Furthermore, it is reflexive: observations, actions, and theoretical concerns of the observers affect and partially shape the very situation they are observing and conceptualizing. The sociology of translation, according to Leydesdorff, correctly emphasizes this aspect of scientific complexity, but it fails in that it doesn't distinguish, on the analytical level, between the underlying dimensions of the knowledge production process, thus failing to address on an empirical basis "the question of how the social and cognitive dimensions co-vary in 'socio-cognitive' (inter-)action at any given moment in time, and the dynamic question of how action shapes and reproduces structure at a next moment."[7]

In Leydesdorff's scheme, the three fundamental dimensions are scientists, cognition, and texts, each with its own peculiar units of analysis at various levels of aggregation. Texts, for instance, are made of words, which belong to sentences organized in paragraphs and sections; scientific articles are packed in journals, which grow into literature archives; scientists aggregate into research groups, which belong to scientific communities; cognitive claims are grounded in scientific theories, which are the building blocks of disciplines. The relationships among the various levels of aggregation, however, are not mechanical, nor do they seem amenable to simple recombinations of elements from bottom up because new, unexpected properties emerge at each level. Word occurrences and co-occurrences, specifically, do not account for the higher levels of text organization, as clearly emerges when statistical analyses of text words in subject-delimited document collections are carried out. In such a case, it turns out that, although the distribution of words over the sections of each individual article seems quite regular, pointing to a predictable pattern of theoretical, methodological, and observational factors, things change dramatically at the level of the document set, where the position as well as the meaning of words over the different sections vary without any clearly identifiable pattern. What counts as a theoretical term in one article, for example, may well be used as an observational or methodological term in another article. It follows that, even within the boundaries of a highly selective corpus, different articles seem incommensurable in terms of word and co-word patterns: co-words are inextricably embedded in changing contexts.[8] The same is true on the part of the observer because analysts using different theoretical frameworks might be expected to generate different relative frequency distributions and co-word clustering algorithms from the same dataset.

Incommensurability at the theoretical level, however, doesn't imply that the results of the measurements are also incomparable, as long as they can be expressed in terms of probability distributions. On the contrary, relying on Shannon's classic information theory and appealing to Bar-Hillel's and Brookes's call for an information calculus, Leydesdorff claims that the expected information content of empirical distributions can be calculated without resorting to the meaning of individual words or to co-occurrences of words. That's why he criticized the clustering algorithm implemented in the software used by co-word analysis pioneers, LEXIMAPPE, whose output didn't seem consistent with the statistical analysis of word distributions over the documents of a given set. Co-word supporters, on the other hand, maintain that the context dependency of word meanings doesn't prevent anyone from using them as mathematical variables or indexes. This opportunity is all the more valuable because changing meanings are exactly what is needed for scientific communication to work properly, at least if science is not regarded as the discovery of an external independent reality but as a complex chain of translations in the spirit of actor-network theory.[9]

5.2. CAN A COMPUTER WRITE HISTORY? ALGORITHMIC HISTORIOGRAPHY AND THE MAP OF INFLUENCES

In the early 1960s, after an unintended suggestion by the geneticist Gordon Allen, Garfield developed the following method for testing the utility of citation indexes in historical research:

1. Take an essay supplying an authoritative and thorough historiographic reconstruction of a significant episode in the history of science, preferably a discovery resulting from the cumulation of many previous small achievements, and draw a map on a sheet of paper with nodes and lines (a graph) standing for the key events and relationships identified by the historian.
2. Perform a literature search to trace the nodal papers and investigators that first reported those key events and relationships.
3. Examine the bibliography of each nodal paper to determine its connections with other nodal papers and draw a second map or, as Garfield

named it, a "historiograph" displaying the network of bibliographic links between the nodal papers.
4. Superimpose the two maps and analyze the extent of their overlapping.
5. Starting from the nodal papers' cited references, perform a thorough citation analysis of each item to evaluate its citation impact and to check whether some new connections overlooked by the historian do eventually emerge.

The matching points in the two maps would uncover the degree of reliability of a history of science written following the thread of citations, a kind of history that computers could substantively help write, and that Garfield's mentors in historiographic matters—John D. Bernal, Derek Price, Chauncey D. Leake, Richard H. Shryock, Robert Merton—encouraged in private communications.

The final report of an experiment following the above blueprint was published in 1964.[10] It focused on the discovery of the DNA code. Isaac Asimov's book, *The Genetic Code* (1963), supplied the official narrative version of the story, while the *Genetics Citation Index* and the first edition of the *SCI* provided citation data for the pool of nodal documents covering the period from Mendel to Nirenberg (1962). The comparative analysis revealed that the two historiographs overlapped for 65 percent of the relationships and that the nodal papers with the highest citation scores were related to the same events counted by Asimov as among the most important contributions to the field. Furthermore, the citation links suggested new relationships between nodes that failed to emerge from the narrative version and whose significance was certified by the historian himself. This confirmed the utility of the historiographs in identifying the most influential documents, along with fresh genealogical relationships among them.

The *HistCite* software, developed by Garfield in collaboration with Alexander Pudovkin and Vladimir Istomin, is the computerized version of the above mapping algorithm.[11] Its purpose is the automatic and interactive construction of historiographs displaying the chronological development of a topic through a quantitative analysis of the bibliographic connections between its core documents. Starting with a bibliography or a collection of papers, each represented by its source record in the *Web of Science*, the software exploits the lists of their cited references to reconstruct and analyze the underlying citation network. A virtual mini-citation index is thus generated out of the initial set. Each record therein is associated with two basic numbers intended to automate the identification of core papers: a "Global Citation Score" (the number of times an item is cited in the *ISI Web of Science*) and a "Local Citation Score" (the number of times an item is cited in the initial bibliography). It takes just a few clicks to activate a variety of analytical tools and sort keys for visualizing directories of entities ranked by total and local citation scores: authors, journals, (title) words, tags attached to individual records by the user, yearly output, document types, languages, institutions, and countries. A "Cited Reference" list, in addition, shows all the references cited in the collection, thereby simplifying the identification of potentially relevant documents missed by the initial bibliography or not covered by the ISI. Once added to the starting bibliography manually or by a *Web of Science* lookup, these "outer" references enrich the analysis by enhancing the complexity and completeness of the network diagram. Eventually, a browseable map, or historiograph, of the chronological citation network hidden in the document collection is generated that highlights, according to a variable citation threshold, the few nodes corresponding to the most cited "core" papers.

Some nontrivial assumptions lie behind the graph-network model of science historiography. The chief one is that the history of science can be read as a chronological sequence of events, or scientific discoveries, each unambiguously referable to a specific publication, and each occupying an almost visualizable point in the Euclidean timeline of scientific evolution. Citations are assumed here to reflect direct genealogical relationships between authors' ideas. They are markers of intellectual influence, and the marking perimeter they let surface is so distinct that, in Garfield's own words, "the bibliographic information contained in a collection of published scientific articles is sufficient for the purpose of recapturing the historiographic structure of the field."[12]

Objections might be raised at various levels. It might be argued that chronology is actually a poor representation of the dynamics of scientific communication as far as it supplies materials for a pure *histoire événementielle* of finite products—the formal representations of scientific concepts and theories in journal papers—as opposed to a deeper investigation of structural connections rooted in social, philosophical, and broadly speaking, cultural grounds. The choice of Asimov's account seems to ratify this argument: the book is a popularization of the subject, based primarily on the author's personal memory, and lacks any form of scholarly apparatus, including the chief actors on the stage (i.e., bibliographic references). More to the point, it might be claimed that the multipurpose concept of influence implicit in the algorithmic model of historiography is not as plainly translatable in bibliographic connections as the software operation would seem to suggest. Professional historians constantly remind us, with good reason, that writing history requires hypotheses, intuition, interpretive skills, and the application of a sound methodological toolkit to a heterogeneous set of sources, both conventional (published documents) and unconventional (unpublished and archival material). In this regard, historians of science agree with historians of other scholarly areas in stressing the ambiguities inherent in the concept of influence.

A classic argument runs as follows. We cannot speak of, say, the influence of author *B* on author *A* without the intervention of a third pole, *X*, made up of the paradigmatic universe of theories, concepts, and methodologies shared by a community of scholars at a certain time.[13] One could directly ask *A* if *B* influenced his or her work. More ambitiously, one could administer a cleverly designed questionnaire to a sample A_1, A_2, . . . , A_n of renowned specialists in a research area, asking them to pinpoint and rate the main attributes of influential papers in that area. What would be the result either way? Unless independent evidence of the conjectured intellectual bonds is supplied, the first analyst would get nothing more than *A*'s personal opinion or conscious misrepresentation of *B*'s relevance, whereas the second would get A_1's, A_2's, . . . , A_n's mixed statements of what presumably makes a document influential across space and time. Given the respondents' personal commitment to the ethos of science, their answer will most likely converge toward such idealized factors as theoretical or practical significance, methodological interest, and the like.[14] Alternatively, one could ask the texts themselves, and specifically the cited references sections of scientific articles, to unravel what authors fail to recognize upon questioning, but the exploitation of formal bibliographic connections between *A* and *B* leaves the analyst in the middle of theoretical quicksand. *A* indeed may decide to cite *B* in a paper, and perhaps not *C*, who was the first to introduce the concept symbolized by the citation, simply because *A* is not informed of *C*'s existence or, if the additional confounding role of the rhetorical use of language is taken into account, because *A* wishes to parade his own awareness of the existence and value of *B* for avoiding any possible future retaliation. It might be the case, therefore, that extant bibliographies are an imperfect mirror of the true genealogical relationships between ideas, and what is commonly referred to as influence escapes any sort of conceivable algorithm.

Chapter 7 delves at greater length into the manifold criminal charges pending on any bibliometric oversimplification of the dynamics of scientific ideas. For the time being, however, a counter-objection is in order. Science differs from other scholarly activities in one basic aspect: for its practitioners to survive as a professional group, the paradigmatic string of constructs lurking behind individual operations has to be exposed to public criticism. Published scientific documents are not simply literary artifacts, but nodes of communicative networks where a conventional consensus among individual scientists about the linguistic (and bibliographic) carving of the details in the common landscape they contemplate (or construct) has to be reached. Scientometricians get the point here in stressing that bibliographic citations are a tiny but not trivial cog in the machine that steadily reshapes the world picture while fostering the self-construction of scientists' professional identity through the adherence to a common set of theoretical and linguistic conventions.

Obviously, no scientometrician would deny the complexity of historical investigation or pretend that algorithmic historiography is in the position of writing the history by itself. What citation analysis can supply, instead, is an additional tool for marking out the perimeter of potentially relevant sources and interpreting the significance of their relationships. After all, before embarking on any complex interpretative job, a historian must perform the preliminary operation of collecting all the relevant literature and trying to figure out the genealogical connections among its units. Citation links are not the exclusive or necessarily the best possible indicators of such relationships but, to the degree that they reflect the self-perceived (or self-constructed) dependence of authors on previous work, they have to be taken into account, both for what they reveal and for what they (intentionally or not) conceal. In addition, an algorithmic history of science redrawn through citation links can facilitate the spotting of the key papers and the identification of the points of bibliographic discontinuity within a stream of research. The next step, linking these modifications to structural changes in the basic units of science—be they disciplines, specialties, research areas, or invisible colleges—became, starting in the mid-1970s, the main task of co-citation analysis.

5.3. THE SPECIALTY STRUCTURE OF SCIENCE: CO-CITATION ANALYSIS

In 1974, Henry Small and Belver Griffith launched the first systematic project of a bibliometric cartography of science based on the analysis of highly co-cited documents, or co-citation analysis. Some of the questions they hoped to answer are, in Henry Small's own words:

> 1. What are the natural, structural units of science? Are they similar to the traditional disciplines such as physics or chemistry, or perhaps smaller units such as specialties or what *Price* called "invisible colleges"? 2. How do these structural units relate to one another? Are they loosely or tightly integrated, or completely isolated from one another? 3. What are the forces which determine these structural units and their interrelations? To what extent is the map of science also a map of scientific knowledge? What role do social factors play in determining structure? 4. How does the structure of science change over time, both on a macro- and a micro-level?"[15]

The next two sections show, respectively, how in practice a bibliometric map is drawn with citation data and what perspectives on the structure and dynamics of science it delivers.

5.3.1. Drawing the Map: Some Glimpses on the Technique and Products of Co-Citation Analysis

The belief that scientific literature can be classified through the analysis of bibliographic links, as opposite or complementary to content analysis, didn't originate with the *SCI*. In the late 1950s, Kessler at Massachusetts Institute of Technology had improved upon a technique formerly envisioned by Robert Fano: bibliographic coupling. Papers are bibliographically coupled when they share one or more references in their bibliographic section. Since the bibliography is one tool authors use to disclose the intellectual background of their work, the number of references in common between two or more papers can be taken as an indicator of their cognitive proximity. Actually, Kessler's test on a corpus of articles issued in thirty-six volumes of the *Physical Review* resulted in a rather fuzzy classification of the coupled papers, which formed classes hardly comparable with any conceivable grouping based on conventional subject headings. In the author's view, nonetheless, the utility of the technique had to be evaluated from the scientists' standpoint insofar as scientific minds work differently from those of laypeople or librarians.[16] Further tests in the context of the Cranfield experiments indicated that Kessler's technique, in combination with other kinds of indexing, could improve the performance of an information retrieval system, so it is not surprising that later databases and digital libraries systematically incorporated some form of bibliographic coupling in the algorithms specifically designed to retrieve papers or web pages similar in content to a given paper or web page. After two decades of quasioblivion, the technique took on new life in the 1990s, when its potential in identifying clusters of subject-related documents or hot research topics on the research front was fully exploited by Wolfgang Glänzel and Hans-Jürgen Czerwon, whose method inspired the recent work of Bo Jarneving on bibliographic coupling.[17]

Relying on the newborn citation index, Garfield's team at the ISI focused on a unit of analysis symmetrical to Kessler's one: pairs of documents cited together (co-cited) by other papers, instead of pairs of papers citing the same documents. This idea was developed simultaneously, though independently, by Henry Small in Philadelphia and Irina Marshakova in Moscow. Clearly, the choice to look forward (citations) instead of backward (references) brings to the foreground the dynamic aspect of the citation process and the evaluative potential of citations, requiring that the analysis be fed with highly or at least fairly cited documents. Bibliographically coupled documents, on the contrary, do not incorporate any vocation for citation-based appraisals, and the only sense in which this represents an advantage over the competing method is that bibliographic coupling doesn't need to wait for citations to accrue to the starting documents. What's more, a common misstatement about co-citation analysis consists of entrusting it with a task that is more likely accomplished by bibliographic coupling, that is, putting together documents that share the same or similar subjects; co-cited documents don't have to be similar in this restricted, library-oriented sense because they may involve entirely different topics or methodologies, and it's exactly in this resilience that citationists see the potential to produce a totally new categorization of scientific literature that outperforms traditional classifications in portraying the actual dynamics of scientific research.

Co-citation analysis rests on the premise that if two documents are cited by a third document, it is likely that some kind of structural relationship between them does exist, the strength of the relationship depending on how many times they are co-cited in a given corpus of literature. Above a certain threshold, structural affinities between the co-cited documents are likely to emerge for the simple reason that, by agreeing on what constitutes the previous significant sources, scientists define, to a certain extent, the intellectual boundaries of their research field. So, if highly cited documents are thought to convey key concepts, methods, and experiments, then the analysis and classification of heavily co-cited documents might well be expected to uncover the patterns of relationships between these key elements, thereby highlighting the sociocognitive structure of a research field, its articulation into subareas and emerging research fronts, and its connections with other specialties.[18] Within certain limits, moreover, the conclusions drawn from co-citation analysis at the document level have been extended to other levels of aggregation, notably authors, journals, and subject categories:

1. *Authors*. In 1979, at Drexel University, Howard White and Belver Griffith started an entirely new strand of research, author co-citation analysis, which combined Small and Griffith's co-citation mapping with a set of online retrieval techniques, formerly perfected by White himself, for extracting co-cited authors from ISI databases. Their proposal centered on a new mapping unit: sets of documents associated with a given group of (first) authors instead of individual documents. That is to say, the more two authors' oeuvres are co-cited, the nearer the authors' position in the mapped region. Maps resulting from the first experiments seemed to convey useful information about the most influential authors, the classification of authors in subgroups or "schools," their location with respect to each other, and their degree of intellectual proximity within a group and across group boundaries. A classic example of the technique's potential is the extensive co-citation analysis of 120 top authors in information science, published in 1996 by White and McCain.[19]

2. *Journals.* Just as authors' oeuvres are, journals are said to be co-cited when at least one article from each journal is listed among the references of a citing article. After McCain's experiment in 1991 on a set of economic journals, this technique's ability to mark out the intellectual structure and subject relatedness of scholarly literature's subsets has been further explored in the fields of genetics, information retrieval, information systems research, urban studies, and semiconductor research.[20]
3. *Subject categories.* Recently, the SCImago group at the University of Granada has implemented a new mapping unit: class and category co-citation. Specifically, they assume the ISI *JCR* categories and the classes of Spain's Agencia Nacional de Evaluacion y Prospective as the high-level recipients of papers being co-cited: the more papers issued by journals belonging to certain categories are cited together, the greater the degree of connectedness of those categories. On this premise, they apply multivariate methods and advanced social network techniques to build maps (or "scientograms") that purport to represent the structure of science as a whole.[21]

The series of standard steps required for drawing a co-citation map has not changed so much since the 1970s. Following is a brief review:[22]

1. *Extraction of the dataset and selection of the appropriate unit of analysis.* The unit of analysis may be located, as we have seen, at the level of documents, journals, authors, and subject categories. The selected units are checked against the ISI databases, and their citation rates are determined. A lower citation threshold is usually fixed, at this stage, to prearrange the sample for the analysis of highly co-cited items. Obviously, the quality of any mapping project is strictly dependent on the accuracy and completeness of the initial dataset. One possible drawback of author co-citation analysis, in particular, is the limitation of ISI citation data to first authors for the greatest part of the indexed documents.
2. *Co-citation measures and normalization.* Co-citation scores are computed and a raw data-matrix is formed, each value in the matrix representing the number of co-citations for a particular pair. If, for instance, the units are authors, the output is a square symmetric matrix where

the cited authors' names are placed in identical order on both rows and columns. The cell corresponding to the intersection of a given row (author *X*) with a given column (author *Y*) contains the co-citation counting for the pair of authors *X-Y*. At this stage, at least in the case of author oeuvres, raw co-citation data undergo systematically some form of normalization to prevent the first level analysis from being distorted by the irregularity of citation practices across different fields. One of the most popular normalization routines is the transformation of the data matrix into a correlation matrix, for instance by replacing the single values with the corresponding Pearson's correlation coefficients.[23]

3. *Data analysis and map generation (ordination).* Since a multiple co-citation analysis involves relationships among multiple variables, whose representation wouldn't fit the two- or three-dimensional surface of either paper or computer screen, it's necessary to apply dimensionality reduction techniques to draw the map. In other words, the complex network of co-citation links has to be replaced by a model that simplifies the reality while retaining its basic features. Well-established methods of multivariate statistics and data mining—such as eigenvalue/eigenvector decomposition, factor analysis, cluster analysis, multidimensional scaling, triangulation, pathfinder network scaling, and self-organizing maps—allow to reduce the number of variables preserving the internal structure of the dataset. In their seminal experiments, for example, Small and Griffith used multidimensional scaling and cluster analysis to translate differences of co-citation strength between pairs of papers into spatial associations. Multidimensional scaling, originally developed in the context of psychophysics and psychometry, starts by calculating what in technical jargon are called the "distances" between each pair of *n* items. In the above example of author co-citation analysis, they are the distances between each pair of author rows in the correlation matrix. Then the analysis seeks to determine *n* points in a reduced number of dimensions, typically two for a bidimensional plot, so that the interpoint distances are as close as possible to the original distances between the rows. After being scaled, the points on the map can be further classified using cluster analysis, a complex set of multidimensional techniques that allow data to group into homogeneous

sets by searching for patterns of similarities between the numerical measures attached to them.

4. *Interpretation and validation of the results.* Whereas the starting correlation matrix could be regarded as the surface of an unknown landscape, a map eventually exists that uncovers its plains, mountains (highly co-cited items), and borders (clusters of subject-related pairs). To make sense, the clusters, factors, or mapped dimensions need to be interpreted in terms of scholarly domains: research fields, specialties, disciplines, and intellectual or institutional ties. Finally, to be valid the proposed interpretation has to be checked against external evidence derived from the qualitative assessment of scientists active in the mapped domain or from the paradigmatic expertise embedded in textbooks, reviews, and authoritative historical reconstructions.

So far, the interpretation of a bibliometric map has appeared not nearly as linear and straightforward as the above algorithm would lead one to expect. In this regard, Small and Griffith's pioneering experiment on a cross-disciplinary sample of papers drawn from the first quarter of 1972 *SCI*, although laying the foundation of much subsequent work in the field, had already raised some long-lasting conceptual and technical issues. The project demonstrated that scientific literature can effectively be subdivided into clusters of highly co-cited documents corresponding, by and large, to subject matter specialties, but the picture was out of focus in many respects. For one thing, fields wherein a lot of papers are published (and cited), for example medicine and related areas, were overrepresented at the expense of research areas with low publication volumes and weaker reference patterns, including mathematics and engineering. In addition, although it seemed relatively easy to spot, within the three major groupings of physics, chemistry, and biomedicine, the specialty structure of areas connoted by a strong consensus on key documents and concepts, such as nuclear or particle physics, it turned out far more difficult to make sense of the fine structure of the biomedical domain. Here, in fact, some frequently cited and co-cited documents, usually papers reporting widely shared techniques and methods, cut across different areas in such a way as to blur the boundaries of virtually different regions, thereby violating the assumption that highly cited documents would be specific to a specialty.[24]

The coexistence of well-delineated local clusters with loosely knitted macro-groups roughly amenable to single specialties, as well as the marked heterogeneity in the size of clusters, degree of connectedness, and age, suggested, in certain cases, a sort of infinite interpoint distance that was inconsistent with Derek Price's trust in the two-dimensionality of science: "It is as though," admitted Small and Griffith, "we had drawn a map and discovered that everything is roughly in order, but that Albany is infinitely distant from Wilmington, even though each is accurately positioned with regard to both New York and Philadelphia."[25] The results also placed a new perspective on the use of citation counts for research performance evaluation: highly cited papers were not all equal, nor did they seem comparable across specialties. Price's old dream of building war maps of immediate use to science managers, therefore, was still far from realization. Maps were clearly a vague approximation of an extremely fluid reality, and just superimposing the annual *SCI*-based cartographies didn't let a single logical thread emerge over time. Research fronts seemed to vary from year to year; new documents continually entered the clusters while others dropped out abruptly; and the partial solidity of some elements at a higher level, such as the comparatively stable boundaries between the disciplines, contrasted with a much greater fluidity at the lowest levels, where many clusters or documents remained isolated.[26]

Notwithstanding the lack of resolution and the fuzziness of early bibliometric maps, from the mid-1970s onward, co-citation analysis became the most popular and widely accepted technique for modeling the sociocognitive structure of scientific communications. The ISI drew on Small and Griffith's work to plan, during the 1970s and 1980s, a systematic clustering of the entire annual files of the *SCI* that culminated in the project, later abandoned, of a world *Atlas of Science*. The *Atlas* should combine specialty-by-specialty co-citation maps, aimed to show the interconnectedness among each specialty's core documents, with bibliographies of the core documents themselves (plus the papers citing them) and mini-reviews of each subject area written by experts whose names appeared in the appropriate clusters.[27] Such a comprehensive mapping of the global structure of science lent itself to wide-ranging speculations on the technical and philosophical potentialities of an entirely new "spatial" information retrieval tool shaped on the model of geographic systems. "In such a system," contended Small,

> the user first views the overall map of science, selects a node on it and then sees the map for this sub-region, perhaps a discipline of science. Nodes on this sub-map may also be viewed as maps of sub-disciplines or specialties. The user may progress down this tree of nested maps until the document level map is reached, for retrieval of specific papers. In addition at each stage the user might have access to textual information describing the disciplinary or specialty network currently displayed, similar to the mini-reviews in ISI's Atlas of Science. . . . Clearly such a system is more than a bibliographic retrieval system; it is more nearly an encyclopedia of science integrated with a bibliographic structure, perhaps a realization of the "World Brain/Memex" combination envisioned by Garfield.[28]

Citations do not link simply documents, they link ideas and arguments, and the connections are both internal to each subject area and transversal to multiple areas. On the internal front, the ability of highly cited documents to symbolize the main concepts at stake led Small to model, in terms of a walk through a co-citation network, the thought process involved in writing down a review of a scientific field.[29] On the external front, given the propensity of citations to reach beyond the boundaries between disciplines and research fields, a global map of the kind envisioned by Garfield and Small could reasonably be expected to reveal, at least in principle, the structure of the argumentative networks connecting scientific theories beyond sectorial languages and practices.[30]

The pioneers' philosophical commitment to the ideal of an encyclopedic information system encompassing the whole network of scientific papers got lost in the next-generation cartographers. Anyway, the task of trawling through the universal citation network to dig out its hidden cognitive structure at the global level of the world science system was taken over, at a lower degree of resolution, by network analyses of journal-journal citation links. On the assumption that journals covering similar subjects also display similar referencing patterns and cite one another in a consistent way, journals cross-citing data have been used, on a route adjacent to co-citation analysis, to group related titles into subject classes. In the late 1960s, economists at the Federal Research Bank, New York, were already examining the inter-citation network of economic journals with multidimensional scaling techniques, under the conviction that treating a journal set as "a system of telephone switchboards, each of which serves a subset of economists," and measuring the switchboard exchanges in terms of frequencies of inter-citations, allowed the visualization of "a discernible structure in the journal network."[31] This research stream was dramatically spurred by the availability, starting in the early 1970s, of prepackaged ISI journal-to-journal citation data via the *Journal Citation Reports* (formerly *Journal Citation Index*). So, after Narin and Carpenter's seminal attempts to derive hierarchical groups of subject-related journals through cluster analysis, a number of authors set out to process *JCR* tabulations with multivariate statistics and graph-analytic techniques.

In 1985, Patrick Doreian and Thomas Fararo applied the sociometric construct of "structural equivalence" to the network of inter-citations in a sample of sociological journals over three distinct time periods so as to detect equivalence classes or "blocks" of journals that related in a similar way to other classes.[32] In 1987, a CWTS team showed how the hidden dimensions of a citation network in a set of astronomy and astrophysics journals could be modeled by means of a newly developed multivariate technique, quasi-correspondence analysis.[33] Nearly a decade later, Juan Campanario adapted a well-known neural network algorithm, self-organizing or Kohonen maps, to four sets of journal-to-journal citation data spanning the fields of chemical physics, communication studies, and sociology.[34] More systematically, Leydesdorff has been applying multivariate statistics and network analytic tools to investigate the fine-grained structure of the journal-journal citation matrix that grows out of the annual *JCR* editions. Besides the traditional objective of figuring out the specialty structure of the sciences, he has been seeking to determine if and how changes in journal citation patterns can be used as indicators of structural changes in the global organization of the sciences over time.[35]

In recent years, two circumstances have arisen that further enhance the resolution and accuracy of scientific maps. One is the dramatic advancement in computer visualization and virtual reality, which, along with a wider range of network analysis and dimensionality reduction techniques, turned maps more and more into fine-grained representations of dynamic scientific territories in the shape of three-dimensional, colorful, interactive, and browseable devices. Interestingly, in perfect line with an elective affinity running throughout the bibliometric tradition, many improvements in this area have been triggered by data mining and information retrieval research, where the design of hypertexts and graphic user interfaces, from the mid-1990s on, profited considerably by the use of three-dimensional maps for the retrieval of pieces of information from vast amounts of data. A striking demonstration of these advancements is the global maps of science at the journal level, worked out, using advanced graph-analytic techniques, by Leydesdorff, I. Samoylenko's team, and Kevin Boyack's team.[36]

The second favorable trend showed up when, benefiting from the array of text mining techniques fostered by the widespread availability of full-text databases, bibliometricians started exploiting hybrid combinations of citation and text analysis to get more accurate pictures of scientific fields or to validate the results obtained by each method separately. In the late 1980s, researchers at the CWTS pioneered this line of investigation. Robert Tijssen and colleagues, for instance, experimented on journal inter-citation networks using a mapping technique based on the joint use of citation data and content profiles derived from subject classification codes. In one case, they mapped the structure of similarities and dissimilarities in the dataset through quasi-correspondence analysis; in another case, they adopted (and adapted from the field of psychometrics) a subclass of the multidimensional scaling models called INDSCAL[37]. Shortly thereafter, Robert Braam and colleagues explored a hybrid approach in which words from the publications citing the documents previously clustered by co-citation analysis were analyzed to check the convergence of both representations and to assess the ability of co-citation methodology to identify all or a major part of the publications in a given specialty.[38] On a parallel track, Michel Zitt, who in the 1990s developed a semiquantitative method of dynamic lexical analysis alternative to the juxtaposition of successive snapshots of synchronic relations for the description of the long-term evolution of a thematic area, has recently tested a combined lexical-citation method for the delineation of emerging fields, such as nanosciences, whose lack of a definite institutional structure is reflected in the fluidity of both the scientific vocabulary and the citation practices.[39] On a smaller scale, Patrick Glenisson and colleagues have utilized full-text analysis—carried out with the classic vector-space model—to classify the complete set of articles issued in 2003 by the journal *Scientometrics*, concomitantly resorting to traditional reference-based citation measures for the refinement of the clusters thus obtained.[40]

It was almost inevitable that, urged by a steadily increasing demand for easy-to-interpret research assessment tools, bibliometricians gradually managed to highlight specific zones of the mapped territories so as to help policymakers in spotting the peaks and the subsidences of the intellectual landscape. In the last analysis, the strategic value of a map is measured by its ability to position the actors (scientists, universities, research institutes) on the stage, to quantify their impact on a research field, and to monitor the variations of their performance over time. Thus, the next logical step was to further enrich the mapping process with information on personal and collective authors, and to develop interfaces enabling the layperson to answer policy-related questions such as, "What does this domain look like? Who are the main actors in this domain? What is their particular expertise? How does this expertise relate to that of others? What are the main developments in a certain period of time? Which actors contribute to these developments? Who may be responsible for a particular change?"[41] To appreciate the shift, one could contrast Garfield's aforementioned project of an *Atlas of Science* with the *Atlas of Science* recently issued online by the SCImago research group at the University of Granada. Here, the descriptive concern with portraying the structure of knowledge domains in the Ibero-American science system is markedly unbalanced by the evaluative concern with inserting, next to each domain, productivity and impact data on the scientific performance of authors, institutions, journals, and countries.[42]

5.3.2. Maps of Science or Maps of Knowledge?

One could legitimately ask, What does a bibliometric map tell us, and how much distortion does it convey into our perception of the social and intellectual organization of the sciences? Is the map simply a snapshot of the conventional borderlines posited, mainly for bureaucratic reasons, by scientists and policymakers between otherwise fluid research areas, or does it effectively dig deeper into true intellectual structures? Answers to these questions can hardly be attempted without a preliminary insight into the way science works or is expected to work as a truth-building enterprise. Stated alternatively, an answer presupposes that, being inspected through theoretical lenses, bibliometric data can be pigeonholed and given voice into a model of scientific growth, regardless of whether the model committed to a cumulative and cognitive-driven or to a discontinuous and socially forced mode of knowledge production. Two classic case studies that exemplify the sharply diverse "philosophical" courts before which citations have been summoned to give evidence are the survey of reverse transcriptase research set forth in Kenneth Studer and Daryl Chubin's *The Cancer Mission* (1980), and the several attempts, recurring since the 1970s, to bibliometrically investigate Kuhn's puzzling notion of scientific paradigm.

Studer and Chubin's work rested on the premise that, for the sociology of science to be relevant to biomedical research policy, a deeper insight into the cognitive dimension of science has to be reached within the framework of a "confluence theory" of scientific development. New discoveries don't emerge in a vacuum, nor is their background circumscribed by a rigidly defined set of disciplinary tools. Science seeks relatedness as much as it seeks truth, and the value of a research field is measured in large part by the contributions it makes and the clarifications it affords to adjacent fields. In biomedicine, too, major advances like the discovery of reverse transcriptase occur when ideas, theories, and techniques from other fields of science (chemistry, physics, etc.) "converge to form a biological problem domain," without the convergence necessarily amounting to the creation of a new discipline or field.[43] Because of the gatekeeping function exercised by scientific journals, journal papers are the most convenient entry point into a problem domain; to the extent that their formal connectedness through bibliographic citations can be analyzed structurally by means of multivariate statistics, they are expected to reveal the connectivity structure of the problem domain. Citations and co-citations, however, cannot be taken at face value as indicators of intellectual cohesiveness, at least not without the support of complementary evidence of a qualitative nature. In confirmation of citations' ambiguous role in the communication system, having applied factor analysis to the co-citation matrix of a core set of articles in the field of reverse transcriptase, Studer and Chubin couldn't help but notice that the similarity of citation profiles behind certain clusters, far from reflecting the progressive subsidence of "hot" ideas and research topics, had a much simpler explanation: the extensive and socially driven inter-citation among members of groups belonging to the same institution. Articles, therefore, clustered not only along cognitive lines but also, and sometimes primarily, along institutional lines as "*there develops a tendency in cocitation analysis to pick off the large laboratories and research groups.*" The authors concluded, therefore, that "we must tread lightly on the implications of cocitation analysis until the parameters that boast visibility and usage, much of which seem artifactual, are better known."[44]

On the same plane upon which Studer and Chubin had tested citations' mapping potential, other bibliometricians projected a Kuhnian theory of scientific change, perpetually dwelling in a limbo between the cognitive and the social space. Kuhn had taught scientometricians that a specialized research community can foster knowledge advancement only through the unconditional and almost dogmatic adherence of its members to a shared set of practices, methods, instruments, values, and theories. He referred variously to these common elements in terms of "paradigm" or "disciplinary matrix,"[45] emphasizing their dependence on some "exemplars" containing "the concrete puzzle-solutions which, employed as models or examples, can replace explicit rules as a blueprint for the solution of the puzzles of normal science."[46] As long as scientists fall into line with a paradigmatic way of thinking and acting, their research activity is not so different from the bricklaying held dear by the positivistic (and the bibliometric) tradition, but true advancements in science occur only when abrupt gestalt switches in the scientists' perception of their research objects are brought about by paradigm shifts marking the transition from the ordinary, basically uncritical, and cumulative puzzle-solving activity of the normal science, to the highly original and philosophically compelling effort of the revolutionary science.

"If I am right," argued Kuhn, "that each scientific revolution alters the historical perspective of the community that experiences it, then that change of perspective should affect the structure of postrevolutionary textbooks and research publications. One such effect—a shift in the distribution of the technical literature cited in the footnotes to research reports—ought to be studied as a possible index to the occurrence of revolutions."[47] Aside from this swerve in referencing patterns, though, no concrete indications were given of how, empirically, a paradigm could be identified, and this vagueness became a matter of serious disputation among sociologists and philosophers of science.[48] In a later contribution, nonetheless, the author opened a window onto information science, conjecturing that the way a scientist learns to recognize a new problem as analogous to a previously solved problem has much in common with the way a child, walking through a zoological garden with his father, learns to distinguish the geese from the ducks and the ducks from the swans, both processes being readily reproducible by a computer. No abstract definitions or correspondence rules are needed; it suffices for the child to see and hear the father saying "look, there is a swan," or "look, there is a goose." Thus, being programmed to recognize what his or her community already knows, the child learns to apply symbolic labels to clusters of natural objects.[49]

Kuhn's insights were interpreted as an invitation to welcome the notion of paradigm within the realm of information science. Notwithstanding his reluctance to look at paradigms through bibliometric and linguistic glasses,[50] a set of frequently co-cited documents had some of the necessary requirements to stand for an empirical approximation of the notion of "exemplar." If scientists involved in the solution of disciplinary puzzles do not apply abstract rules, but rather learn, by means of shared examples of standard practice, to perceive the similarities between a new problem and the problems previously solved by their colleagues and predecessors, then a core set of key documents incorporating such exemplary models of practice may well be supposed to exist. And if exemplary documents do actually inform scientific practices, it might also be the case that not only do they receive more co-citations than average, but the significance attached to those citations is surprisingly regular. Even more, it might be construed that the substitution, in a specialized domain, of a set of frequently co-cited documents with a different set provides a rough indicator of a paradigm shift or scientific revolution.

Small has gone a long way toward handling in a bibliometric style the ambiguities inherent in Kuhn's construct. As early as 1972, he changed his mind on how a conceptual map of the history of nuclear physics could be drawn. Instead of an intensive reading of the key papers and the figuring out of their conceptual relationships, the formal connections among textual and bibliographic elements—i.e., the co-occurrences of words and co-citations—could work as well, with an additional saving of time and effort. In this framework, the clusters of highly cited documents extracted from ISI citation data through co-citation analysis appeared a rough but fairly acceptable translation of Kuhn's "exemplars," while the sudden disappearing of key documents from a cluster followed by their replacement with a distinct set closely resembled the drastic shift in research orientation implicit in Kuhn's notion of revolutionary change. A 1977 longitudinal study of the specialty structure of collagen research substantiated Small's approach, showing not only that the clusters of highly cited documents corresponded to groups of significant documents in the eyes of the specialists, but also that "citations provided, in effect, a physical and measurable manifestation of a collective mental switch from a static to a dynamic conception of collagen."[51]

In the wake of Small's original project, and taking advantage of increasingly refined visualization tools and multivariate analysis techniques, later authors took over and elaborated on the idea that not only the structure of scientific disciplines, subdisciplines, and research areas, but also their inner dynamics, their going through paradigm shifts and scientific revolutions, can be adequately portrayed in a two- or three-dimensional scale. Perhaps the most impressive demonstration in this sense is offered by Chaomei Chen's *Mapping Scientific Frontiers* (2003), in which, reviewing previous research on the subject, the author describes and displays several three-dimensional models aimed at capturing and visually browsing, in the literary space of many scientific specialties, the salient features of the growth and development of competing paradigms as reflected in co-word and co-citation patterns.[52]

Of course, one might promptly object that to see is not the same as to believe. And to believe is even more difficult if both the limitations of the mapping methodology and the complexity of the philosophical target are taken into due consideration. Citation data, as chapter 7 argues in greater detail, do not possess the same degree of stability of geographic mapping units, and the lenses through which they are inspected—i.e., the multivariate statistics toolkit—are not capable of returning a picture comparable to the output of a Geographical Information System handling geographically referenced information. In factor analysis, for instance, the initial variables (citation or co-citation frequencies) are reduced to (linear) combinations of a lower number of "latent" variables called "factors," typically interpreted in terms of specialties or research areas. The factors are deemed able to explain the correlations among the initial variables, but a number of reasons make this assumption controversial. Unlike the real variables, factors can't be measured or observed, the determination of their number is not univocal, and there are many methods of extracting them out of the correlation matrix and many techniques of rotating factor loadings to achieve a simpler and more interpretable structure.

As to the target, it has to be recognized that a Kuhnian paradigm contains much more in terms of tacit knowledge and social negotiations than what shines through the brilliant surface of published literature. Kuhn himself, after all, would hardly agree on the bibliometric translation of his construct. In his later view of scientific theories as complex language systems, he maintained that comparisons among such languages cannot be achieved by measuring them against external, allegedly "objective" criteria, but rather involve a complex learning and evaluation process. And also before the "linguistic turn," the only place he found for citation analysis was in helping sociologists break a potentially vicious circle. A paradigm or disciplinary matrix is what the members of a scientific community share and vice versa, a scientific community consists of those who share a common paradigm; thus, to escape the circularity, it is necessary that the scientific community be defined without resorting in advance to the concept of paradigm. This is a comparatively easy task for well-delineated or institutionalized professional groups; affiliation to university departments, membership in professional associations, and the particular set of journals read are usually sufficient to distinguish a physicist from a chemist, a solid-state physicist from a high-energy physicist, an organic chemist from an inorganic chemist. At a slightly lower level, instead, where an area of ultra-specialized research topics is being formed around some core interests, practical and conceptual difficulties do actually emerge. Here, Kuhn himself acknowledged the potential utility of citation analysis, underlining that, for a specialty group to surface prior to its public acclaim, "one must have recourse to attendance at summer institutes and special conferences, to preprint distribution lists, and above all to formal and informal communication networks, including the linkages among citations."[53] Significantly, the footnote references supporting Kuhn's statement point to Garfield's 1964 work on algorithmic historiography, Derek Price's 1965 article "Networks of Scientific Paper," and one of Myer Kessler's reports on bibliographic coupling.

NOTES

1. The "jigsaw puzzle" model is introduced, in connection with Small and Griffith's co-citation maps, in Derek Price, "Toward a Model for Science

Indicators," in *Toward a Metric of Science: The Advent of Science Indicators*, ed. Yehuda Elkana, Joshua Lederberg, Robert K. Merton, Arnold Thackray, and Harriet Zuckerman, 69–95 (New York: John Wiley & Sons, 1979).

2. The study of geographic maps as social products, texts to deconstruct rather then true representations of a landscape, was conducted, toward the end of the 1980s, by John B. Harley in the wake of Foucault, Derrida, and Barthes. See Harley, "Maps, Knowledge and Power," in *The Iconography of Landscape: Essays on the Symbolic Representation, Design and Use of Past Environments*, ed. Denis Cosgrove and Stephen Daniels, 277–312 (Cambridge, Mass.: Cambridge University Press, 1988); Harley, *The New Nature of Maps: Essays in the History of Cartography* (Baltimore: The Johns Hopkins University Press, 2001).

3. A key source on the theoretical foundations and early applications of co-word analysis is Michel Callon, John Law, and Arie Rip, eds., *Mapping the Dynamics of Science and Technology: Sociology of Science in the Real World* (Houndmills: Macmillan, 1986). For an introductory survey on co-word analysis techniques and conceptually related issues, see Oin He, "Knowledge Discovery Through Co-Word Analysis," *Library Trends* 48, no. 1 (1999): 133–59, findarticles.com/p/articles/mi_m1387/is_1_48/ai_57046530.

4. A clear statement of the assumptions behind co-word analysis and a comparative assessment of keyword and title word analysis in relation to a concrete case study is in John Whittaker, Jean Pierre Courtial, and John Law, "Creativity and Conformity in Science: Titles, Keywords and Co-Word Analysis," *Social Studies of Science* 19, no. 3 (1989): 473–96. The use of title words instead of keywords on the score of the "indexer effect" limiting the latter's usefulness is advocated by Leydesdorff, "Words and Co-Words as Indicators of Intellectual Organization," *Research Policy* 18, no. 4 (1989): 209–23.

5. For an up-to-date introduction to actor-network theory, see Latour, *Reassembling the Social: An Introduction to Actor-Network-Theory* (Oxford: Oxford University Press, 2005). Two more concise but extremely valuable outlines are in Law, "Notes on the Theory of the Actor-Network: Ordering, Strategy, and Heterogeneity," *Systemic Practice and Action Research* 5, no. 4 (1992): 379–93, www.lancs.ac.uk/fass/sociology/papers/law-notes-on-ant.pdf, and Callon et al., "From Translation to Problematic Networks: An Introduction to Coword Analysis," *Social Science Information* 22, no. 2 (1983): 191–235.

6. Latour and Woolgar, *Laboratory Life* (Princeton, N.J.: Princeton University Press, 1986).

7. Leydesdorff, *The Challenge of Scientometrics: The Development, Measurement, and Self-Organization of Scientific Communications* (Parkland, Fla.: Universal Publishers, 2001), 7.

8. Leydesdorff, "Why Words and Co-Words Cannot Map the Development of the Sciences," *Journal of the American Society for Information Science* 48, no. 5 (1997): 418–27, users.fmg.uva.nl/lleydesdorff/cowords/index.htm.

9. Along with its relatives and heirs, such as CANDIDE (the Macintosh version of LEXIMAPPE), SDOC, SAMPLER, and SEMIOMAP, LEXIMAPPE has been used in the past decades to map the intellectual space of several scientific and technological fields. The LEXIMAPPE debate on the pages of *Scientometrics* is worth following, as it involves some core issues for the understanding of the role of co-words as scientometric indicators. Leydesdorff, "A Validation Study of Leximappe," *Scientometrics* 25, no. 2 (1992): 295–312; Jean Pierre Courtial, "Comments on Leydesdorff's 'A Validation Study of Leximappe'," *Scientometrics* 25, no. 2 (1992): 313–16; Leydesdorff, "A Reply to Courtial's Comments," *Scientometrics* 25, no. 2 (1992): 317–19; Courtial, "Comments on Leydesdorff's Article," *Journal of the American Society for Information Science* 49, no. 1 (1998): 98.

10. The experiment is reported in Eugene Garfield, Irving H. Sher, and Richard J. Thorpie, *The Use of Citation Data in Writing the History of Science* (Philadelphia: Institute for Scientific Information, 1964), www.garfield.library.upenn.edu/papers/useofcitdatawritinghistofsci.pdf. By processing the same data with the method of critical path analysis, similar conclusions on the structure of the DNA research network have been drawn by Norman P. Hummon and Patrick Doreian, "Connectivity in a Citation Network: The Development of DNA Theory," *Social Networks* 11, no. 1 (1989): 39–63, www.garfield.library.upenn.edu/papers/hummondoreian1989.pdf.

11. The software, commercially distributed at www.histcite.com, was first presented in Garfield, "From Computational Linguistics to Algorithmic Historiography" (paper presented at the Lazerow Lecture held in conjunction with panel on Knowledge and Language: Building Large-Scale Knowledge Bases for Intelligent Applications, University of Pittsburgh, September 19, 2001), www.garfield.library.upenn.edu/papers/pittsburgh92001.pdf; and Garfield, "From Bibliographic Coupling to Co-Citation Analysis via Algorithmic Historio-Bibliography: A Citationist's Tribute to Belver C. Griffith" (paper presented at Drexel University, Philadelphia, Pa., November 27, 2001), www.garfield.library.upenn.edu/papers/drexelbelvergriffith92001.pdf; then jointly by the three authors, in Eugene Garfield, Alexander Pudovkin, and Vladimir S. Istomin, "Algorithmic Citation-Linked Historiography—Mapping the Literature of Science" (paper presented at the ASIS&T 2002: Information, Connections and Community. 65th Annual Meeting of ASIST, Philadelphia, Pa., November 18–21, 2002), www.garfield.library.upenn.edu/papers/asis2002/asis2002presentation.html. Several examples

of applications of the *HistCite* algorithm to specific case studies can be accessed at www.garfield.library.upenn.edu/histcomp/.

12. Garfield, Pudovkin, and Istomin, "Why Do We Need Algorithmic Historiography?" *Journal of the American Society for Information Science and Technology* 54, no. 5 (2003): 400, www.garfield.library.upenn.edu/papers/jasist54(5)400y2003.pdf.

13. For an insightful literary perspective on the concept of influence, see Umberto Eco, *On Literature* (Orlando, Fla.: Harcourt, 2004), 118–35.

14. Such, for instance, is the outcome of a multivariate statistical analysis of 252 American psychologists' answers to a questionnaire on the drivers of articles' impact in psychology reported by Robert J. Sternberg and Tamara Gordeeva, "The Anatomy of Impact: What Makes an Article Influential?" *Psychological Science* 7, no. 2 (1996): 69–75.

15. Small, Sweeney, and Grenlee, "Clustering the Science Citation Index Using Co-Citations, 2: Mapping Science," *Scientometrics* 8, nos. 5–6 (1985): 322.

16. Kessler, "Bibliographic Coupling Between Scientific Papers," *American Documentation* 14, no. 1 (1963): 10–25; Kessler, "Bibliographic Coupling Extended in Time: Ten Case Histories," *Information Storage & Retrieval* 1, no. 4 (1963): 169–87.

17. Glänzel and Czerwon, "A New Methodological Approach to Bibliographic Coupling and Its Application to the National, Regional and Institutional Level," *Scientometrics* 37, no. 2 (1996): 195–221; Jarneving, "Bibliographic Coupling and Its Application to Research-Front and Other Core Documents," *Journal of Informetrics* 1, no. 4 (2007): 287–307. The latter article also contains a synthetic review of previous research on the subject.

18. Small, "Co-Citation in the Scientific Literature: A New Measure of the Relationship Between Two Documents," *Journal of the American Society for Information Science* 24, no. 4 (1973): 265–69, www.garfield.library.upenn.edu/essays/v2p028y1974–76.pdf.

19. The technique was introduced in White and Griffith, "Author Cocitation: A Literature Measure of Intellectual Structure," *Journal of the American Society for Information Science* 32, no. 3 (1981): 163–71, garfield.library.upenn.edu/hwhite/whitejasist1981.pdf, and subsequently codified in McCain, "Mapping Authors in Intellectual Space: A Technical Overview," *Journal of the American Society for Information Science* 41, no. 6 (1990): 433–43. The experiment on information scientists cited above is reported in White and McCain, "Visualizing a Discipline: An Author Co-Citation Analysis of Information Science, 1972–1995," *Journal of the American Society for Information Science* 49, no. 4 (1998): 327–55, www.asis.org/Publications/JASIS/Best_Jasist/1998WhiteandMcCain.pdf.

20. McCain, "Mapping Economics Through the Journal Literature: An Experiment in Journal Cocitation Analysis," *Journal of the American Society for Information Science* 42, no. 4 (1991): 290–96; McCain, "Core Journal Networks and Cocitation Maps: New Bibliometric Tools for Serials Research and Management," *Library Quarterly* 61, no. 3 (1991): 311–36; McCain, "Neural Networks Research in Context: A Longitudinal Journal Cocitation Analysis of an Emerging Interdisciplinary Field," *Scientometrics* 41, no. 3 (1998): 389–410; Ying Ding, Gobinda G. Chowdhury, and Schubert Foo, "Journals as Markers of Intellectual Space: Journal Co-Citation Analysis of Information Retrieval Area, 1987–1997," *Scientometrics* 47, no. 1 (2000): 55–73; Ming-Yueh Tsay, Hong Xu, and Chia-Wen Wu, "Journal Co-Citation Analysis of Semiconductor Literature," *Scientometrics* 57, no. 1 (2003): 7–25; Zao Liu, "Visualizing the Intellectual Structure in Urban Studies: A Journal Co-Citation Analysis (1992–2002)," *Scientometrics* 62, no. 3 (2005): 385–402; and Linda S. Marion, Concepción S. Wilson, and Mari Davis, "Intellectual Structure and Subject Themes in Information Systems Research: A Journal Cocitation Study," *Proceedings of the American Society for Information Science and Technology* 42, no. 1 (2005) eprints.rclis.org/archive/00005217/.

21. Benjamin Vargas-Quesada and Félix Móya-Anegon, *Visualizing the Structure of Science* (Berlin: Springer, 2007).

22. The evolution of mapping techniques over the past decade may be followed through the ARIST reviews published so far: White and McCain, "Visualization of Literatures," *Annual Review of Information Science and Technology* 32 (1997): 99–168; Katy Börner, Chaomei Chen, and Kevin W. Boyack, "Visualizing Knowledge Domains," *Annual Review of Information Science and Technology* 37 (2003): 179–255, www.cs.sandia.gov/projects/VxInsight/pubs/arist03.pdf; Steven A. Morris and Betsy Van der Veer Martens, "Mapping Research Specialties," *Annual Review of Information Science and Technology* 42 (2008): 213–95. A comprehensive overview of the technical and conceptual issues involved in the drawing and validation of several types of scientific maps obtained with multidimensional scaling methods can be found in Robert J. W. Tijssen, *Cartography of Science: Scientometric Mapping with Multidimensional Scaling Methods* (Leiden: DSWO Press, 1992).

23. The use of Pearson's r as a similarity measure at this stage of author co-citation analysis, canonized by White and McCain, stirred up an ongoing debate on its statistical meaning, its effects on the mapping process, and the pros and cons of the possible alternatives. The row was sparked off by Per Ahlgren, Bo Jarneving, and Ronald Rousseau, "Requirements for a Cocitation Similarity Measure, with Special Reference to Pearson's Correlation Coefficient," *Journal of the American Society for Information Science and Technology* 54, no. 6 (2003): 550–60.

24. Small and Griffith, "The Structure of Scientific Literature. I: Identifying and Graphing Specialties," *Science Studies* 4, no. 1 (1974): 17–40.

25. Griffith et al., "The Structure of Scientific Literature. II: Toward a Macro- and Microstructure for Science," *Science Studies* 4, no. 4 (1974): 363. To cope with these objections, ISI cartographers tested manifold standardization procedures aimed at improving the quality of the maps, including: the use of co-citation normalization formulas (Jaccard and Salton's cosine), fractional counting of citations (the value of a citation was fractioned according to the number of references contained in the citing article), variable level clustering (the level of co-citation for cluster formation varies according to a preestablished rule based on cluster size), and the iteration of the clustering routine at each level of aggregation (first level, clusters of documents; second level, clusters of first level clusters; third level, clusters of second level clusters; and so on). These enhancements are documented in Small and Sweeney, "Clustering the Science Citation Index Using Co-Citations, 1: A Comparison of Methods," *Scientometrics* 7, nos. 3–6 (1985): 391–409, and Small, Sweeney, and Grenlee, "Clustering the Science Citation Index Using Co-Citations. 2," 321–40.

26. In the second half of the 1980s, doubts about the validity of co-citation maps and criticisms of the methodological and statistical weaknesses of cluster analyses performed by early cartographers were raised, among others, by Diana Hicks, "Limitations of Co-Citation Analysis as a Tool for Science Policy," *Social Studies of Science* 17, no. 2 (1987): 295–316; Leydesdorff, "Various Methods for the Mapping of Science," *Scientometrics* 11, nos. 5–6 (1987): 295–324, and J. E. J. Oberski, "Mapping of Science: Possibilities and Limitations," in *Handbook of Quantitative Studies of Science and Technology*, ed. Anthony F. J. van Raan, 431–62 (Amsterdam: Elsevier, 1988).

27. Garfield, "Introducing the ISI Atlas of Science: Biochemistry and Molecular Biology, 1978/80," in *Essays of an Information Scientist 1981–1982*, 279–87 (Philadelphia: ISI Press, 1983), www.garfield.library.upenn.edu/essays/v5p279y1981-82.pdf.

28. Small, Sweeney, and Grenlee, "Clustering the Science Citation Index Using Co-Citations. 2," 339.

29. Small, "The Synthesis of Specialty Narratives from Co-Citation Clusters," *Journal of the American Society for Information Science* 37, no. 3 (1986): 97–110, www.asis.org/Publications/JASIS/Best_Jasist/1987Small.pdf. The automatic generation of review articles on the basis of the identification and classification of citation contexts has been pursued by Hidetsugu Nanba and Manabu Okumura, "Towards Multi-Paper Summarization Using Reference Information," in *Proceedings of the 16th International Joint Conferences on Artificial Intelligence* [(IJCAI-99), Stockolm, Sweden], ed. Thomas Dean, vol. 2, 926–31 (San

Francisco: Morgan Kaufmann, 1999), citeseer.ist.psu.edu/204343.html, while Small's conjecture of a specialty narrative has been recently taken up to test whether word similarity metrics can be used to automatically select the citation context to be used as a document summary, by Jeff Hand, "Feasibility of Using Citations as Document Summaries" (thesis in partial fulfillment of the requirements for the degree of doctor of philosophy, Drexel University, December 2003), dspace.library.drexel.edu/bitstream/1860/288/8/hand_jeff_thesis.pdf.

30. On the cross-disciplinary virtues of citations, see Small, "Citations and Consilience in Science," *Scientometrics* 43, no. 1 (1998): 143–48; Small, "A Passage Through Science: Crossing Disciplinary Boundaries," *Library Trends* 48, no. 1 (1999): 72–108, findarticles.com/p/articles/mi_m1387/is_1_48/ai_57046528.

31. Robert V. Eagly, "Economic Journals as a Communications Network," *Journal of Economic Literature* 13, no. 3 (1975): 887–88.

32. Doreian and Fararo, "Structural Equivalence in a Journal Network," *Journal of the American Society for Information Science* 36, no. 1 (1985): 28–37.

33. Tijssen, Robert J. W., Jan De Leeuw, and Anthony F. J.van Raan, "Quasi-Correspondence Analysis on Scientometric Transaction Matrices," *Scientometrics* 11, no. 5–6 (1987): 351–66.

34. Campanario, "Using Neural Networks to Study Networks of Scientific Journals," *Scientometrics* 33, no. 1 (1995): 23–40.

35. Two methods employed by Leydesdorff for building indicators of scientific change through the analysis of journal citation patterns are the spotting of central tendency journals and the measure of probabilistic entropy. The former is exemplified in Leydesdorff and Cozzens, "The Delineation of Specialties in Terms of Journals Using the Dynamic Journal Set of the *SCI*," *Scientometrics* 26, no. 1 (1993): 135–56; for the latter see Leydesdorff, "Indicators of Structural Change in the Dynamics of Science: Entropy Statistics of the *SCI Journal Citation Reports*," *Scientometrics* 53, no. 1 (2002): 131–59, users.fmg.uva.nl/lleydesdorff/jcr/index.htm. For recent applications of factor analysis and advanced graph-analytic techniques to the clustering of the *JCR*, see Leydesdorff, "Can Scientific Journals Be Classified in Terms of Aggregated Journal-Journal Citation Relations Using the Journal Citation Reports?" *Journal of the American Society for Information Science and Technology* 57, no. 5 (2006): 601–13, www.leydesdorff.net/classif03/index.htm; and Leydesdorff, "Clusters and Maps of Science Journals Based on Bi-Connected Graphs in the *Journal Citation Reports*," *Journal of Documentation* 60, no. 4 (2004): 371–427, users.fmg.uva.nl/lleydesdorff/jcr01/art/index.htm.

36. Boyack, Klavans, and Börner, "Mapping the Backbone of Science," *Scientometrics* 64, no. 3 (2005): 351–74, ivl.slis.indiana.edu/km/pub/2005-boyack-mapbckbn.pdf (for the map go to www.mapofscience.com/); I. Samoylenko et

al., "Visualizing the Scientific World and Its Evolution," *Journal of the American Society for Information Science and Technology* 57, no. 11 (2006): 1461-69, phy.ntnu.edu.tw/~cchen/pdf/jasist1461.pdf; Leydesdorff and Rafols, *A Global Map of Science Based on the ISI Subject Categories* (2008), users.fmg.uva.nl/lleydesdorff/map06/texts/index.htm (the map is available at www.leydesdorff.net/map06/index.htm).

37. Tijssen, De Leeuw, and van Raan, "A Method for Mapping Bibliometric Relations Based on Field-Classifications and Citations of Articles," in *Informetrics 87/88: Select Proceedings of the First International Conference on Bibliometrics and Theoretical Aspects of Information Retrieval*, ed. Leo Egghe and Ronald Rousseau, 279–92 (Amsterdam: Elsevier, 1988), hdl.handle.net/1942/847; Tijssen et al., "Integrating Multiple Sources of Information in Literature-Based Maps of Science," *Journal of Information Science* 16, no. 4 (1990): 217–27.

38. Braam, Moed, and van Raan, "Mapping of Science by Combined Co-Citation and Word Analysis. I: Structural Aspects," *Journal of the American Society for Information Science* 42, no. 4 (1991): 233–51; Braam, Moed, and van Raan, "Mapping of Science by Combined Co-Citation and Word Analysis. II: Dynamical Aspects," *Journal of the American Society for Information Science* 42, no. 4 (1991): 252–66.

39. Zitt and Bassecoulard, "Delineating Complex Scientific Fields by an Hybrid Lexical-Citation Method: An Application to Nanosciences," *Information Processing & Management* 42, no. 6 (2006): 1513–31, www.nanodistrict.org/scientific-output/papers/delinnano-mz-eb.PDF.

40. Glenisson et al., "Combining Full Text and Bibliometric Information in Mapping Scientific Disciplines," *Information Processing & Management* 41, no. 6 (2005): 1548–72, ftp.esat.kuleuven.ac.be/sista/ida/reports/04-205.pdf.

41. Ed Noyons, "Science Maps within a Science Policy Context," in *Handbook of Quantitative Science and Technology Research: The Use of Publication and Patent Statistics in Studies of S&T Systems*, ed. Henk F. Moed, Wolfgang Glänzel and Ulrich Schmoch, 247 (Dordrecht: Kluwer Academic, 2004). An example of a knowledge domain visualization tool explicitly aimed at supporting competitive intelligence in a science management perspective is described in Boyack, Wylie, and Davidson, "Domain Visualization Using Vxinsight® for Science and Technology Management," *Journal of the American Society for Information Science and Technology* 53, no. 9 (2002): 764–74.

42. The *Atlas* is accessible, after a free registration, at www.atlasofscience.net.

43. Studer and Chubin, *The Cancer Mission: Social Contexts of Biomedical Research* (Beverly Hills, Calif.: Sage, 1980), 17.

44. Studer and Chubin, *The Cancer Mission*, 206 (emphasis in the original).

45. Kuhn, *The Structure of Scientific Revolutions*, 2nd ed. (Chicago: University of Chicago Press, 1970), 174–210. The multifaceted concept of "paradigm"

is at the core of Kuhn's theory of scientific development in the 1962 edition of this book, whereas the postscript to the second edition introduces the notion of "disciplinary matrix."

46. Kuhn, *The Structure*, 175.

47. Kuhn, *The Structure*, ix.

48. The main reactions to Kuhn's model of scientific development and the criticisms of the vagueness of his construct of paradigm are documented by the symposium on Kuhn's work held in London in 1965, which is the origin of one of the most important books in the historiography and philosophy of science, Imre Lakatos and Alan Musgrave, eds., *Criticism and the Growth of Knowledge: Proceedings of the International Colloquium in the Philosophy of Science, London, 1965* (Cambridge: Cambridge University Press, 1970).

49. Kuhn, "Second Thoughts on Paradigms," in *The Essential Tension: Selected Studies in Scientific Tradition and Change*, 293–319 (Chicago and London: University of Chicago Press, 1977). This is a reprint of a 1974 essay.

50. Kuhn's reluctance to explain paradigms in terms of citation analysis is complained about by Henry Small:

> Kuhn's reaction to our initial papers on the cocitation structure of science was one of puzzlement. He wondered why we had focused on highly cited papers and authors, rather than defining the total community of researchers involved with the topic, representing the complete paradigm-sharing community. Any notion that highly cited papers might stand for exemplars or other paradigmatic constructs was not on his radar. Presumably, Kuhn saw bibliometric methods only as a means of performing a social inventory of a specialty, not as a way to define the paradigm itself. Therefore, it is unlikely that Kuhn would have accepted any of our bibliometric reinterpretations of his theory.

Small, "Paradigms, Citations and Maps of Science: A Personal History," *Journal of the American Society for Information Science and Technology* 54, no. 5 (2003): 395–396.

51. Small, "A Co-Citation Model of a Scientific Specialty: A Longitudinal Study of Collagen Research," *Social Studies of Science* 7, no. 2 (1977): 159.

52. Chen, *Mapping Scientific Frontiers: The Quest for Knowledge Visualization* (London: Springer, 2003). The entire issue no. 5 of volume 54 (2003) of the *Journal of the American Society for Information Science and Technology* is devoted to the visualization of scientific paradigms.

53. Kuhn, "Second Thoughts," 296. On the evolution of Kuhn's concept of paradigm, see Erich von Dietze, *Paradigms Explained: Rethinking Thomas Kuhn's Philosophy of Science* (Westport, Conn.: Praeger, 2001).

Chapter Six

Impact Factor and the Evaluation of Scientists: Bibliographic Citations at the Service of Science Policy and Management

Constrained on the dissecting table of sociologists, science delivers a rather disenchanted picture of its truth-chasing mission: a professional work organization in which the pursuit of collective goals in the advancement of knowledge about nature and society is inseparable as much from the allocation of scientific reputation among colleagues/competitors as from the publish-or-perish game pending on the individual's desire to change the course of history for the better. Scientific progress, in fact, rests on the recognition and legitimization of individual contributions by a research community sharing a set of notions, methodologies, practices, and values relative to the field wherein personal skills are exercised. Hence a scientist is properly a scientist only as part of a community whose members, in addition to being oriented toward the resolution of a class of shared problems, are in the position to evaluate and reward specific research results.

The recognition of scientific quality, which basically amounts to the assessment of the novelty of a contribution, is a motivating drive for the researcher, to the point that those who don't attain it are generally diverted to other activities, such as teaching or administrative duties. From a science policy standpoint, it is also a useful guide in the allocation of the financial and human resources necessary for the machinery of science to work properly. Before taking the tangible form of university tenure or promotion, funds granting, or prizes, recognition is awarded through the evaluation of scientific publications. The usual method for assessing the quality of scientific publications or research projects competing for funding is the critical examination of their content by a group of acknowledged experts (peer reviewing). The peer-review system, born in seventeenth-century Europe to counteract the knowledge fragmentation caused by the branching of scientific specialties, is a basic component of the collective, cumulative, and self-corrective character of science because, through the competent judgment of peers, scientists rely confidently on the work of colleagues and predecessors without having to reinvent the wheel each time.

The reliability of peer reviewing as a quality detector tool requires that the evaluation procedure be as encoded and impersonal as possible, and that there be a spontaneous convergence of judgment criteria from different experts on a given piece of research. In the best possible world, both conditions might be thought of as the ultimate fulfillment of a sort of mission, in which the judges temporarily abandon their human condition and reappear as pure, rational spirits who manage to perform an objective appraisal in accordance with a set of universal rules. Scientists themselves, though, have gradually realized, at their own expense, that such a spiritual distillation is difficult, if not impossible, to achieve. As extensively documented by the history of science and by the less honorable history of successful frauds, the peer-reviewing system is biased by prejudices, inconsistencies, and conflicts of interest. Sociologists, on their part, have promptly traced this weakness back to the conflictual dynamics that govern scientific fields, while philosophers echoed them by demolishing the myth of a single, universal scientific method that should warrant the application of objective peer assessment criteria. Among the criticisms, one stands out as the mainstay of many others in undermining the peer-reviewing system's alleged vocation of promoting "pure" knowledge. In Kuhn's perspective, scientists operating within the frame of a consolidated paradigm maintain a rigid and conservative attitude toward the body of current scientific theories so as to discourage alternative, potentially destabilizing ideas and hypotheses. This resistance to change takes strategic advantage of the direct or indirect social control of the main distribution channels of scientific ideas exercised by means of the peer assessment procedures so vital to the scholarly publication system. Resistance to change, one might add, neatly dovetails with the heated reactions stirred up by the application of external (i.e., quantitative) evaluation criteria to scientists' work, something they are likely to regard as an unwarranted meddling with their semiprivate dialogue with nature. Nevertheless, in a Kuhnian universe, conservatism is also the staple of scientific change: whenever the power and precision of paradigmatic science result in the emergence of unmanageable technical difficulties, or "anomalies" that would make sense only in light of new paradigmatic theories, the stage is set for a scientific revolution whereby a paradigm shift takes place. Throughout the time spans separating one revolution from another, then, research work is "normal science," mere routine or puzzle solving, a kind of activity that "does not aim at novelties of fact or theory and, when successful, finds none."[1] Thus, a peer review conducted during a period of normal science is suspected of hindering the introduction of significant theoretical innovations through a sort of "iron law" of disciplinary oligarchy.[2]

The peer review crisis triggered off the search for alternative or complementary evaluation methods more detached from subjective judgments. So, at least from the 1970s, the use of citation analysis to produce indicators of scientific performance in terms of citation scores began to receive unprecedented attention from politicians and science managers, clearly attracted by the several virtues of a numerical index: convenient, quickly understood, easily applied, and, above all, easy to calculate thanks to ISI databases. Citation scores and, in general, bibliometric indicators display an obvious, congenital defect: they are never the first to arrive on the crime scene. The unit under evaluation, be it a beginner scientist on the verge of publishing his or her first paper, or a paper on the verge of being cited, cannot indeed be imbued with bibliometric life until, respectively, the publication process is completed and the citation received. The quality potential of aspiring scientists or new entries in the publishing arena, as well as the quality potential latent in a paper before someone notices (and cites) it, are thus irremediably shut out of the bibliometric window, their appraisal being relegated to qualitative judgment of an irremediably local—at worst parochial—character. Of course, it could reasonably be contended in the case of newbies that, if not for their direct recruitment, bibliometric appraisals can at least help select their better teachers and evaluators, but being a high-impact evaluator doesn't automatically warrant the interiorization of the ethical requirements involved in any fair evaluation. By contrast, after the recruitment has taken place and the scientist is aboard, so deeply involved in the publication and citation game that productivity and impact patterns become discernible and eventually measurable, the question of whether quantity and quality fit together takes on more distinct contours.

Scientometricians have constantly adopted a dual strategy to corroborate the equation quality = citation impact implicit in bibliometric performance indicators, explaining why the bibliometric medicine is supposed to work and, at the same time, enhancing its preventive power against potential rejections:

1. *Why the medicine is supposed to work:* because a sizable amount of statistical evidence has been generated, every now and then, on the positive correlation existing between citation scores and independent appraisals of individual performance, above all peer ratings. Likewise, as the long series of Garfield's *Citation Classic Commentaries* testifies, if one turns to the twentieth-century history of science, it doesn't take too much to realize that the most prominent scientists, those who have seen their excellence punctually rewarded by prizes and honors (the Nobel prize in the first place), have managed to publish, more often than not, citation classics.[3]
2. *How to avoid rejections:* simply anticipating, by a good deal of statistical evidence, some of the objections raised against the significance of citation counts, two of the most recurrent being the ambiguous role of self-citations and the phenomenon of delayed recognition.

Scientists usually cite their own earlier contributions upon which ongoing work is built, thereby reinforcing in the audience the sense of continuity between present and past accomplishments. For a long time the ISI has been interested in finding out what the meaning of self-citations is in the case of journals, while a 1977 paper by Renata Tagliacozzo laid down the basis for a series of subsequent critical analyses of individual authors' self-citation practices, addressing such relevant issues as the relative proportion of self-citations in different research fields, their frequency of occurrence and age within selected groups of papers, and their dependence on other critical bibliometric features, including the number and productivity of coauthors and the size of bibliographies.[4] Self-citations are not an evil in themselves. Yet, beyond a certain threshold, they are suspected of deceitfully inflating the citation impact of the unit under assessment (paper, scientist, journal, institution). To neutralize this suspicion, bibliometricians have sought to demonstrate that, though at the micro-level of analysis self-citations may effectively represent a problem and ought to be removed before any comparative evaluation, when macro-units come into play and large aggregates of citations are analyzed over an adequate number of years, not only do they lose their deviating weight and not need to be excluded from the evaluation, but they work, to a certain extent, as a size-dependent "impact-reinforcing mechanism" by triggering a "chain reaction" that ultimately brings about an increase in external citations.[5]

Another potential threat to many bibliometric studies making use of short time-windows for citation tallying is the historical recurrence of premature discoveries, that is, important contributions completely ignored for several years because of an excess of originality that puts them ahead of time, as happened during the second half of the 1800s to Mendel's work. Once again, this is not a problem for evaluative bibliometrics insofar as it can be demonstrated that, from a statistical viewpoint, delayed recognition is an exception to the general rule that "a paper not yet cited, say, ten years after publications has little chance to become even fairly cited in the future."[6]

The use (and abuse) of citation indexes in the evaluation of scientific performance has taken two main avenues. The first is a shortcut, leading to the association of the quality of a paper with the same index expressing, in statistical terms, the impact of the journal in which it appeared (the impact factor); the second is a more tiring road, in great part still under construction, which concentrates on citation counting at the individual or collective level and, without denying the perils of the journey, makes its own way through a wide range of methodological considerations on the limits and conditions of applicability of bibliometric indicators.

6.1. THE SHORTCUT: THE JOURNAL IMPACT FACTOR

The impact factor (IF) is a journal citation measure devised in the early 1960s by Eugene Garfield and Irving Sher to help select journals for the *Current Contents* and the *SCI*. Since 1975, it has been regularly supplied by the *Journal Citation Reports* (*JCR*), an annual ISI publication covering both the natural and social sciences in separate editions. The *JCR* is a byproduct of the main ISI indexes: processing data from the *SCI* and *SSCI*, it categorizes and compares over 9,100 international scholarly journals with respect to various metrics, the chief one, or at least the most popular, widely used and heatedly debated, being the IF.

The IF of a journal is an estimation of its average article's citation score over a relatively short time span. It is computed for a given year through a simple arithmetical division between a numerator and a denominator. The numerator is the number of citations received, in the processing year, by (all) the items published in that journal during the previous two years; the denominator is the overall number of "citable" items (research articles, reviews, and notes) issued by the journal during the same two years. If, for instance, in 2006 the items issued in 2005 and 2004 by journal *X* were cited 100 and 150 times, respectively, and if the total number of citable items published in that journal in 2005 and 2006 was 70, then the 2006 IF of journal *X* would be

$$\mathrm{IF(X)} = \frac{100 + 150}{70} = 3.57$$

The use of a ratio, with a portion of published items summed up as the denominator, in lieu of a crude count of citations to all the journal backfiles, serves the purpose of leveling out the size gap between journals publishing many items and those carrying few but potentially influential contributions. Likewise, the adoption of a short time window allows discounting of the skewing effect of overcited articles published by older journals. One could reasonably argue, incidentally, that a total citation count relative to all articles published by a journal throughout its life would ensure a better historical perspective on its impact than a two-year interval. In effect, Garfield himself used a total citation count to derive his Law of Concentration from Bradford's Law. But having delimited the set of multidisciplinary research journals forming the core of the entire science communication system, he needed a more flexible analytical tool for further extending the pool of valuable sources. The choice of a two-year window, in particular, stemmed from an early observation by Garfield that "the typical cited article is most heavily cited during the two years after its year of publication."[7] Across disciplinary domains, instead, the chronological distribution of cited references varies considerably, thereby explaining the numerical difference of IF ranges between journals of different disciplines. For this reason, the *JCR* classifies journals into subject categories and ranks them by decreasing IF within each category, delivering also additional measures useful to estimate the time-dependence of citation and reference patterns: the immediacy index and the cited half-life.[8]

Even though librarians have long been aware of the multidimensionality of scientific journals' value and the irreplaceable role of local use surveys in planning acquisition policies, the IF and related indexes have been used extensively, over the last thirty years, to support informed decisions in library collection management. A major source of controversy originated when, in the context of a general revision of research quality control criteria, science administrators started employing the IF as a tool of decision making in science policy issues involving individual scientists, research groups, departments, institutions, and countries, instead of journal titles. The higher the IF of the journals listed in the publication records of the units under assessment, the greater the candidate's chance of outperforming all the other applicants in a competition for promotion, tenure, or funding allocation. In European countries, especially, the practice of associating the citation impact (and the individual merit) of an author with what originally designated just the impact of a journal has become endemic, to the point of conditioning even the scientists' behavior in the manuscript submission process, which is strongly biased toward the titles scoring higher in the bibliometric top ten. Journal editors, on their part, immediately grasped the IF's strategic potential, taking advantage of its alleged linkage with scientific quality in advertising editorials like "Journal *X* receives Impact Factor," "Journal *X* Impact Factor sets another record," "Journal *X* Impact Factor increases by n%," and the like. As in the case of television audience ratings, estimated by the number of people tuned in to a channel regardless of how much they actually enjoyed the program, the proportional increase in the IF score became a prelude to marketing success, to an increased commercial and symbolic visibility, and occasionally to a more profitable sale of advertising spaces, making room also for fraudulent practices, such as the inflation of the IF by journal editors who keep a diamond lane for articles firmly embedded in a network of reciprocal references.[9]

Over the last decade, plenty of criticism and blame has piled up against IF-based evaluation methods, so much so that it would be impossible, perhaps even useless given the repetitiveness of many arguments, to review all or even a small part of the papers setting them forth.[10] Curiously enough, since professional bibliometricians do agree with their alleged critics about the inadequacy of the IF as a valid measure of scientific achievement, what the critics criticize is not a theory or an argument backed up by this or that author but, as usually happens when the stakes go beyond a sterile methodological contention, an ideal "pro-IF argument" created or remodeled from scratch so as to counter it more effectively. IF-addicted evaluation procedures are not theorized by leading lights in the bibliometric field; they are simply "planned" and "executed" here or there as the umpteenth manifestation of the whole panoply of local micro-powers that get a share of the spoils in the academic policy game. Objections to the IF, nevertheless, are of extreme theoretical interest because they run the gamut from the technical limitations of ISI databases in coverage and accuracy to the structural inadequacy of citations as building blocks of performance indicators. Two points deserve particular attention here: the skewness of citation distribution and the several conceptual and technical limitations bearing upon the significance of the IF score.

Detractors of citation analysis censure the widespread association of the IF with a virtual object that hardly exists, the average article. A journal, even though boasting a high IF, contains good articles next to mediocre ones and articles with a strong impact next to others with little or no impact, and the distribution of citedness is extremely skewed, so that the few papers attracting the most citations are quite systematically outnumbered by the share of low cited or totally neglected ones. But if high-impact titles owe their primacy to frequent citation of only a limited number of papers, then the journal IF is definitely inappropriate as surrogate of the impact of an individual article, all the more so because the poor statistical correlation existing between the citedness of individual articles and the IF of the journals wherein they appeared is well documented in bibliometric literature.[11] Still, one might ask to what degree the poor correlation at the individual article level justifies general conclusions about the probabilistic deficiency of the IF. Is it possible to assess the validity and reliability of journal citation scores? What kind of information, if any, does the IF ultimately provide on the relative standing of a journal? A clue to partially answering such questions may be found at the statistical reasoning crossroads where different bibliometric research routes usually meet.

When confronted with league tables of social significance, be they institutional, individual, or documental comparisons based on simple numbers, professional statisticians resolutely stress the importance of data integrity, the necessity to take into account potentially confounding factors, and the crucial issue of specifying an appropriate statistical model enabling the analyst to estimate the uncertainty inherent in results along with the degree of reliance that may eventually be placed on them.[12] For citation league tables, nonetheless, each of the above prerequisites is not that easy to comply with: ISI data integrity is questioned, every now and then, on the ground of the deforming filter imposed on their processing routines by a private company.[13] Furthermore, the confounding role of external variables, such as differences in citation practices among research fields or differences in prestige among journals, is not amenable to an immediate translation into equation parameters, while the choice of a suitable stochastic model necessary to ascertain which differences are, and which are not, significant, is hindered as much by the perpetual skewness of citation scores as by the equally satisfactory way in which different models can fit the same dataset.

Early concerns about the degree of stability and reliability associated with citation rankings were expressed during the 1970s both within the American and the European (specifically Dutch) bibliometric subculture, as testified, respectively, by Narin's *Evaluative Bibliometrics* and Dennis Dieks and Hans Chang's 1976 paper.[14] The latter featured a probabilistic model of the citation process in which the total number of citations received by a paper in a given year is a stochastic variable assumed to follow the Poisson distribution. At around the same time, in view of improving comparisons among authors in similar fields by means of appropriate hypothesis tests, Nancy Geller and colleagues envisioned the even more challenging possibility of estimating, under a series of suppositions about the regularity of citation patterns and the growth rate of scientific literature, the lifetime citation rate of a paper.[15] Their model incorporated the assumption that the variable expressing the number of citations settled on a paper over a forty-year lifetime follows a negative binomial distribution (a deviation from the Poisson arising in contagious processes of the "St. Matthew" type). A similar hypothesis informed Allison's 1980 attempt to provide a scale-invariant measure of inequality in publication/citation counts as well as Schubert and Glänzel's 1983 design of a reliability test for determining the statistical significance of the observed differences in citation impact. Noticeably, the latter authors concluded that their simple test "proved to be suitable for comparisons of journal impact factors or citation rates for similar source and citation periods" while "in other situations (e.g., lifetime citation rates or citations to all previously published papers) the distribution of citations may fundamentally differ in character and even the applicability of classical Gaussian statistics can rightfully be questioned."[16]

Carrying on the above line of statistical modeling, Bensman has recently performed an exploratory investigation of the 2005 *JCR*'s probabilistic structure. Here the frequency distribution of journal IFs across *SCI*'s and *SSCI*'s subject categories is tested against the theoretical probability distributions traditionally holding the field in information science: the Poisson distribution and the negative binomial.[17] Under specific assumptions, the test results indicate that, besides the well-known IF bias toward big and well-funded fields such as biomedicine, a structural cleavage comes into sight from a bird's-eye view. Indeed, as compared to the almost random and Poisson-compliant IF distribution of the *SSCI* journals, the IF distribution of the *SCI* titles is much more positively skewed and conforms to a contagious pattern of the negative binomial type, both effects being amenable to the larger prevalence of highly cited review journals in the *SCI* class. Such a probabilistic unbalance is nothing more than a further corroboration of Garfield's early discoveries relative to the centrality of review papers in scientific communication and the surprisingly low mean citation rates of the vast majority of published scientific literature. Needless to say, it concurs to advise against any acritical use of IF-based measures for research evaluation purposes, at least as far as journals at the lower end of the IF's range are concerned. Exceeding variability, indeed, causes the small differences observed in rank order among journals to be scarcely significant, a probable effect of mere random variation.[18] This is not, however, tantamount to completely dismissing the IF's utility in evaluative settings.

Turning back to the source itself, Bensman has also sought to demonstrate, on a sample of 120 chemistry journals, the validity of Garfield's journal citation measures (total citation counts and IF) as indicators of journal importance against peer ratings and usage rates.[19] In perfect accordance with the tradition of advanced mathematical treatment of information processes outlined in chapter 4, he posits a compound Poisson distribution as the most suitable stochastic model for explaining the frequency distributions of journals across the four measures. Such a complex model is necessary to account for two concurrent sources of variability: the structural ambiguity injected by Bradford's Law into the scattering of chemical articles across journals and the postulate of a success-breeds-success mechanism at work behind the skewness of frequency patterns. The relationships among the four measures of journal importance, then, are gauged by means of parametric and nonparametric techniques, leading to a substantial confirmation of Garfield's findings: 1) total citation count and IF capture different facets of journal importance; 2) the former is better than the latter as a global measure of importance, but the gap narrows if only a better classification is introduced in the sample sorting out review journals from research journals; and 3) both measures are surprisingly stable over time at the higher level of citation rankings. Hence the inappropriateness of the IF in accounting for the relative standing of individual articles doesn't automatically compromise its ability to capture one facet of journals' quality: the higher propensity of high IF journals to publish heavily cited articles compared to low IF journals. Some go even further in claiming that, if a journal set is deemed to correspond roughly with the documentary base of a discipline, the IF is no longer to be intended as the mean citedness of the papers published in a journal, but rather as their "chance for citedness" resulting from the relative contribution of the journal to the overall impact of the entire set of journals devoted to the discipline.[20]

Probabilistic considerations aside, many objections have been raised against the IF's ability to catch journal prominence in any meaningful way, partly because of an ambiguous definition of citable items in the IF ratio, and partly because of a series of external or contingent factors bearing on the significance of citation counts at the journal level. Let's briefly review these limitations.[21]

1. *Classification of citable items.* By definition, the numerator and the denominator of the IF formula are incongruent, as the units that contribute to augmenting the first do not enter the second: the number of citable items at the denominator does not take into account letters, editorials, or conference abstracts, in other words, all the contributions not classified as original research articles, notes, or reviews. This gap involves the definition of strict but necessarily conventional classification criteria for distinguishing citable from noncitable units. As emphasized by

many critics, moreover, even noncitable materials can be cited and do indeed enter the overall counting of citations at the numerator, thereby spuriously inflating the score for journals that publish many trivial items.[22]

2. *Accuracy issues*. From a bibliographic and cataloging point of view, journals are complex entities, often with a turbulent editorial life exposing them to title changes, splits, mergers, and supersessions, which can dramatically affect citation counts. Nonetheless, the ISI doesn't combine citation data on the basis of lineage, nor for sections of the same journal. This choice has occasionally raised controversies over the accuracy of IF scores even for internationally prominent titles, such as the dispute dating back to the 1980s on the double-counting of citations spuriously inflating *Angewandte Chemie*'s IF. It turned out that the miscounts were due to several authors citing both the English and German versions of the same paper, whereas only the former had been indexed by the ISI.[23]
3. *Density and age of cited references*. Since, intuitively, the more one cites, the more can be cited, the average number of bibliographic references per article in a set of subject-related journals is roughly proportional to the citation impact of the journals themselves. Recent statistical analyses confirm, in this respect, that the IF is sensitive both to variation in the average number of references per paper and, more deceptively, to self-citation rates.[24] The age distribution of those references, too, strongly affects the IF calculation, because it weighs upon the number of citations falling in the two-year window of Garfield's measure. Density and age of cited references emphasize the variability of citation cultures among disciplines and research fields. It is well known, for example, that mathematicians cite less than biochemists and that older literature is more significant for them than for life scientists. Most of mathematicians' citations, in fact, fall beyond the two-year span rewarded by IF-based statistics. As peremptorily pointed out by an official report on citation statistics recently sponsored by, among others, the International Mathematical Society, the temporal short sight testifies for the structural inadequacy of such coarse measures of journal value to research evaluation, especially if they are not supplemented with other kinds of substantive information.[25] A similar variability also exists among research fields belonging to the same dis-

cipline; in the biomedical domain, for instance, basic research journals tend to be "consumed" (and cited) more quickly than applicative journals, so that a higher number of citations concentrates on more recent years, especially on the last two involved in the IF calculation. Such aging patterns are hardly manageable in bibliometric terms because each research field (or even subfield) seems to contain a specific mix of journals with different aging characteristics. That's why, to facilitate fair comparisons, efforts have been made to identify and classify the different kinds of aging processes starting from the distinction between a maturing and a declining phase in the journals' citation life.[26]

4. *Journal format and article type*. Once a journal gets published, its format and scope affect the IF via the speed and intensity with which different types of articles attract citations. A rapid communications title, for instance a physics letter journal, is usually cited more quickly than a full paper journal, but for a shorter period of time; hence, its IF will be comparatively higher. Likewise, given the well-known propensity of review articles to attract many citations on the score of their bibliographic vocation to collapse entire research traditions, journals that publish many review articles are likely to boost their IF. That's why the *JCR* itself recommends processing journal source data by document type.

Whereas accuracy issues can in principle be tackled by means of a careful cleaning of ISI data, the other structural pitfalls require effective adjustments of the original IF formulation or, alternatively, the design of more sensitive measures of journal impact.[27] A readily available adjustment is the modification of the time window for either the cited or the citing years. In the first case, as noted above, the counting of citations to articles older than two years marks out the aging differences between journals of various disciplines and research fields. Yet evidence also exists that, if comparisons of journals' citation records are drawn within each subject category, IF-based rankings do not change significantly.[28] The second choice entails the computation of a "diachronous" in lieu of a "synchronous" IF. Instead of counting citations from a single citing year to documents that appeared in two or more previous publication years, citations are tracked from two or more citing years (hence from several *JCR* editions) to documents issued in a fixed publication year. Being more sensitive to the time evolution of citation patterns, a diachronous impact indicator better reflects the actual impact of a set of papers; hence, it is particularly suitable for evaluative purposes. It has been implemented, for example, in the evaluation methodology of the Hungarian Academy of the Sciences as well as by the Leiden group at the CWTS.[29]

Another method for reducing IF biases and enhancing its cross-field comparability is the creation of a normalized measure, which takes into account the main variables at stake: (sub)field citation practices, types of documents published by the journal, and age of cited papers. One early example of normalization is Graeme Hirst's discipline impact factor, dating back to the late 1970s, which measured the number of times a journal is cited by the core literature of a single subfield rather than by the complete set of ISI journals.[30] At about the same time, by far the most interesting proposal came from Pinski and Narin, whose Google-like algorithm of journal ranking is examined more carefully in subsection 6.2.2. A third remarkable example of normalization is the "Journal to Field Impact Score" recently introduced by van Leeuwen and Moed. The new index seeks to overcome the IF's pitfalls on various fronts. It counts the same items (original articles, letters, notes, reviews) both at the numerator and denominator; it is field-specific, in the sense that the impact of the individual journal is compared to the world citation average in the corresponding research fields; it differentiates the normalized impact for the various document types (reviews, research articles, etc.); and finally, it employs variable citation and publication windows for the count depending on the communicative patterns of the research field under evaluation.[31]

So far, none of the revised versions or substitutes of ISI IF has gained general acceptance beyond its proponents, probably because the alleged alternatives lack the degree of interpretability of the original measure. Furthermore, their reproducibility is seriously hindered by the absence of an ISI-like institution maintaining an updated database of comparable solidity. Even so, whatever the limitations and broken promises of journal impact measures, they are vital to the survival of scientometrics because the IF and its descendants play a key part in the standardization of citation measures for the purpose of comparative assessment. In fact, anyone seeking to establish a scale of measurement for citedness—that is, a standard against which a fair judgment of "how much" the citation score of a given set of papers is "low" or "high" can be formulated—needs an accurate preliminary estimation of the expected number of citations for the research field encompassing the papers under evaluation. A widely practiced solution, though not necessarily the best, is to set the standard at the level of the scientific journals. The rationale for this choice is simple: by virtue of the high degree of specialization and the strict quality control exercised at the content level by the "gatekeepers" of science (editors and referees), scientific journals are supposed to contain "coherent sets of papers both in contents and in professional standards."[32] The mean IF of the journals covering a specific research area is then assumed, in many research evaluation models, as a rough estimation of the number of times a paper falling into that research area could have been cited.

The reader should be aware that the above operation is by no means a neutral one. In fact, any attempt to normalize a citation score through journal measures must necessarily assume a preexisting, clear-cut delimitation of disciplinary and subdisciplinary boundaries between sets of scientific journals. In other terms, the normalization has to rely on a journal-compliant classification of scientific specialties, which is expected to accomplish in advance the complex task of "cutting the cake" so as to predispose each slice to the assignment of a field-specific standard score. As discussed in section 5.3, the application of multivariate statistics and graph-analytic techniques to journal-journal citation patterns can provide the classifier with assistance in assigning journals to different subject specialties. But one general lesson can also be drawn from those experiments, namely that no conclusive "structural" classification is achievable by quantitative methods alone, at least as long as substantive assumptions are required for a multivariate or network analytical model to fit the complex patterns of interrelationship hidden in ISI citation data. Thus, unless one simply trusts *JCR*'s subject categories for practical reasons, an unambiguous classification of research fields to be used as a baseline for scientometric evaluations is severely hindered by the fluidity of disciplinary boundaries and the high degree of variability in subject coverage across scientific journals (think of Bradford's Law). At a deeper layer, it might even be construed that such a classification is thwarted by the way in which modern scientific disciplines have evolved following the pattern of historical developments outlined in section 1.3: as the system of knowledge production transcended university departments, the degree of mutual dependence among scientific fields grew accordingly under the effect of cognitive and social factors; university-based disciplines, with their clear-cut subdivisions conventionally implemented in bibliographic (and bibliometric) information systems, consequently ceased to represent the basic unit of social organization for knowledge production.[33] Some authors have also moved forward in claiming that, from the mid-twentieth century onward, a new form of dynamic, context-driven, problem-focused, and interdisciplinary knowledge production mode took hold in contrast to the academic, discipline-based organization,[34] a knowledge production mode in which, to put it succinctly, any preconceived classification of research fields has to come to grips with the disturbing prospect that the intellectual and institutional organizations of the sciences no longer coincide to any considerable degree.

6.2. THE LONG AND WINDING ROAD: DESIGN AND APPLICATION OF ADVANCED SCIENTOMETRIC INDICATORS

Hardly anyone would dispute that scientific research is a complex, multidimensional activity, which cannot be flattened onto an evaluation scale centered on a single output or publication type. Scientists do not spend all their time writing; they teach, plan, experiment, organize, and report provisional findings. Furthermore, when they write, they don't write only journal papers, but also books, technical reports, manuals, conference papers, and software, materials that, though citable and cited to a certain extent by journal papers, live and exercise an impact often outside the formal communication channels of journal literature. In computer science, for example, conference proceedings play an outstanding role in the timely communication of cutting-edge knowledge and cannot be reduced to a provisional form of scholarly transaction to be replaced, sooner or later, by the full-fledged journal paper. Even in the biomedical field, where journal papers are the standard currency of research activity, the proper criteria for evaluating the impact of the findings being communicated reach far beyond the crude consensus manifested by formal citations since the ultimate goal of biomedical research, namely health improvement and the promotion of patient care, lies well outside the boundaries of the academic audit process.

On the other hand, given the lack, outside the ISI products, of comprehensive databases containing structured information on the various types of output for all individual scientists and institutions worldwide, the assumption that ISI data provide a good sample of global and local scientific standing is quite well accepted, at least for the natural and life sciences. Insofar as publishing papers in peer-reviewed international journals comes to be regarded as the main business of scientists, therefore, ISI citation data offer the raw materials for building indicators of scientific performance. And even when, as in the above examples of computer science and biomedicine, their poor representativeness is acknowledged outright, it is generally agreed that further inquiry into the conditions under which they can be integrated with complementary information sources is worthwhile. For instance, Leiden bibliometricians at the CWTS completed a pilot study in 2006 for expanding ISI indexes with source papers from refereed proceedings of computer science international conferences in view of developing field-specific bibliometric indicators.[35] Similarly, Grant Lewison has envisioned the expansion of the set of analytical tools necessary to trace the multiple routes along which biomedical research influences health decisions through the bibliometric analysis of nonjournal citation linkages, such as citations to biomedical articles from patents, clinical guidelines, and newspapers.[36]

Over the past decades, the debate on research evaluation methods and standards has produced an impressive amount of literature, making room for a wide variety of hypotheses and solutions. At one extreme, the opponents of quantitative methods, although recognizing the need for change, get rid of all bibliometric indicators after a superficial trial because, in the last analysis, nothing appears as reliable as an accurate peer review. At the opposite extreme, some have placed so much trust in quantitative analysis as to claim that, at least to a certain extent, properly weighted indicators can and should be implemented by expert systems and computer-assisted procedures to help determine career progression and university chair assignment.[37] Bibliometricians, on their part, have gone beyond simple schematizations: none would deny the centrality of an accurate peer review for the appraisal of scientific merit; all the same, they purport to complement and render more accurate the peer-reviewing system through additional, less subjective analytical tools. To this end, they set out to thoroughly analyze the limits and conditions of applicability of citation-based indicators at whatever level of aggregation: individuals, research groups, institutions, or nations.

6.2.1. Evaluation of Individual Scientists: From Citation Counting to the Hirsch Index

Early on in the political life of citation indexes, the citation score of individual scientists appeared a readily available candidate for the role of performance indicator. Yet since the beginning, this apparently straightforward measure met with some basic issues of validity for the simple reason that, despite the distributional justice mechanism inherent in the act of citing, when the page is turned upside down and the perspective of the cited item comes into focus, citations become goods as valuable as precious metals on the score of a similar property: rarity. It didn't take too much, indeed, to realize that rarity is a structural property of the citation network seen from below. After the launching of the *SCI*, early statistical surveys revealed that the ratio between references processed each year and the number of unique items cited by those references was nearly constant and approximately equal to 1.7, a sort of "magical" number christened "Garfield's constant" by its discoverer. Stated alternatively, in a single year, each paper was cited on average only 1.7 times, and 25 percent of the papers were never cited even once. Subsequent surveys basically confirmed the above trend, showing that, regardless of the exponential growth of scientific literature, the corresponding increase in the value of Garfield's constant was minimal.[38]

As discussed in chapter 4, the average doesn't tell a true story in bibliometric distributions because of their structural skewness: few documents are likely to live in the spotlight, while a much greater number are expected to fall into the spacious dustbin of the uncited. Far from being of merely theoretical interest, this marked inequality raises some fundamental questions of immediate practical value: How can a comparison among average or low citation scores improve the evaluation of individual scientists' work? If the candidates, for instance, have not been publishing long enough to acquire a meaningful number of citations, are the differences statistically significant to the point of justifying the conclusions drawn with the help of citation analysis? To what extent can individual research productivity be correlated with the number of publications and citations received? And to what extent do publication and citation rates depend on the particular subject area in which the candidate's competence is exercised?

Garfield, although recognizing the ease with which citation data for individual researchers can be "misinterpreted or inadvertently manipulated for improper purposes,"[39] maintains that, if properly handled, they can substantively support the evaluation process, supplying it with an objective measure of the usefulness of an individual scientist's work to other scholars. Obviously, ISI-driven citation analysis is a fair evaluation tool in a given field only insofar as publication in scholarly journals is the field's primary vehicle of communication. Comparative citation-based assessments, moreover, are meaningful only if "like" is compared with "like," so that each scientist is first ranked with peers (the practitioners of the same specialty area), and then a relative comparison among rankings is carried out. Citation data are not intended to replace informed peer review and, to be correctly interpreted, ought to be adjusted by taking into consideration the wide variability of citation practices across research fields and disciplines. In addition, a thorough examination of the content and context of citations is required for determining the exact position of the cited items in the communication network, so as to enable the evaluator to answer the question, Who are the citers, and why did they cite a particular document?[40]

Perfectly aware of these caveats, scientometricians seek for a precise formulation of the conditions under which a theoretically informed comparative evaluation can be carried out. As expected, they are faced with several difficulties at each step and level of aggregation: individuals, groups, institutions, and countries. But it is with respect to the individual scientist that things become especially complicated, because of the ambiguities inherent in the definition of objective evaluation standards and the irreparable damage that would otherwise result from an uncritical interpretation of bibliometric data. By way of an example, let's take the simplest case of two applicants working in the same specialty area, under similar organizations and funding facilities, and writing papers on similar subjects in the same journals, who challenge each other in a competition for tenure or promotion. How large should their publication outputs be to obtain statistically reliable indicators? Which differences between citation scores should be considered significant with a certain probability, and which amenable to mere chance? How can the comparative value of their work be assessed in quantitative terms if, due to a difference in age, they exhibit a different publication (and hence citation) score? And what if one boasts a better citation record simply by virtue of a single early publication, compared to which his or her recent work appears to be almost totally ignored? It is obvious that the use of crude citation counts, even though extended to the life work of both candidates, can lead to completely unreliable results, while a longitudinal approach seeking for (and comparing) trend variations in citation rates could be a more fruitful support to the qualitative evaluation of the publications' content.[41] Things get even harder if the simplifying hypothesis of an equal starting point is discarded and replaced by the case of applicants from various research fields who compete for recognition. A major source of problems is now from the same substantial cross-field variability of publication and citation cultures that undermines the evaluative potential of the journal IF. It obviously doesn't suffice to count publications or citations and record their trends over four, ten, or more years; rather, a cross-field normalization is necessary to determine how much a score is low, average, or high with respect to the scores of people working in different areas.

Normalization is usually attained by relating the citedness of a set of papers to a conventional standard that may be either relative or absolute. A relative standard is the citation score of a "control group" of papers allegedly similar to those under evaluation. An absolute standard is the expected number of citations per paper in the research (sub)field encompassing the papers under scrutiny. Relative comparisons, as suggested by Garfield, can take advantage of bibliometric mapping techniques, such as co-citation analysis or bibliographic coupling, in the selection of thematically similar papers. Yet the criterion for defining similarity necessarily entails a good deal of arbitrariness in the choice of the control group, thereby making quite impossible any cross-field evaluation.[42] Absolute comparisons, instead, put evaluation into a broader perspective in an effort to compare the actual citation score of a document with the ideal number of times it could have been cited. The computation of such expected citedness, clearly, takes different forms depending on whether one chooses to zoom in (a narrow research area) or out (large disciplinary conglomerates) on the shaky ground of science neighborhoods. In addition, as emphasized in section 6.1, it cannot help but rely on a plausible classification of scientific documents into relatively stable cognitive units, a task accomplished, at simplest, by drawing on the bibliometric properties and classificatory compliance of the most familiar documentary units situated at a higher level of aggregation than the individual paper: scientific journals. Schubert and Braun, for example, introduced a relative citation rate indicator for papers published in the same journal that relates the number of citations actually settled on them to the mean citation rate of all papers appearing in that journal. Other important methodologies for relative and absolute comparisons between countries, institutions, and research groups, moreover, are reviewed in the next section. Sophisticated as they may be, though, there's no room for absolute certainty in the resulting output because the choice of the level of normalization and classification scheme inevitably injects some variability into the final ranking.[43] Without necessarily pointing to a generalized pyrrhonism in research evaluation—the no-judgment's land where, being in the position to select the most appropriate normalization strategy for ranking highest among the peers, every researcher is basically the measure of himself or herself—the cross-field and cross-scale instability of citation rankings warns against any simplistic interpretation of ready-made league tables.

Science administrators do not usually take the time to carry out complex calculations, nor do they linger over the subtle specifications that a complex standardization method would require. In a "conspiracy hunting" scenario, one could even speculate that their being satisfied with "quick and dirty" estimations of individual performance is the best way to keep local academic equilibria from too fine-knit networks of power-constraining checkpoints. The universally blamed faults manifested by the journal IF when applied to research evaluation, however, pushes for the adoption of equally handy but hopefully less biased indicators of individual research achievement. A current candidate for such a role is the *h*-index proposed in 2005 by the physicist Jorge Hirsch.

In the spirit of Derek Price's commitment to the elective affinities between quantity and quality, Hirsch's measure is meant to provide a joint characterization of both productivity and cumulative research impact through a simple number that can be easily picked up by ISI (or similar citation-driven) databases and is defined as follows: a scientist has index h if h of the papers he or she has (co)authored have at least h citations each, while the rest have fewer than h citations each.[44] A scientist with h

equal to fifty, for example, has contributed fifty papers with at least fifty citations each, while the remaining papers are cited fewer than fifty times each. A scientist with *h* equal to zero has published, at best, a heap of papers devoid of any impact on the formal level of citations. The subset of medium-highly cited papers bearing on the calculation of the *h* has been dubbed "*h* core" by Rousseau. Implicit in its definition is the insensitivity to extreme values, a sort of fail-safe device against the well-known troubles induced by the structural skewness of bibliometric distributions. The author of many low-cited papers will get as weak an *h*-index as the one who publishes only a few "blockbusters," the chance being real for both of them to increase their own *h* on condition that new citations breathe fresh life into past (or future) dormant papers. Enduring performance blessed by medium-high citation rates is thus rewarded at the expense of occasional exploits and less regular patterns of contribution. On top of all this, according to its inventor, the *h* value is a better predictor of future individual achievements than traditional indicators made up of total citation count, mean citations per paper, and number of papers.

The recent increase in interest in *h*-type indexes, exemplified by a 2007 special issue entirely devoted to the subject in the neonate *Journal of Informetrics* (volume 1, issue 3), is worth recalling here not so much for the (still debatable) indicator's practical value in research evaluation as for its being fairly representative of the bibliometric way of coming to grips with the whole set of theoretical opportunities hidden in a straight number; once again, an attitude quite compliant with Derek Price's invitation to "stretch the method to the full and examine critically any benefits which might thereby accrue." Indeed, as with other performance indicators and perhaps at an even greater pace, bibliometricians are covering the two highways leading off the plain Hirsch number. On the one hand, mathematically inspired researchers cope with a formal foundation of its properties by the same tools and assumptions used for modeling classic bibliometric distributions and growth/aging patterns of scientific literature in either a probabilistic or deterministic fashion: Poisson processes and the Gamma distribution (Burrell), Lotkaian informetrics (Egghe, Rousseau), and the Paretian family of probability distributions (Glänzel).[45]

On a parallel track, the ability of the new measure to discriminate between different levels of scientific performance, its advantages with respect to competing indicators of prominence, and its convergence with independent appraisals of research quality are being investigated on empirical grounds.[46] The flexibility of the index has also encouraged its extension to entirely different target units, including topics and compounds, research groups, journals, and countries. The *h* of a topic or compound, for example, which is expected to provide an indication of its degree of "hotness," is obtained by simply replacing the "author search" with a "topic or compound search" in the citation database and then applying the original definition to the results ranked by citation frequency.[47] Yet the conceptual weaknesses of this umpteenth one-dimensional ranking tool have been easily identified.

First of all, *h* values cannot exceed the number of a scientist's publications and don't decrease for those who give up publishing or don't get citations from a certain point on, so young researchers are unfairly put at a disadvantage due to their relatively skimpy publication record, whereas scientists who have authored many papers in the past can increase their *h* even if intellectually retired in the present, simply resting on their laurels. This is why Hirsch's seminal paper suggested also, as a first rough adjustment, dividing *h* by the years of academic activity. For almost the same reason, the index is scarcely sensitive to significant variations in performance levels, so a small corpus of highly cited, potentially outstanding papers might fail to receive an adequate visibility. *H*'s crude definition, moreover, overlooks the effects of factors notoriously influencing the significance of individual citation scores, such as publication type and age, citation age, self-citation rate, and number of coauthors. Even worse, as repeatedly advocated by anti-bibliometricians and "advanced bibliometricians" as well, publication and citation rates often vary considerably, not only across disciplines and research fields but also within a given research field, in relation to the subject area, thereby making extremely difficult the definition of a proper standard for comparative assessments. If to this is added the growing availability, over the last decade, of non-*ISI* online sources for citation grabbing (*Scopus* and *Google Scholar* above all), then the task of gaining a deeper insight into *h*'s power and range of applicability appears even more challenging.

The above shortcomings are currently keeping scientometricians busy in the search for corrections, supplements, complements, or alternatives to the original formulation. On the whole, although giving the impression that it is just a matter of time until someone new wakes up in the morning inspired by a fresh variant or an ad hoc corrective factor, the latest profusion of such creative efforts reinforces the conviction that Hirsch's proposal incorporates some of the most desirable and long-overdue properties of an indicator of research performance:[48]

a. A series of variations on the theme of improving *h*'s sensitivity to the number and quality of citations in the *h*-core—i.e., to high levels of citation performance—have taken as many forms as Egghe's *g*-index; BiHui Jin's *A*, *R*, and *AR*-indexes; Marek Kosmulski's *h(2)*-index; Egghe and Rousseau's (citation weighted) h_w-index; Thomas Anderson and colleagues' "tapered *h*-index"; and Qiang Wu's *w*-index.
b. Lutz Bornmann and colleagues' *b*-index seeks to determine a field-specific baseline or reference standard useful for marking off the most productive core of a scientist's publication record, thereby facilitating the calculation of *h*-values more sensitive to world class excellence.
c. Mark Sanderson's h_{mx} is a simple correction for disparities in *h*-value registered across different databases (*Web of Science*, *Scopus*, *Google Scholar*).
d. Pablo Batista and colleagues' h_i-index, Michael Schreiber's h_m-index, and Jin-Kun Wan and colleagues' h_p-index correct individual scores for the number of coauthors, while a further adjustment proposed by Schreiber formalizes the effect of self-citations.

The need for an adequate representation of the effects of time patterns and aging processes on research performance, already partially fulfilled by Jin's *AR*-index, has been explicitly addressed by a number of authors along with the likewise thorny issue of comparing *h*-values for scientists active in areas ruled by different citation practices:

e. Antonis Sidiropoulos and colleagues' "contemporary" and "trend" *h*-indexes refine the straight number of citations by taking into account, respectively, the publication year of an article and the age of each citation settled on it; their proposal has been taken up and further elaborated by Chen's team at Drexel University in view of developing an effective measure of the timeliness of the citation impact of scientific discoveries.

f. Jonas Lundberg's *citation z-score* builds on the work of Leiden bibliometricians (see subsection 6.2.2) to attain a normalization of citation impact at the level of the individual publication, which is supposed to allow a better control over the variability of citation rates across research fields.
g. Filippo Radicchi and colleagues' "generalized" *h*-index corrects individual articles' citation rates for field variation by means of a reference standard defined as the average number of citations received by all articles published in the same year by the journals belonging to the same field (according to ISI subject categories).
h. In order to single out significant variations in individual scientists' citation patterns across different research domains, their *h*-'s rate of change over an extended period of time has been modeled through technical solutions as different as Liming Liang's "*h*-index sequences and matrices," Burrell's "Hirsch rate," Frances Ruane and Richard Tol's "rational" *h*-index, and Egghe's dynamic *h*-index.
i. Juan Iglesias and Carlos Pecharromán introduce—and formally derive from exponential and power law models of citation distribution—a correction factor for *h*-values that accounts for both the number of papers published by each researcher and the world average number of citations per paper in his or her research field, as reported by ISI *Essential Science Indicators*.
j. Kalervo Järvelin and Olle Person's *DCI* (*Discounted Cumulated Impact*)*-Index* seeks to enhance the measure's sensitivity to both citation age (devaluing old citations with respect to more recent ones) and quality of citing articles in a Google-like fashion (more weight given to citations from highly cited publications).
k. John Antonakis and Rafael Lalive's *IQP* (*Index of Quality and Productivity*) corrects straight citation counts for scholarly productivity, author's academic age, and field-specific citation habits with reference to an expected citation rate, defined as the global number of citations a scholar's work would receive if it were of average quality in its field.

As the next section discusses in greater detail, the problem of finding a suitable standard for the normalization and cross-field comparison of straight citation counts, far from being limited to individual comparisons, is integral to any advanced bibliometric theory at whatever scale of analysis. Crucial as it may seem, however, even the most out-and-out scientometrician couldn't help but recognize that, in pursuit of true scientific quality, several indicators are needed in combination with an accurate peer review for detecting critical nuances in individual performance levels. No single number, indeed, can tell the whole story about research performance, inasmuch as, taken alone, it "crashes the multidimensional space of bibliometrics into one single dimension."[49]

6.2.2. Evaluations of Countries, Institutions, and Research Groups

A classic objection to citation-based evaluations of research performance is that the low probability of getting cited and the skewness of citation distribution irremediably compromise the significance of citation counts at the individual level. The simplest way to go around the obstacle, then, is choosing the evaluation unit at an aggregation level higher than the individual scientist: the research group, the institution (department, faculty, university), or the country. With all the necessary caution imposed on the analyst by the several traps scattered across the mathematically skewed universe depicted in chapter 4, there exists a widespread belief among bibliometricians that citation analysis applied to the corpus of publications (oeuvre) produced over a certain period of time by the members of a collective entity deals with a number of items large enough to allow, at least in principle, a fairly safe application of standard statistical tools.

Nowadays it is easy to compile hit parades of institutions and individual scientists ranked by publication or citation scores. Since 2001, the *Web of Science* portal has been equipped with an analytical tool called *Essential Science Indicators*, which draws upon citation data to rank scientists, institutions, countries, and journals in twenty-two predefined research fields. The most frequently cited scientists also make the best of themselves on the *ISI HighlyCited.com* website, where biographical and bibliographic records, as well as information on professional profiles, awards, honors, memberships, and past and present professional positions are available for each author. It takes just a few steps to discover how far any of us is from becoming a citation classic. Following a similar trend, academic rankings of world universities and champion leagues of research institutions have proliferated in recent years. Such compilations, though often welcomed by an increasingly market-oriented audience of science managers, nevertheless run the risk of making an out-of-context use of citation data against which militant bibliometricians have occasionally raised their voices.[50] While gaining in statistical tractability, indeed, any evaluation exercise at the institutional level comes to grips with the fundamental issue of comparing output and impact data of research organizations with sharply dissimilar, if not incommensurable, organizational profiles, missions, managerial culture, financial resources, and research facilities. Finding a suitable standard for cross-country and cross-field comparisons and applying it to concrete situations is no less challenging here than in the evaluation of individuals, inasmuch as technical and methodological difficulties show up all over the place.

On the technical side, a well-known obstacle to the reliability of any *SCI*-based assessment stems from ISI citation databases being full of noise, which is amenable partly to random errors, such as misprints and spelling errors, incorrect attributions of papers to authors, and erroneous identification of journal titles (mainly on the score of the great variety of journal abbreviations), and partly to systematic errors dictated by ISI editorial policy. Typical random errors affecting large-scale studies are those produced by inaccurate or ambiguous attribution of publications to specific institutions through authors' addresses. Many universities, for example, appear with variant names in the byline of the papers authored by their members without any proper unification taking place in the citation index; similarly, the allocation of distinct research groups to one main organization, as in the case of medical schools belonging to the same alma mater, is often missed.[51] Systematic errors, in turn, depend on important peculiarities of the indexing process, notably 1) the limitation of citation indexing to the first author in the case of multiauthored papers; 2) the decision, already mentioned above, to not provide unified citation counts for journals undergoing complex editorial changes; and 3) the criteria applied to the selection of source journals, which penalize, in the name of Bradford's Law, peripheral scientific areas and specialized literature in non-English languages. Source selection criteria are regarded by many critics as a primary cause of systematic error, insofar as they do not adequately take into account the actual dynamics of scientific communication and peer judgment in the various subject areas. Even more, they are charged with paying too little attention to non–Anglo-American journals, non-English-language journals, and nonjournal materials, in particular conference proceedings, which provide a vital communication channel in engineering and computer sciences, and books, which, apart from being irreplaceable vectors of new findings in the social sciences and the humanities, play a substantial role in authoritative synthesis of ratified knowledge and techniques in many paper-centered scientific areas.

Bibliometricians maintain that both random and systematic errors can be minimized on condition that a preliminary careful work of cleaning, correction, and integration of the original ISI datafiles predisposes the system to accurate matching and to more context-sensitive appraisals of scientific output. A solution envisaged by leading scientometric agencies since the 1970s has been the creation of local databases and in-house software for storing and processing ISI citation data in view of their correction and enrichment with all pieces of information indispensable to fine-grained bibliometric analyses, including unified authors' addresses, detailed citation data for the different types of articles (research articles, reviews, letters, etc.), normalized measures of impact, and citation data for ISI journals classified into scholarly subfields. Three outstanding examples of analytic versions of the *SCI* are the National Science Foundation's *Science Literature Indicators Database*, compiled for NSF by ipiQ (formerly CHI Research); the database of the ISSRU at the Library of the Hungarian Academy of Sciences; and the database of the CWTS at Leiden University. The CHI database is currently used by, among others, the U.S. National Science Board, for compiling the biennial *Science Indicator Reports* on science, engineering, and technology. The Hungarian datafile is the origin of Tibor Braun and colleagues' pioneering work of design, systematic collection, and updating of a comprehensive set of publication output and citation impact indicators for cross-field and cross-country comparisons on a worldwide scale. Starting from a thirty-two-country comparative evaluation for the period 1976–1980, the project continued during the 1980s with a supplement of "facts and figures" issued under the *World Flash on Basic Research* series in the journal *Scientometrics*, and culminated in the *Scientometric Datafiles*, a comprehensive set of indicators on 2,646 journals and ninety-six countries in all major science fields and subfields for the five-year period 1981–1985.[52] The Leiden database, finally, is the main source for the "advanced bibliometric methodology" developed by van Raan's team and currently employed in Dutch scientometric studies focused primarily on European research groups and institutions.[53]

On the methodological side, the construction, application, and validation of citation-based indicators is the true challenge taken up by current bibliometricians and the Holy Grail sought by science policymakers. It is well known that bibliometric measures are inherently problematic and cannot be taken at face value. Citation distributions are notoriously skewed, and the average citedness of a research group, an institution, or a country, just like the IF of a journal, is decidedly influenced by the share of a few highly cited papers that, being subject to large annual variations, are hardly representative of "normal" performance levels. One could take the chance here of killing two birds with one stone and correct jointly for the size of scientific fields and the non-normality of citation scores, simply by shifting the focus from the total number of citations to the number of highly cited papers produced by the research unit under assessment; a solution that, at any rate, doesn't exempt the analyst from delivering a preliminary unbiased definition of the threshold value for the identification of highly cited papers as such.[54] Theoretical weaknesses notwithstanding, however, it is generally agreed that for practical purposes, 1) skewed distributions can be equally informative of a collective unit's research quality level only if the evaluation is carried out, on a large enough sample of bibliometric data, through a battery of carefully designed indicators; and 2) field-dependent publication and citation practices have to be taken into account before any comparative evaluation. Peer judgment, needless to say, still plays a pivotal role in the overall process, acting as both the high court judge and the indictee. As a judge, it secures the yardsticks for checking out the validity of bibliometric measures against external, strictly qualitative criteria; as an indictee, it ends up on trial each time its response disagrees with bibliometric scores, to such a degree that a rethinking of the entire assessment exercise might seem advisable. Different views, however, exist about how field-specific and reliable bibliometric measures ought to be defined, applied to complex real-life situations, and validated. Let's take a closer look at the most authoritative research programs developed so far.

a. *"Googling" citation networks: Pinski and Narin.* In the mid-1970s, Pinski and Narin developed an advanced analytical technique that bypassed the statistical and conceptual weaknesses of individual citation counts by focusing on the structural properties of citation networks. The "influence methodology" is a general scheme for pinpointing the position occupied in the network by any institutional aggregate of publications undergoing qualitative assessment, such as journals or collections of papers, referable to research fields and subfields, research groups, institutions, and countries. Its practical value in science evaluation, however, is closely related to the statistical properties of the most firmly established point of contact between the intellectual and social organizational units of the sciences: scholarly journals. Indeed, a central assumption in the methodology is that, although individual article counts can vary considerably within a single journal, "aggregates of publications can be characterized by the influence measures of the journals in which they appear."[55] Traditional influence measures, such as total number of citations or Garfield's IF, are unsatisfactory in many respects: total citation rates are size-dependent and cannot be ranked on an absolute scale, while journal IF doesn't correct for article lengths (to review journals' advantage), citation practices (to the detriment of fields with low citation density and slower literature aging), and quality of citations (to the detriment of journals cited by more prestigious sources). To escape these pitfalls, the influence methodology introduced a major breakthrough in the form of a revolutionary journal ranking algorithm inspired by a basic principle of social networking: citations, just like social connections, are not all equal, their weight being adjustable as a function of the prestige of the citers, that is, of the ability of the citers themselves to attract citations from other highly cited sources. On this premise, assuming that journals do not change in size relative to each other and represent a constant subject area, the "influence weight" is defined as a size-independent measure of the weighted number of citations a journal receives—higher weights being given to citations from more highly cited journals—normalized by the number of references it gives to other journals. At the article level, accordingly, the "influence per publication" for a journal is defined as the weighted number of citations each of its articles receives from other journals. Of course, as noted elsewhere, every normalization of journal measures that aspires to be of practical utility in research evaluation has to come to grips with the fuzziness of research fields' boundaries as reflected in the deforming mirror of journal literature. The influence methodology makes no exception, as "influence weights are meaningful only for subfields which are distinct or self contained with respect to the journal literature."[56] Applicative issues aside, Pinski and Narin's algorithm, whose rationale had been foreshadowed by Manfred Kochen's 1974 book on information retrieval, didn't enjoy much popularity for a long time, but the rediscovery of its value occurred punctually in connection with all subsequent attempts to normalize citation counts on the citing side and to implement iterative algorithms that take full advantage of the "qualitative" information hidden in citation networks' structure: from Google's celebrated *PageRank* to the *Eigenfactor* algorithm for journal ranking developed at the University of Washington, all the way to more recently proposed variants of the journal IF.[57]

b. *"Big science" bibliometrics: Martin and Irvine.* A milestone in the history of evaluative bibliometrics is Ben Martin's and John Irvine's "methodology of converging partial indicators," worked out in the late 1970s, when both researchers joined the Big Science Project at the Science Policy Research Unit (SPRU) in Brighton. The authors' seminal 1983 paper on basic research performance in four radio astronomy observatories paved the way for many subsequent institutional-oriented analyses. Its evaluative framework, focused on the basic scientific performance of groups annexed to large research facilities, such as radio telescopes and high-energy physics accelerators, appealed to a combination of several bibliometric and non-bibliometric indicators, including publication counts, citation analysis, and an extensive form of peer review fed by direct interviews with scientists.[58] Whatever the indicator, obviously no absolute appraisal is allowed, but only relative or comparative ones. Most important, none of the measures ensures conclusive evidence on the relative contribution to scientific progress of a research unit; rather, each serves the purpose of building, at best, a partial indicator of scientific output or performance. Each measure, in fact, is affected by inescapable sources of errors. Publication counts, for example, are clearly limited in scope by the inequality of value of individual contributions (quantity doesn't amount to quality); likewise, citation analysis is marred by the technical limitations of citation indexes as well as by the ambiguous functions of citations in scientific communication. Peer review, in turn, suffers from the subjectivity of peer judgment, often diverted by social and political factors. Even so, if comparisons are drawn between similar research units—i.e., groups working in the same internationally homogeneous research fields, taking advantage of the same research facilities, publishing in the same peer reviewed international journals, and so on—and if convergent results turn up, then it can be safely assumed that the disturbing variables were minimized and bibliometric scores convey a reasonable estimate of the groups' relative contribution to scientific progress.

c. *The Hungarian way.* Scientometricians at the Hungarian Academy of Sciences devised, during the 1980s, a set of relative indicators of publication output and citation impact that allow, under certain conditions, cross-field comparisons among countries, research institutes, departments, and scientific societies in a mathematically sound fashion. The first attempt to test the applicability of the newly developed measures came in conjunction with a comprehensive statistical evaluation of the literature of analytical chemistry, a favorite piece of early post–World War II quantitative investigation of science.[59] Provided the community under study has produced a large enough number of publications to enable a statistically robust treatment and a reliable error estimation, the indicators are defined as follows for countries:

   1. The Activity Index (AI), characterizing the relative strength of a country's effort in a given research field, is the ratio between the country's share in the world's publication output in the field and the country's share in the world's publication output in all science fields combined.
   2. The Attractivity Index (AAI), qualifying the relative impact of a country in a field, is the ratio between the country's share in citations attracted by publications in the field and the country's share in citations attracted by publications in all science fields combined.
   3. The Relative Citation Rate (RCR) is the ratio between a summation of observed values and a summation of expected values for all the papers published by a country in a given research field:

$$\mathrm{RCR} = \frac{\Sigma \text{ Observed citation rate}}{\Sigma \text{ Expected citation rate}}$$

The observed citation rate for a paper is simply the actual number of citations accrued to it, while the expected citation rate is the average citation rate of the journal in which that paper was published. As already emphasized, one conventional but critical assumption here is that journals are evenly distributed across scientific specialties. In Schubert and Braun's words, this amounts to saying that science journals "encompass definite research areas (frequently a single 'paradigm') and also a certain standard of quality is guaranteed by the editorial 'gatekeeping' process. Therefore, it seems justified to assign a set of papers to subject fields on the basis of the field classification of journals. Moreover, the average citation rate of the journal in which a paper was published is a valid standard to which its citation rate can be matched."[60]

d. *Detecting excellence among research groups: The Leiden school.* A fourth breeding ground for new ideas and practices in research evaluation is van Raan and colleagues' bibliometric research program at the CWTS of Leiden University. Unlike the Hungarian approach, the Leiden "advanced bibliometric methodology" disregards the analysis at the macro-level of the country, charged with being too generic to characterize research performance in a politically relevant fashion, and traces the roots of scientific excellence explicitly to the meso-level of the institution, notably the university and its operative units (faculties, departments, institutes), where small or large research groups find their natural habitat. The hallmark of scientific excellence, here as in the entire *SCI*-centered tradition of evaluative bibliometrics, is international scientific influence or impact, which is a measurable aspect of scientific quality as long as two basic assumptions hold: 1) "scientists who have something important to say publish their findings vigorously in the open, international journal (serial) literature";[61] and 2) even though not necessarily vehicles of breakthrough or cutting-edge science, highly cited papers in the international scientific literature are "statistically valid proxy measures of academic scientific excellence."[62] Indicators range from the conventional (average) number of citations per publication and the percentage of noncited papers to more sophisticated ratios between local and global impact measures intended to control for some of the most critical variables affecting the validity of crude citation counts, namely type of article (research paper, review, letter, etc.), incidence of self-citations, publishing years (with the corresponding annual

> fluctuations in journal citation impact), and field-related differences in baseline levels of citation performance. Not all indicators, however, are equal or equally informative because their relative position is markedly unbalanced by the prominence of a "crown" number aimed to estimate the impact of the research group or institution on the international scientific community over a relatively long time period. For a corpus of publications by a given research group or institute, it is defined as the ratio between the average number of citations per publication (corrected for self-citations) and a field-specific world average based on the citation rate of all papers appearing in all journals belonging to the same field in which the unit under evaluation has been publishing (in case of more than one field, a weighted average value is calculated). The world average defines, exactly as the expected citation rate did in the Hungarian methodology, a worldwide reference level whereby an estimate of how much the impact of a given local group is above or below the international standard can be obtained. In further confirmation of the new indicator's significance, a statistical analysis of the results from two large-scale evaluation studies covering, respectively, the publication and citation performance of all university chemistry research groups in the Netherlands during the period 1991–2000 and all medical research groups at Leiden university during the interval 1990–2001, supports van Raan's claim that, despite the disruptive skewness of bibliometric distributions, conventional statistical tools keep preserving their validity at the chosen level of aggregation, insofar as, due to the magic of the central limit theorem, the distribution of values for the crown indicator and similar normalized measures is very close to normal. Hence "meaningful comparisons (benchmarking) between groups can be made in terms of reference values based on mean values and variances."[63]

Within the orbit of the SPRU, J. Sylvan Katz has challenged the validity of performance indicators built, in the Hungarian or Leiden style, on the ratio between the impact exercised by the set of papers produced by a group or institution in a given field and the impact of all publications in the field. Such relative indicators, in fact, derive their alleged validity from the false assumption that the number of citations accrued to a collection of documents is independent of its size. Accurate data and improved counting techniques show, on the contrary, that a power law relationship exists between the number of citations received by a group, institution, or nation and its publishing size, the exponent of the power law being relatively independent of the nationality and size of the science system under observation. Within certain limits, of course, size is an aspect of performance: successful research attracts funds and researchers, thereby triggering an expansion in the thematic richness (and potential impact) of a group's research portfolio. But if the impact of a community increases as a power law of its publishing size, and if this sort of rich-get-richer or Matthew Effect is scale-independent—i.e., holding from the small size group to the world level—then larger citation impact is mostly a scale-dependent effect integral to the self-similar system of science, and its meaning as performance indicator approaches the status of an illusion. New indicators, therefore, ought to be developed that compensate for the size effects in comparative evaluations so as to make rewards proportional to size. It takes just a few steps, for example, to demonstrate that an adjusted relative citation impact (ARCI) of the kind proposed by Katz reshuffles the cards in world science hierarchies, removing the United States from the highest ranking in many fields.[64]

How sharply different outcomes can follow upon the application of competing methodologies to the same dataset is testified by the heated debate during the 1980s on the decline of British science, sparked off by a series of Martin's and Irvine's papers. Drawing on the National Science Foundation's *Science Literature Indicators Data-Base*, compiled by CHI Research with *SCI* data, the two British authors and their colleagues employed cross-national publication and citation figures to support the claim that the trajectory of British science between 1973 and 1982 had registered a steady, long-term decline, with just a little slowing in the downward trend during the first half of the 1980s.[65] Immediately afterward, other scientometricians tackled the same issue, resorting to alternative versions of ISI databases, each equipped with specific data cleaning techniques. Quite surprisingly, the conclusions they drew were completely different from Martin's and Irvine's. Relying on simple online searches of the ISI database, Leydesdorff found evidence for a relative stability of British science during the 1970s, followed by "a remarkable increase from 1981 onwards."[66] Braun, Glänzel, and Schubert, simultaneously, exploited the Hungarian version of the *SCI* and, having tested a series of publication productivity indicators, eventually dismissed both the previous verdicts, at least as far the early 1980s pattern of British science was concerned, on the ground that "the annual change of any of the indicators considered had no statistical significance, and no trend distinguishable from the effect of random fluctuations could be observed."[67] It soon became clear that, in trying to spot the same object through the glasses of similar telescopes, each observer perceived fairly different details of the landscape due to mutually exclusive assumptions about the way raw data should be handled. It turned out, interestingly, that the generation of such contrasting figures depended on critical choices about 1) using a fixed journal set or a dynamic set (reflecting annual changes in journal coverage) to define the world total and the national percentage shares of the world total; 2) computing annual publication totals on the basis of tape-years, i.e., the date a publication entered the *SCI*, or on the basis of the date it was published; 3) limiting, as CHI database did, countable output to specific publication types, namely research articles, notes, and reviews, or taking into account also letters, editorials, and other materials; and 4) adopting a fractional author (in this case country) count in the case of multiauthored papers or a full count assigning a full point to each country on the basis of institutional addresses. Each of the above choices is questionable, to a certain extent, on either empirical or theoretical grounds, and section 7.3 demonstrates why this holds especially true in light of the complex issue of multiauthorship. It is not surprising, therefore, that the Hungarian attempt to test multiple time-series of indicators obtained from different combinations of the foregoing assumptions met further methodological objections by Martin and Irvine, thereby leaving to posterity any conclusive judgment on the initial question: Did British science experience a real decline during the 1980s?[68]

On the eve of the British science affair, a serious methodological scrutiny of Martin's and Irvine's work had already animated the August 1985 issue of *Social Studies of Science*, involving scholars from a variety of professional backgrounds. Charged with writing up the conclusions and the possible moral of the story, Michael Moravcsik advocated the multidimensional nature of scientific and technological systems and the consequent need to clarify, before any evaluation exercise, such vital factors as the goals and the specific subset of the science and technology system that is the object of the analysis; the uses to which the results of the evaluation will be put; the appropriate set of indicators addressing the system's components; and the way to measure them and to interpret the measures, along with the limitations inherent in such measures. What the debate ultimately reaffirmed is the impossibility of obtaining fixed and unambiguous rankings out of a complex set of variables. At best, concluded Moravcsik, a study based on these premises "does not give a methodology for assessment. Instead, it gives a methodology for finding a methodology for assessment."[69]

6.3. CITATIONS OF PATENTS BETWEEN SCIENCE, TECHNOLOGY, AND LAW

The realization that knowledge plays an important role in the economy is not a recent discovery. In the last decade of the twentieth century, however, gaining momentum from the revolution in information and communication technologies brought about by the Internet and the World Wide Web, it came to the surface with unprecedented strength. Its concrete manifestations were interpreted, accordingly, as signs of the emergence of a new system, a global and "knowledge-based" economy, wherein critical business competition is exercised not only, as in the past centuries, through the control of natural resources, commodity markets, and low-cost manpower, but also, if not primarily, through the production, distribution, and deployment of investable intellectual capital.

Different views exist about the validity of the very construct of knowledge-based economy, ranging from critical arguments on its rhetorical status of "umbrella concept" or "buzzword" to attempts at making analytically distinguishable and measurable the knowledge-base dimension of the economy against the background of the complex dynamics taking place in modern, self-organizing social systems.[70] It can hardly be denied, in any case, that knowledge-driven innovation is now integral to the commercial success of firms, and that a strategic asset to many businesses is the ability to create, or simply to trace and absorb, commercially valuable ideas. At the forefront of corporate value creation and shareholder wealth, therefore, one is not surprised to find the main artifact designed to meet the twofold paradoxical requirement of stimulating innovation while inhibiting diffusion though intellectual property restrictions: patents.

A patent is a legal document issued by a governmental agency that, in exchange for the public disclosure of the technical details of an invention, grants the inventor, or any person or organization to whom the inventor's prerogatives have been transferred, the monopoly on its production and commercial exploitation. The right holds, as long as certain fees are paid, within the boundaries of the issuing agency's country. U.S. patents, by way of illustration, are granted by the United States Patent and Trademark Office (USPTO) for a period of time that begins at the date of issue and ends twenty years from the date the application was filed. Since most inventions are, in large part, enhancements built upon previous objects or techniques, the ultimate verification of patentability calls for an in-depth analysis of the invention's technical specifications by a skilled examiner. Prior art disclosed in scientific literature and in earlier subject-related patents is especially relevant to the examination process insofar as the occurrence of a similar idea or invention, anywhere and at any prior time, is the basis for the final judgment on either accepting or rejecting the applicant's claims.

A typical U.S. patent is composed of three basic sections:

1. A title page, containing bibliographic data and practical information useful to identify the document unambiguously: title, abstract, classification number, name and address of both the inventor and the assignee, date, application number, and patent number. When a U.S. patent is granted, the title page also contains a list of bibliographic references supplied by the patent examiner; they are the building blocks of most patent citation analyses.
2. The description of the invention, explaining how to make and use it. This includes drawings, technical specifications, and, scattered throughout the text, the references to prior relevant literature supplied by the inventor.
3. The claims defining the scope or boundaries of the patent, that is, the specific features of the invention for which legal protection is being requested. Possible patent infringements are checked against the statements made here.

In drafting sections 2 and 3, the contribution of a patent attorney is often crucial. Choosing the right words in the right places is a basic prerequisite for successful patent applications, and even if many science analysts think the same holds true for scientific discourse construction, here this basic rule of the game is overtly acknowledged at the outset. It is worth noting that only a small fraction of research output is ever patented; public domain technologies and company or trade secrets like Coca-Cola are not. Hence patents are just a partial clue to technological innovation. To be patentable, moreover, an invention must be novel, nontrivial, and commercially exploitable, and checking for compliance with such requisites, and above all with that of novelty, is not easy. In fact, the mix of technical and juridical jargon resulting from the multilevel job at the origin of most patents, the non-word character of many patent specifications (drawings, diagrams, formulas), and the strategy intentionally followed by some companies not to use descriptive words that would be obvious to a researcher in their field so as to limit potentially fraudulent uses, pose serious problems to the patent searcher striving to verify the originality of an invention. If the language barrier to comprehensive searches and the limits of both the international and the national classification schemes are also taken into account, the reason why patent searches are usually carried out by expert consultants on commercial, added-value databases, like the ISI-Thompson *Derwent Innovations Index*, can easily be grasped.

Patents are tough to manage, not only because of the intrinsic diversity in scope, structure, language, and publication history from journal papers, but also because the extent of their content's dependence on scientific knowledge, bearing on the basic issue of the relationships between technology and science, is one of the most controversial topics in science studies. At the most basic level, one could rely on commonsense and argue that, unlike journal papers, a patent does not create or aspire to creating new knowledge, but deploys an already existing set of notions and techniques to assemble a new object satisfying certain requirements of social usefulness and industrial reproducibility. Such a clear distinction is plausible and familiar to many people. It is stumbled upon, for example, in science popularizations or in educational projects, beginning with primary schools, whose methodology in teaching scientific concepts makes full use of practical examples drawn from everyday life objects that are supposed to "incorporate" those concepts. Carried to the extreme, this line of reasoning assumes a linear relationship among the three domains: economic development depends on innovative technologies that, in turn, build on scientific knowledge. Or, in the reverse: scientists do research and elaborate theories tested against empirical evidence; the results are communicated to a wider audience and then increasingly codified as manuals and higher education programs; the corpus of acquired theoretical knowledge is manipulated by the inventor to obtain a patentable device or technique; and the invention, once it has been manufactured on a large scale, is put on the market and thus contributes to economic advancement. This pattern essentially reduces technology to applied science. It can be contrasted with the theory, referable to Aristotle, that depicts science and technology as two radically different forms of knowledge: the first purely speculative and curiosity-driven, and the second of a practical nature and motivated by strict socioeconomic interests.

The history of science shows extensively that both the commonsense and the Aristotelian views are oversimplified and completely misleading models of the complex of intimate relationships existing among science, technology, and economy. They neglect, for example, that there are (or have been) technologies capable of developing almost independently of speculative knowledge; that many sectors of modern science incorporate, from the very first steps of empirical data collection and recording, a strong technological component (think of high-energy physics experiments); and that science is no less sensitive than technology to social and economic influences in the selection of its research goals. What bibliometrics is asked for, in the above scenario, is not the empirical foundation of a speculative theory on science/technology relationships, but the provision of factual evidence, still to be confronted with evidence of diverse origins, on the opportunity to extend to technological documents the same analytical techniques applied to scientific literature both for quality assessment purposes and for mapping the formal connections between scientific and technological research areas.

Patents, like many other human artifacts, have long been of interest to economists concerned with the output of scientific and technological research and its correlation with standard indicators of economic performance, such as the Gross Domestic Product.[71] Indicators based on output measures, however, are of limited use for the assessment of the actual value of patented inventions: simply counting and classifying patents doesn't tell anything about the weight of each patent's contribution to economic and technological advancement. So, following the same trajectory of scientific performance indicators, traditional econometric statistics based on output measures were supplemented, at a certain point, by patent citation analysis. To be accomplished at best, such extension should have been driven by a catalyst for raw data harvesting comparable to the *SCI*, with patents as source documents instead of scientific papers. The initial lack of a standard tool, by contrast, caused patent citation analysis to split its empirical base into different local datafiles occasionally growing out of the two parallel, seemingly unrelated research traditions of bibliometrics and econometrics: the former's initial input came from information retrieval, but soon found its way to an evaluative arena firmly rooted in the long-established conceptual and methodological framework of scientometrics, whereas the latter simply took for granted what many bibliometricians still consider a puzzle to be solved, namely the significance of citations as quality indicators.

The idea to use citations as an aid to effective patent searches alternative (or complementary) to subject-based classification codes was circulating among American patent attorneys as early as the 1940s. A decade later it was put into concrete form by Garfield who, inspired by a proposal of Arthur H. Seidel, tested a patent citation index to 4,000 chemical patents in 1957. Its official version, published in the 1964 and 1965 editions of the *SCI*, included as sources all U.S. patents, but was soon dropped for lack of financial support. Anyway, it demonstrated the feasibility of the project on a large scale.[72] At almost the same time, Phyllis Reisner, at the IBM Research Division, tested a machine-readable citation index to patents as a tool for monitoring the performance of classification systems.[73] Somewhat predictably, it turned out that, in connecting patents, just as in connecting papers, citations had far more to say than the sheer subject affinity exploitable for retrieval purposes. After Merton, Price, and Garfield, they incorporated some form of use and endorsement of the cited document: a patent *B* citing an earlier patent *A* is likely to build upon its scientific and technological content; inversely, the citation of a patent *A* by many subsequent patents *B*, *C*, *D*, . . . indicates that *A* has probably set the ground for many subsequent inventions or, as economists are wont to say, has generated significant "technological spillovers." Hence restoring the network of patent citations opened a window on the promised land of (technological) research evaluation. Once again, the bibliographic mission of citations found itself deeply interlocked with the critical task of supplying quantitative evidence of a complex character in matters of impact, quality, and knowledge diffusion.

Early experiments on patent citation networks, dating back to the mid-1970s, were performed by Charles Oppenheim's team on samples of U.S. patents. They exploited Garfield's historiographs to identify key patents in well-defined subject areas.[74] A formal recognition of the evaluative potential of patent citations came from major U.S. government agencies: the Office of Technology Assessment and Forecast's 1976 report suggested interpreting the number of citations accrued to a patent as an index of its technological significance. In the same vein, the National Science Foundation sponsored a study, carried out by CHI Research in the late 1970s, to consider whether patent citations should be added to the bulk of science indicators in the national *Science Indicators* series (having obtained positive results, the addition actually took place, and the reports were renamed *Science and Engineering Reports*). Ever since, the interest in patent citation analysis has flourished, receiving further impetus during the 1980s, when large-scale computerized patent data became increasingly available for automatic processing. At that time, both the bibliometric and econometric citation crunching machines mentioned above were definitively set in motion, each sustained by the creation of unique U.S. patent databases with added-value citation data. Narin's team at CHI Research systematically extended the core of bibliometric techniques to technology indicator construction and, taking additional advantage of ISI citation data, delved into the thorny issue of the relationships between technology and science; on a parallel route, inspired by Simon Kuznets's visionary proposals and Jacob Schmookler's pioneering use of patent statistics in the 1960s, Adam Jaffe's and Manuel Trajtenberg's research group carried on the work initiated by the Harvard economist Zvi Griliches under the National Bureau of Economic Research Program, employing patent citations to quantify the market "value" of patents and the flows of technological knowledge at the heart of the complex dynamics of economic growth.

The statistical analysis and classification of patent citation data is currently being used to design a wide range of indicators of technological prominence and diffusion. Some hot research areas can be summarized under four main headings:[75]

1. *Knowledge diffusion.* A lot of attention has been devoted to the geography of knowledge spillovers, seeking to build bibliometric maps of knowledge transfer processes between technological fields, institutions, and countries. One notable finding here is the local character of many patent citation patterns. Citations to domestic patents are more likely to be domestic, coming preferably from patents of the same state and metropolitan area; the inventor's citations, in particular, are comparatively more parochial than the examiners' ones. Though the effect fades with time, international borders seem to present a persistent barrier to spillovers.[76]
2. *Technology and science.* A classic topic in patent bibliometrics is the multiple lines of knitting between scientific and technological literature: patent-to-paper citations, along with patent-to-patent citations, are the clue to tracing knowledge transfer processes from science (in particular from university-related basic research) to technology, from technology to technology, and from the defense to the civil fields. The share of patent-to-paper citations, specifically, is deemed worth computing, as it supposedly allows estimating the extent to which, building upon cutting edge science, technology keeps pace with the latest developments in basic research.[77]
3. *Evaluation studies.* A good deal of studies build on patent citation counts to carry out the comparative evaluation of firms', institutions', and countries' competitiveness and technological power according to their citation-weighted patent stocks. Much work of this kind has been performed, on behalf of government agencies, by Narin and colleagues.[78] As in journal-based citation analysis, the number of citations accrued to a set of patents over a given time span is assumed to be an indicator of technological impact/quality, while the median age of the referenced patents is used to estimate the speed of innovation. Of course, as the reader might expect after the discussion in section 6.2, any evaluation exercise requires a normalization against a proper standard, and any standard is built upon a series of nontrivial assumptions about the classification of the units composing the domain under evaluation (whether scientific disciplines or industry groups).
4. *Business intelligence.* In a business intelligence perspective, the patent citation score is used to establish the value of patent portfolios for

> licensing purposes or to select publicly traded companies for inclusion in a stock market portfolio. The underlying assumption is that a company whose portfolio contains highly cited and science-intensive patents is likely to be generating leading-edge technologies and to strengthen its market position, causing investors to look favorably at its stock. Critical developments here, dating back to the 1990s, still revolve around Narin's work, which has found its natural way to a patented method based on bibliometric indicators for choosing a stock portfolio called *Tech-Lin*.[79]

Each history, of course, has a counter-history questioning its assumptions, and the strong commitment to economics inherent in patent citation analysis makes even more critical here than in journal-based bibliometrics the issues of validity naturally arising from the foregoing research lines. Do patent citations actually measure what they are supposed to measure in the dynamics of technological innovation? What are the differences between patent and journal paper citations, and what kinds of information are hidden in patent citation networks?

Journal papers and patents have been found to behave similarly in many respects. Narin has supplied a good deal of evidence in support of the thesis that, in high-tech and fast-moving areas, there is a striking similarity between the referencing cycles of scientific articles and those of patents; that is to say, the amount of time it takes for scientific articles to cite earlier articles is almost the same as that observed for patents citing earlier patents and for patents citing earlier scientific articles. This might be interpreted in terms of convergence of basic research and technological innovation research processes, as manifested by the material overlapping of the two activities. "The inventor," notes Narin with regard to biotechnology, "works in the university or the government lab in the morning in the United States, and he works at or consults with a private company in the afternoon, and the time lag between his academic research and his private inventive activity is lunch!"[80] Patents and research papers are also similar in three other important respects: the tendency to cite more quickly and heavily documents belonging to the same country of origin, the inherent diversity of citation patterns across fields (a patent in electronics, for example, cites on average a higher number of recent documents than a patent in naval technologies), and the skewness of citation distribution. Regarding the last point, just like scientific papers, and with similar effects on any would-be mathematical framework for scientometric assessment, patents have been found enmeshed in scale-free citation networks governed by a power law distribution that imposes an uneven allocation of symbolic wealth among units of supposedly different caliber. In each applicative area, then, there are a limited number of frequently cited patents, which take the lion's share in the knowledge diffusion process and are concomitantly held in higher esteem by experts, surrounded by a much higher number of patents whose impact almost disappears into thin air.[81]

In spite of the above affinities, if the bibliographic structure is taken into account, patents and journal papers differ strongly, thereby suggesting the necessity to be extremely cautious in generalizing standard bibliometric techniques to patent literature.[82] Patent references, unlike those from papers, are the result of a social process involving at least three actors with different perspectives and quite incomparable interests. The inventor, in the first place, supplies the raw materials in the form of descriptions, drawings, and technical specifications along with a personal version of which bibliographic references are worth including. Concomitantly, the patent attorney or agent, by virtue of extensive legal skills, assists the inventor in drafting the text and choosing suitable bibliographic references for the application to be successful. Finally the patent examiner, under an official mandate, conducts a supplemental literature search to verify and expand upon the bibliographic background with references that the inventor has overlooked or willingly ignored. Their scope is to delimit or disallow the applicant's claims for the invention's novelty and usefulness. The heterogeneity of origins and goals of bibliographic references inevitably bears upon the quality of bibliometric analysis. The inventor's citations, usually scattered in the text of the patent specification, are expected to document the theoretical and technological background of the invention. More realistically, they simply make visible what the inventor, motivated primarily by the wish to stress the novelty of the proposal, endeavors to present as a theoretical or technological foundation. Thus their function, like that of many academic citations, is rhetorical, at least to a certain degree. The examiner's citations, on the other hand, are in principle devoid of rhetorical or hidden motivations. Under the hypothesis that the examination process is being conducted carefully and impartially, they should allow assessing the "objective" technological relevance of the patent.

Upon closer inspection, however, the supposed higher usefulness of the patent examiner's citations rests on shaky ground. In the United States, for instance, where the provision of bibliographic references to document the state of the art is mandatory for the inventor, many of the examiner's references to earlier patents or scientific papers are identical to those supplied by the inventor, who endeavors to anticipate any objection of inadequate bibliographic coverage by citing even the minimally relevant pieces of prior art. The examiner's search report has consequently a broader scope than mere patentability, being more similar to a legally obliged archival reconstruction of the (bibliographic)-historical context of the invention. As such, it probably says very little about the true channels of influence and intellectual spillovers. That's one of the reasons why bibliometricians markedly disagree about the value of examiners' citations, which are sometimes regarded as inappropriate and misleading, to the point of suggesting their exclusion from the analysis and their replacement with citations from the applicant, whose reasons to cite more closely resemble those of the authors of scientific articles.[83]

As in classic citation analysis, the entire building of patent citation analysis rests on the assumption that a highly cited patent contains a technological innovation of particularly high impact/quality/importance. One of the main concerns of early patent citation analysts, accordingly, has been to validate the equivalence quality = citation impact against external criteria, above all the opinion of experts on the technical importance of the cited invention.[84] To some extent, such equivalence is undeniably marred as much by the inefficiency of the patent examination system as by our ignorance of the roles played by bibliographic citations in the construction of the scientific and technological discourse at the individual level. But if, here as in journal-centric bibliometrics, citations are taken collectively instead of individually, the trust in the validity of their statistical analysis shifts abruptly. Punctually, then, the argument shows up again that, although the uniformity and transparency of patent citation data and their mechanical adherence to the Mertonian paradigm can't reasonably be asserted at the individual level, "in large aggregates the idiosyncratic is overwhelmed, and the overall nature of the aggregate dominates."[85]

NOTES

1. Kuhn, *The Structure of Scientific Revolutions,* 2nd ed. (Chicago: University of Chicago Press, 1970), 52.

2. Thomas J. Scheff, *Peer Review: An Iron Law of Disciplines?* (2002), www.soc.ucsb.edu/faculty/scheff/23.html. An empirical test of the low degree of inter-referee consensus manifested by peer-reviewing procedures in a variety of disciplines is reported by Domenic V. Cicchetti, "The Reliability of Peer Review for Manuscript and Grant Submissions: A Cross Disciplinary Investigation," *Behavioral and Brain Sciences* 14, no. 1 (1991): 119–86. The article is followed by a series of open peer commentaries discussing the significance of Cicchetti's findings, the causes of the peer review crisis, and the ways to improve the quality of peer evaluation criteria.

3. A series of empirical studies dealing with the correlation between citation counts and peer ratings across several disciplines within the frame of the U.K. Research Assessment Exercise have been undertaken by Charles Oppenheim's team since the mid-1990s. The latest product of this effort, from which the references to previous relevant work can be quickly retrieved, is Oppenheim and Summers, "Citation Counts and the Research Assessment Exercise, Part VI: Unit of Assessment 67 (Music)," *Information Research* 13, no. 2 (2008), informationr.net/ir/13-2/paper342.html. Going back in time, a pre-ISI concern with the correlation between qualitative and quantitative performance measures, most notably citation counts, can be traced back to Kenneth E. Clark, "The APA Study of Psychologists," *American Psychologists* 9, no. 3 (1954): 117–20. In the post-ISI era, chapter V and chapter X of Narin's *Evaluative Bibliometrics* review early correlational studies of research performance, while other useful points of departure may be found, in chronological order, in Cole and Cole, "Scientific Output and Recognition: A Study in the Operation of the Reward System in Science," *American Sociological Review* 32, no. 3 (1967): 377–90; Cole and Cole, *Social Stratification in Science* (Chicago: University of Chicago Press, 1973); Julie A. Virgo, "A Statistical Procedure for Evaluating the Importance of Scientific Papers," *Library Quarterly* 47, no. 4 (1977): 415–30; Michael E. D. Koenig, "Bibliometric Indicators Versus Expert Opinion in Assessing Research Performance," *Journal of the American Society for Information Science* 34, no. 2 (1983): 136–45; Garfield, "Do Nobel Prize Winners Write Citation Classics?" in *Essays of an Information Scientist 1986*, 182–87 (Philadelphia: ISI Press, 1988), www.garfield.library.upenn.edu/essays/v9p182y1986.pdf. More recently, besides the papers stemming from the growing interest in the Hirsch index's ability to predict peer evaluation (see subsection 6.2.1), the problem of validation of citation studies is discussed in Moed, *Citation Analysis*, chapter 18. For a critical view

on RAE-based correlational approaches, see Julian Warner, "A Critical Review of the Application of Citation Studies to the Research Assessment Exercises," *Journal of Information Science* 26, no. 6 (2000): 453–60.

4. Tagliacozzo, "Self-Citations in Scientific Literature," *Journal of Documentation* 33, no. 4 (1977): 251–65. For a recent overview of the literature on self-citations, see Glänzel, Thijs, and Schubert, "A Concise Review on the Role of Author Self-Citations in Information Science, Bibliometrics and Science Policy," *Scientometrics* 67, no. 2 (2006): 263–77, www.cais-acsi.ca/proceedings/2005/glanzel_2_2005.pdf.

5. van Raan, "Self-Citation as an Impact-Reinforcing Mechanism in the Science System," *Journal of the American Society for Information Science and Technology* 59, no. 10 (2008): 1631–43, arxiv.org/abs/0801.0524.

6. Glänzel, Schlemmer, and Thijs, "Better Late Than Never? On the Chance to Become Highly Cited Only Beyond the Standard Bibliometric Time Horizon," *Scientometrics* 58, no. 3 (2003): 578.

7. Garfield, "Citation Analysis as a Tool in Journal Evaluation," *Science* 178, no. 4060 (1972): 472, www.garfield.library.upenn.edu/essays/V1p527y1962-73.pdf.

8. Here the "immediacy index" (not to be confused with the concept of immediacy introduced in section 3.4) is the ratio between the number of citations a journal receives in a given year and the number of articles it issues during the same year. It is interpreted as a measure of the speed with which the journal's average article gets cited. The "cited half-life" (not to be confused with the half-life of scientific literature discussed in section 4.6) is the number of journal publication years going back from the current year that account for 50 percent of the citations received in that year, while the "citing half-life" is the number of journal publication years accounting for 50 percent of the references given in that year. The former roughly estimates the average age of the journal's cited articles, whereas the latter points to a similar property for the journal's referenced articles. On these two measures see *Journal Citation Reports® on the Web V. 4.0*, (Philadelphia: The Thomson Corporation, 2005), scientific.thomson.com/media/scpdf/jcr4_sem_0305.pdf.

9. This and similar issues of manipulability of the IF are examined, on empirical grounds, in Jan Reedijk and Henk F. Moed, "Is the Impact of Journal Impact Factor Decreasing?" *Journal of Documentation* 64, no. 2 (2008): 183–92.

10. Sound arguments against the use of ISI IF in research evaluation are advanced, with the support of fresh empirical evidence, in Per O. Seglen, "Why the Impact Factor of Journals Should Not Be Used for Evaluating Research," *BMJ* 314, no. 7079 (1997): 498–502, www.bmj.com/cgi/content/full/314/7079/497, and Seglen, "The Skewness of Science," *Journal of the American Society for*

Information Science 43, no. 9 (1999): 628–38. For a recent (freely accessible) entry point to the debate, see the articles issued in Howard I. Browman and Konstantinos I. Stergiou, "The Use and Misuse of Bibliometric Indices in Evaluating Scholarly Performance," *Ethics in Science and Environmental Politics* [Theme Section] 8, no. 1 (2008), www.int-res.com/abstracts/esep/v8/n1/.

11. See, on this point, the case made in Seglen, "Causal Relationship Between Article Citedness and Journal Impact," *Journal of the American Society for Information Science* 45, no. 1 (1994): 1–11.

12. Harvey Goldstein and David J. Spiegelhalter, "League Tables and Their Limitations: Statistical Issues in Comparisons of Institutional Performance," *Journal of the Royal Statistical Society, Series A* 159, no. 3 (1996): 385–443, www.cmm.bristol.ac.uk/team/HG_Personal/limitations-of-league-tables.pdf (Preprint devoid of figures).

13. Mike Rossner, Heather Van Hepps, and Emma Hill, "Show Me the Data," *The Journal of Cell Biology* 179, no. 6 (2007): 1091–92, www.jcb.org/cgi/content/full/179/6/1091.

14. Chapter 9 of Narin's *Evaluative Bibliometrics* is entirely devoted to the reliability of his newly developed influence measure. On the European side see Dieks and Chang, "Differences in Impact of Scientific Publications: Some Indices Derived from a Citation Analysis," *Social Studies of Science* 6, no. 2 (1976): 247–67.

15. Geller, De Cani, and Davies, "Lifetime-Citation Rates: A Mathematical Model to Compare Scientists' Work," *Journal of the American Society for Information Science* 32, no. 1 (1981): 345–65. This is a refinement of a 1978 *Social Science Research* paper on the same subject.

16. Schubert and Glänzel, "Statistical Reliability of Comparisons Based on the Citation Impact of Scientific Publications," *Scientometrics* 5, no. 1 (1983): 65. Their model is similar to the one proposed in Allison, "Inequality and Scientific Productivity," *Social Studies of Science* 10, no. 2 (1980): 163–79.

17. Bensman, "Distributional Differences of the Impact Factor in the Sciences Versus the Social Sciences: An Analysis of the Probabilistic Structure of the 2005 Journal Citation Reports," *Journal of the American Society for Information Science and Technology* 59, no. 9 (2008): 1366–82, garfield.library.upenn.edu/bensman/bensman072008.pdf.

18. This is the conclusion drawn, upon an analysis of citations to experimental medicine journals based on the 2005 *JCR*, by Darren C. Greenwood, "Reliability of Journal Impact Factor Rankings," *BMC Medical Research Methodology* 7, no. 48 (2007), www.biomedcentral.com/1471–2288/7/48. The study features Bayesian Markov chain Monte Carlo methods and hypothesizes a Poisson distribution for the empirical frequency of citations to a journal over the two years contributing to the IF calculation.

19. Bensman, *Garfield and the Impact Factor: The Creation, Utilization, and Validation of a Citation Measure. Part 2: The Probabilistic, Statistical, and Sociological Bases of the Measure* (2007), garfield.library.upenn.edu/bensman/bensmanegif22007.pdf.

20. See, for example, the interpretation of the IF given by Peter Vinkler, "Characterization of the Impact of Sets of Scientific Papers: The Garfield (Impact) Factor," *Journal of the American Society for Information Science and Technology* 55, no. 5 (2004): 431–35.

21. The main issues involved in the use of citation data for evaluation purposes, with a particular emphasis on the weaknesses of the journal IF, have been recently reviewed in Leydesdorff, "*Caveats* for the Use of Citation Indicators in Research and Journal Evaluations," *Journal of the American Society for Information Science and Technology* 59, no. 2 (2008): 278–87, www.leydesdorff.net/cit_indicators/cit_indicators.pdf.

22. An account of the inaccuracies in *JCR* journal ranking caused by an inappropriate definition of "citable items" is given by Moed and van Leeuwen, "Improving the Accuracy of Institute for Scientific Information's Journal Impact Factors," *Journal of the American Society for Information Science* 46, no. 6 (1995): 461–67.

23. The issue was raised by Braun and Glänzel, "The Sweet and Sour of Journal Citation Rates," *The Chemical Intelligencer* 1 (1995): 31–32, and further discussed in Moed, van Leeuwen, and Reedijk, "A Critical Analysis of the Journal Impact Factors of Angewandte Chemie and the Journal of the American Chemical Society Inaccuracies in Published Impact Factors Based on Overall Citations Only," *Scientometrics* 37, no. 1 (1996): 105–16.

24. Mohammad Hossein Biglu, "The Influence of References per Paper in the SCI to Impact Factors and the Matthew Effect," *Scientometrics* 74, no. 3 (2008): 453–70, eprints.rclis.org/archive/00011935/.

25. Robert Adler, John Ewing, and Peter Taylor, *Citation Statistics: A Report from the International Mathematical Union (IMU) in Cooperation with the International Council of Industrial and Applied Mathematics (ICIAM) and the Institute of Mathematical Statistics (IMS)* (2008), www.mathunion.org/fileadmin/IMU/Report/CitationStatistics.pdf.

26. A classification of journal aging patterns functional to the refinement of impact measures has been developed in Moed, van Leeuwen, and Reedijk, "A New Classification System to Describe the Ageing of Scientific Journals and Their Impact Factors," *Journal of Documentation* 54, no. 4 (1998): 387–419. The authors build on a scheme formerly worked out in Glänzel and Schoepflin, "A Bibliometric Study on Ageing and Reception Processes of Scientific Literature in the Science," *Journal of Information Science* 21, no. 1 (1995): 37–53.

27. For a review of the various indexes proposed as alternatives to Garfield's *IF* see Glänzel and Moed, "Journal Impact Measures in Bibliometric Research," *Scientometrics* 53, no. 2 (2002): 171–93, www.steunpuntoos.be/WG_Papers/Scientometrics_53_2_171.pdf.

28. This observation stems from Garfield's study on a sample of 200 ISI journals, reported in Garfield, "Long-Term vs. Short-Term Journal Impact: Does It Matter?" *The Scientist* 12, no. 3 (1998): 10–12, www.garfield.library.upenn.edu/commentaries/tsv12(03)p10y19980202.pdf, and Garfield, "Long-Term vs. Short-Term Journal Impact, Part II: Cumulative Impact Factors," *The Scientist* 12, no. 14 (1998): 12–13, www.garfield.library.upenn.edu/commentaries/tsv12(14)p12y19980706.pdf.

29. Synchronous and diachronous impact factors, with their respective advantages and disadvantages, are briefly reviewed in Peter Ingwersen et al., "The Publication-Citation Matrix and Its Derived Quantities," *Chinese Science Bulletin* 46, no. 6 (2001): 524–28, 219.238.6.200/getfile?category=article&code=00WC0565&file-name=01ky0524.pdf.

30. Hirst, "Discipline Impact Factors: A Method for Determining Core Journal Lists," *Journal of the American Society for Information Science* 29, no. 4 (1978): 171–72.

31. van Leeuwen and Moed, "Development and Application of Journal Impact Measures in the Dutch Science System," *Scientometrics* 53, no. 2 (2002): 249–66.

32. Schubert and Braun, "Reference Standards for Citation Based Assessments," *Scientometrics* 26, no. 1 (1993): 22.

33. A classic interpretation of the causes of this gap may be found in Richard Whitley, *The Intellectual and Social Organization of the Sciences* (Oxford: Clarendon Press, 1984).

34. It is the "mode 2" theorized in Michael Gibbons et al., *The New Production of Knowledge: The Dynamics of Science and Research in Contemporary Societies* (London: Sage, 1994).

35. The final report of the pilot project is in Moed and Visser, *Developing Bibliometric Indicators of Research Performance in Computer Science: An Exploratory Study* (CWTS Report 2007–01) (Leiden: CWTS, 2007), www.socialsciences.leidenuniv.nl/cwts/publications/nwo_indcompsciv.jsp.

36. Lewison, "Beyond Outputs: New Measures of Biomedical Research Impact," *ASLIB Proceedings* 55, nos. 1/2 (2003): 32–42.

37. See, for example, Claudio Nicolini et al., "Can the Assignment of University Chairs Be Automated?" *Scientometrics* 32, no. 2 (1995): 93–107. A critical analysis of this paper appeared in the same issue of the journal: Vinkler, "Some Aspects of the Evaluation of Scientific and Related Performances of Individuals," *Scientometrics* 32, no. 2 (1995): 109–16.

38. Garfield, "Is the Ratio Between Number of Citations and Publications Cited a True Constant?" in *Essays of an Information Scientist 1974–1976*, 419–25 (Philadelphia: ISI Press, 1977), www.garfield.library.upenn.edu/essays/v2p419y1974-76.pdf. In the late 1960s, M. C. Gomperts had noticed a similar pattern on more than 150 years of literature in the field of vibrating plates and made a case for the existence of a "citation law" that caused average citation rates to be independent of the size of the citing literature: Gomperts, "The Law of Constant Citation for Scientific Literature," *Journal of Documentation* 24, no. 2 (1968): 113–17.

39. Garfield, "How to Use Citation Analysis for Faculty Evaluations, and When Is It Relevant? Part 1," in *Essays of an Information Scientist 1983*, 354–55 (Philadelphia: ISI Press, 1984), www.garfield.library.upenn.edu/essays/v6p354y1983.pdf.

40. Garfield, "How to Use Citation Analysis for Faculty Evaluations, and When Is It Relevant? Part 2," in *Essays of an Information Scientist 1983,* 363 (Philadelphia: ISI Press, 1984), www.garfield.library.upenn.edu/essays/v6p363y1983.pdf.

41. See, for example, Anton J. Nederhof, "Evaluating Research Output Through Life Work Citation Counts," *Scientometrics* 7, nos. 1–2 (1985): 23–28.

42. The difficulties arising in control groups selection for relative comparisons are examined by Ronald N. Kostoff, "Citation Analysis of Research Performer Quality," *Scientometrics* 53, no. 1 (2002): 49–71. On the crucial role of journal citation rates for standardization purposes, see Schubert and Braun, "Cross-Field Normalization of Scientometric Indicators," *Scientometrics* 36, no. 3 (1986): 311–24.

43. The instability of citation rankings associated with the choice of the level at which articles are aggregated (journal, field, discipline, etc.) in normalization exercises has been recently discussed, on the basis of a four-year citation analysis of papers published in 1998 and classified according to ISI subject categories, by Zitt, Ramanana-Rahary, and Bassecoulard, "Relativity of Citation Performance and Excellence Measures: From Cross-Field to Cross-Scale Effects of Field-Normalisation," *Scientometrics* 63, no. 2 (2005): 373–401. A further test of the same variability in the relative citation performance of articles submitted by three UK research units to the 2001 Research Assessment Exercise is reported by Jonathan Adams, Karen Gurney, and Louise Jackson, "Calibrating the Zoom: A Test of Zitt's Hypothesis," *Scientometrics* 75, no. 1 (2008): 81–95.

44. Hirsch, "An Index to Quantify an Individual's Scientific Research Output," *Proceedings of the National Academy of Sciences* 102, no. 46 (2005): 16569–72, arxiv.org/abs/physics/0508025; Hirsch, "Does the *h* Index Have Predictive Power?" *Proceedings of the National Academy of Sciences* 104, no. 49 (2007): 19193–98, arxiv.org/abs/0708.0646.

45. Burrell, "Hirsch's *h*-Index: A Stochastic Model," *Journal of Informetrics* 1, no. 1 (2007): 16–25; Egghe and Rousseau, "An Informetric Model for the Hirsch-Index," *Scientometrics* 69, no. 1 (2006): 121–29, hdl.handle.net/1942/943; Glänzel, "On the *h*-Index—a Mathematical Approach to a New Measure of Publication Activity and Citation Impact," *Scientometrics* 67, no. 2 (2006): 315–21.

46. Experiments confirming the substantial convergence between *h*-values and other bibliometric indicators as well as peer assessment in different research areas are reported in, among others, Cronin and Meho, "Using the *h*-Index to Rank Influential Information Scientists," *Journal of the American Society for Information Science and Technology* 57, no. 9 (2006): 1275–78, dlist.sir.arizona.edu/1717/; van Raan, "Comparison of the Hirsch-Index with Standard Bibliometric Indicators and with Peer Judgment for 147 Chemistry Research Groups," *Scientometrics* 67, no. 3 (2006): 491–502, front.math.ucdavis.edu/0511.7206; Bornmann and Daniel, "Convergent Validation of Peer Review Decisions Using the *h* Index: Extent of and Reasons for Type I and Type II Errors," *Journal of Informetrics* 1, no. 3 (2007): 204–13, www.lutz-bornmann.de/icons/BornmannDanielRevised.pdf; Bornmann, Wallon, and Ledin, "Is the *h* Index Related to (Standard) Bibliometric Measures and to the Assessments by Peers? An Investigation of the *h* Index by Using Molecular Life Sciences Data," *Research Evaluation* 17, no. 2 (2008): 149–56, www.lutz-bornmann.de/icons/BornmannWallonLedinRevised.pdf. On the other hand, *h*'s lack of precision in discriminating between scientists of different stature is asserted, on the ground of a Bayesian analysis of a sample of citation data from the SPIRES database, by Sune Lehmann, Andrew D. Jackson, and Benny E. Lautrup, "Measures for Measures," *Nature* 444, no. 7122 (2006): 1003–4. Perplexities on *h*'s biases due to its size-dependence and variability across disciplines and research fields have been expressed also, on the basis of comparisons with other normalized measures of performance, by Rodrigo Costas and Maria Bordons, "The *h*-Index: Advantages, Limitations and Its Relation with Other Bibliometric Indicators at the Micro Level," *Journal of Informetrics* 1, no. 3 (2007): 193–203, and Thed N. van Leeuwen, "Testing the Validity of the Hirsch-Index for Research Assessment Purposes," *Research Evaluation* 17, no. 2 (2008): 157–60. A fine-tuned analysis of the different dimensions of research performance measured by *h*-type indexes can be found, in connection with an empirical investigation of nine *h*-variants' ability to predict peer judgment on a sample of postdoctoral applicants for long-term fellowship at the Boehringer Ingelheim Fonds, in Bornmann, Mutz, and Daniel, "Are There Better Indices for Evaluation Purposes Than the *h* Index? A Comparison of Nine Different Variants of the *h* Index Using Data From Biomedicine," *Journal of the American Society for Information Science and Technology* 59, no. 5 (2008): 830–37, www.lutz-bornmann.de/icons/BornmannMutzDanielFinal.pdf.

47. Michael G. Banks, "An Extension of the Hirsch Index: Indexing Scientific Topics and Compounds," *Scientometrics* 69, no. 1 (2006): 161–68, arxiv.org/abs/physics/0604216.

48. The references corresponding to each variant of the *h*-index mentioned above are given here cumulatively in the same order as they are listed in the text:

a. Egghe, "Theory and Practise of the *g*-Index," *Scientometrics* 69, no. 1 (2006): 131–52, hdl.handle.net/1942/4536; Jin et al., "The *R*- and *AR*-Indices: Complementing the *h*-Index," *Chinese Science Bulletin* 52, no. 6 (2007): 855–63, users.pandora.be/ronald.rousseau/CSB_Jin_et_al.pdf; Kosmulski, "A New Hirsch-Type Index Saves Time and Works Equally Well as the Original *h*-Index," *ISSI Newsletter* 2, no. 3 (2006): 4–6; Egghe and Rousseau, "An *h*-Index Weighted by Citation Impact," *Information Processing & Management* 44, no. 2 (2008): 770–80; Anderson, Hankin, and Killworth, "Beyond the Durfee Square: Enhancing the *h*-Index to Score Total Publication Output," *Scientometrics* 76, no. 3 (2008): 577–78, www.noc.soton.ac.uk/GDD/people/tra/durfee.pdf; Wu, *The w-Index: A Significant Improvement of the h-Index* (2008), arxiv.org/abs/0805.4650. The superiority of the *g* and *R* indexes over the simple *h* in discriminating between different citation patterns is asserted, on the basis of a citation analysis of twenty-six physicists, by Schreiber, "An Empirical Investigation of the *g*-Index for 26 Physicists in Comparison with the *h*-Index, the *A*-Index, and the *R*-Index," *Journal of the American Society for Information Science and Technology* 59, no. 9 (2008): 1513–22, arxiv.org/abs/0802.1820.
b. Bornmann, Mutz, and Daniel, "The *b* Index as a Measure of Scientific Excellence: A Promising Supplement to the *h* Index," *Cybermetrics* 11, no. 1, paper 6, www.cindoc.csic.es/cybermetrics/articles/v11i1p6.html.
c. Sanderson, "Revisiting *h* Measured on UK LIS and IR Academics," *Journal of the American Society for Information Science and Technology* 59, no. 7 (2008): 1184–90, dis.shef.ac.uk/mark/publications/my_papers/2008-JASIST.pdf.
d. Batista et al., "Is It Possible to Compare Researchers with Different Scientific Interests?" *Scientometrics* 68, no. 1 (2006): 179–89, arxiv.org/pdf/physics/0509048; Schreiber, "Self-Citation Corrections for the Hirsch Index," *Europhysics Letters* 78, no. 3 (2007): 30002, arxiv.org/abs/physics/0701231; Wan, Hua, and Rousseau, *The Pure h-Index: Calculating an Author's h-Index by Taking Co-Authors into Account* (2007), eprints.rclis.org/archive/00011401/; Schreiber, "To Share the Fame in a Fair Way, h_m Modifies *h* for Multi-Authored Papers," *New Journal of Physics* 10, no. 4 (2008): 040201. iopscience.iop.org/1367-2630/10/4/040201/fulltext.

e. Sidiropoulos, Katsaros, and Manolopoulos, "Generalized Hirsch *h*-Index for Disclosing Latent Facts in Citation Networks," *Scientometrics* 72, no. 2 (2007): 253–80, arxiv.org/abs/cs/0607066; Chaomei Chen et al., "Delineating the Citation Impact of Scientific Discoveries" (paper presented at the IEEE/ACM Joint Conference on Digital Libraries [JCDL 2007], Vancouver, BC, Canada, June 17–22, 2007), hdl.handle.net/1860/1885.
f. Lundberg, "Lifting the Crown—*Citation z-score*," *Journal of Informetrics* 1, no. 2 (2007): 145–54.
g. Radicchi, Fortunato, and Castellano, *Universality of Citation Distributions: Towards an Objective Measure of Scientific Impact* (2008), arxiv.org/abs/0806.0974.
h. Liang, "*h*-Index Sequence and *h*-Index Matrix: Constructions and Applications," *Scientometrics* 69, no. 1 (2006): 153–59; Burrell, "Hirsch Index or Hirsch Rate? Some Thoughts Arising from Liang's Data," *Scientometrics* 73, no. 1 (2007): 19–28; Ruane and Tol, *Refined (Successive) h-Indices: An Application to Economics in the Republic of Ireland* (2007), ideas.repec.org/p/sgc/wpaper/130.html; Egghe, "Dynamic *h*-Index: The Hirsch Index in Function of Time," *Journal of the American Society for Information Science and Technology* 58, no. 3 (2007): 452–54, hdl.handle.net/1942/980.
i. Iglesias and Pecharromán, "Scaling the *h*-Index for Different Scientific ISI Fields," *Scientometrics* 73, no. 3 (2007): 303–20, arxiv.org/abs/physics/0607224.
j. Järvelin and Persson, "The DCI Index: Discounted Cumulated Impact-Based Research Evaluation," *Journal of the American Society for Information Science and Technology* 59, no. 9 (2008): 1433–40, www.info.uta.fi/tutkimus/fire/archive/2008/DCF-JASIST-08.pdf.
k. Antonakis and Lalive, "Quantifying Scholarly Impact: IQp Versus the Hirsch *h*," *Journal of the American Society for Information Science and Technology* 59, no. 6 (2008): 956–69.

49. Glänzel, "On the *h*-Index," 320.

50. This is what happened upon the publication of the 2003 *Academic Ranking of World University* by the Jiao Tong University in Shanghai, accessible online at ed.sjtu.edu.cn/ranking.htm. The Shangai ranking methodology is fully criticized in van Raan, "Fatal Attraction: Conceptual and Methodological Problems in the Ranking of Universities by Bibliometric Methods," *Scientometrics* 62, no. 1 (2005): 133–43, www.cwts.nl/TvR/documents/AvR-RankingScientom.pdf. It can be contrasted with the ranking of world universities based on normalized measures recently developed by the Leiden group and available online at www.cwts.nl/cwts/LeidenRankingWebSite.html. The Leiden ranking's theoretical

background is extensively explained in Moed, *Bibliometric Ranking of World Universities (CWTS Report 2006–01)* (Leiden: CWTS, 2006), www.cwts.nl/hm/bibl_rnk_wrld_univ_full.pdf.

51. An outcry about a counting error arose, for instance, when the amazingly low number of citations recorded for the landmark *Nature* article (February 2001) announcing the International Human Genome Sequencing Consortium's draft sequence of the human genome led ISI staff to a substantial revision of the initial statistics. The error plainly resulted from two distinct ways of addressing the *Nature* report: some papers specified the full consortium as author, while others, referring to the report's first listed name, cited simply "E. S. Lander and colleagues." On this episode, see "Crime Against Nature?" *Science Watch* 13, no. 1 (2002), www.sciencewatch.com/jan-feb2002/index.html. An early empirical analysis of data inaccuracies marring citation counts is given by Moed and Vriens, "Possible Inaccuracies Occurring in Citation Analysis," *Journal of Information Science* 15, no. 2 (1989): 95–107. The description of a similar, more recent test on a sample of chemistry articles, along with a concise review of previous literature on citation errors, can be found in Robert A. Buchanan, "Accuracy of Cited References: The Role of Citation Databases," *College & Research Libraries* 67, no. 4 (2006): 292–303, www.ala.org/ala/acrl/acrlpubs/crljournal/backissues2006a/julycrl/Buchanan06.pdf.

52. Schubert, Glänzel, and Braun, "World Flash on Basic Research: Scientometric Datafiles; A Comprehensive Set of Indicators on 2649 Journals and 96 Countries in All Major Science Fields and Subfields 1981–1985," *Scientometrics* 16, no. 1–6 (1989): 3–478.

53. For the database description, see Moed, De Bruin, and van Leeuwen, "New Bibliometric Tools for the Assessment of National Research Performance: Database Description," *Scientometrics* 33, no. 3 (1995), 384–94.

54. The number of highly cited papers scored by highly cited (senior) scientists has been considered "a much more reliable predictor of a research group's scientific performance than the number of short-time citations of the group." Reinier Plomp, "The Highly Cited Papers of Professors as an Indicator of a Research Groups Scientific Performance," *Scientometrics* 29, no. 3 (1994): 391. See also Plomp, "The Significance of the Number of Highly Cited Papers as an Indicator of Scientific Prolificacy," *Scientometrics* 19, no. 3 (1990): 185–97. The distortion of impact measures at the country level caused by the skewed distribution of citedness is discussed in Dag W. Aksnes and Gunnar Siversten, "The Effect of Highly Cited Papers on National Citation Indicators," *Scientometrics* 59, no. 2 (2004): 213–24, english.nifustep.no/content/download/2864/28408/file/Scientometrics-paper.pdf.

55. Narin, *Evaluative Bibliometrics*, 183. For a synthesizing description of the methodology, see Pinski and Narin, "Citation Influence for Journal Aggregates of Scientific Publications," *Information Processing & Management* 12, no. 5 (1976): 297–312.

56. Narin, *Evaluative Bibliometrics*, 202.

57. An anticipation of the influence algorithm can be found in Kochen, *Principles of Information Retrieval* (Los Angeles, Calif.: Melville Publishing, 1974), 79–84. A freely accessible database for retrieving the Eigenfactors of various categories of scholarly (and nonscholarly) materials, including all the journals listed in ISI *JCR*, is available at www.eigenfactor.org. On the algorithm and the rationale of the ranking system, see Bergstrom, "Eigenfactor: Measuring the Value and Prestige of Scholarly Journals," *C&RL News* 68, no. 5 (2007), www.ala.org/ala/acrl/acrlpubs/crlnews/backissues2007/may07/eigenfactor.cfm. Two recent examples of journal IF weighting in the Pinski/Narin style are reported in Farrokh Habibzadeh and Mahboobeh Yadollahie, "Journal Weighted Impact Factor: A Proposal," *Journal of Informetrics* 2, no. 2 (2008): 164–72, and Zitt and Small, "Modifying the Journal Impact Factor by Fractional Citation Weighting: The Audience Factor," *Journal of the American Society for Information Science and Technology* 59, no. 11 (2008): 1856–60.

58. The methodology of converging partial indicators is explained in Martin and Irvine, "Assessing Basic Research: Some Partial Indicators of Scientific Progress in Radio Astronomy," *Research Policy* 12, no. 2 (1983): 61–90. The paper's results were later updated in Irvine et al., "Assessing Basic Research: Reappraisal and Update of an Evaluation of Four Radio Astronomy Observatories," *Research Policy* 16, no. 2–4 (1987): 213–27. Three classic implementations of the proposed methodology to the evaluation of big science are Irvine and Martin, "Assessing Basic Research: The Case of the Isaac Newton Telescope," *Social Studies of Science* 13, no. 1 (1983): 49–86; Irvine and Martin, "Evaluating Big Science: CERN's Past Performance and Future Prospects," *Scientometrics* 7, no. 3–6 (1985): 281–308; and Irvine and Martin, "Basic Research in the East and West: A Comparison of the Scientific Performance of High-Energy Physics Accelerators," *Social Studies of Science* 15, no. 2 (1985): 293–341.

59. Braun, Bujdosó, and Schubert, *Literature of Analytical Chemistry: A Scientometric Evaluation* (Boca Raton, Fla.: CRC Press, 1987), chapter 10.

60. Schubert and Braun, "Relative Indicators and Relational Charts for Comparative Assessment of Publication Output and Citation Impact," *Scientometrics* 9, nos. 5–6 (1986): 282.

61. van Raan, "The Pandora's Box of Citation Analysis: Measuring Scientific Excellence—The Last Evil?" in *The Web of Knowledge: A Festschrift in Honor*

of Eugene Garfield, ed. Blaise Cronin and Helen Barsky Atkins, 304 (Medford, N.J.: Information Today Inc., 2000).

62. Tijssen, Visser, and van Leeuwen, "Benchmarking International Scientific Excellence: Are Highly Cited Research Papers an Appropriate Frame of Reference?" *Scientometrics* 54, no. 3 (2002): 386.

63. van Raan, "Statistical Properties of Bibliometric Indicators: Research Group Indicator Distributions and Correlations," *Journal of the American Society for Information Science and Technology* 57, no. 3 (2006): 429. For a nontechnical overview of the methodology, see van Raan, "The Use of Bibliometric Analysis in Research Performance Assessment and Monitoring of Interdisciplinary Scientific Developments," *Technikfolgenabschätzung: Theorie und Praxis* 12, no. 1 (2003): 20–29, www.itas.fzk.de/tatup/031/raan03a.htm#[1].

64. Katz, "The Self-Similar Science System," *Research Policy* 28, no. 5 (1999): 501–17, www.sussex.ac.uk/Users/sylvank/pubs/SSSS.pdf; Katz, "Scale Independent Indicators and Research Evaluation," *Science and Public Policy* 27, no. 1 (2000): 23–36, www.sussex.ac.uk/Users/sylvank/pubs/SI-RE.pdf. In the above cited van Raan, "Statistical Properties," the author seeks to demonstrate that Leiden normalized bibliometric indicators are immune from the scaling behavior described by Katz.

65. The dispute started with Martin, Irvine, and Turner, "The Writing on the Wall for British Science," *New Scientist* 104 (1984): 25–29. The initial statistics were updated in several subsequent papers containing additional figures: Irvine et al., "Charting the Decline in British Science," *Nature* 316, no. 6029 (1985): 587–90; David C. Smith et al., "National Performance in Basic Research," *Nature* 323, no. 6090 (1986): 681–84; Martin et al., "The Continuing Decline of British Science," *Nature* 330, no. 6144 (1987): 123–26; Martin, "British Science in the 1980s—Has the Relative Decline Continued?" *Scientometrics* 29, no. 1 (1994): 27–56. An overview of the early phase of the dispute is in Leydesdorff, "The Science Citation Index and the Measurement of National Performance in Terms of Numbers of Scientific Publications," *Scientometrics* 17, no. 1 (1989): 111–20.

66. Leydesdorff, "Problems with the 'Measurement' of National Scientific Performance," *Science and Public Policy* 15, no. 3 (1988): 149.

67. Braun, Glänzel, and Schubert, "Assessing Assessments of British Science: Some Facts and Figures to Accept or Decline," *Scientometrics* 15, nos. 3–4 (1989): 170.

68. Martin, "The Bibliometric Assessment of UK Scientific Performance: Reply to Braun, Glänzel and Schubert," *Scientometrics* 20, no. 2 (1991): 333–57. See also the comments to this reply posted in Braun, Glänzel, and Schubert, "The Bibliometric Assessment of UK Scientific Performance—Some Comments on

Martin's 'Reply'," *Scientometrics* 20, no. 2 (1991): 359–62, and the additional criticisms in Leydesdorff, "On the 'Scientometric Decline' of British Science: One Additional Graph in Reply to Ben Martin," *Scientometrics* 20, no. 2 (1991): 363–67, and Terence Kealey, "Government-Funded Academic Science Is a Consumer Good, Not a Producer Good: A Comparative Reassessment of Britain's Scientific and Technological Achievements Since 1794 and a Comment on the Bibliometry of B. Martin and J. Irvine," *Scientometrics* 20, no. 2 (1991): 369–94. A further reply to these papers is in the above-cited Martin, "British Science in the 1980s," and also in Martin and Irvine, "The Position of British Science," *Nature* 355, no. 6363 (1992): 760. More recently, publication and citation data on UK science for the period 1985–2003 have been reexamined and reinterpreted in terms of the scientists' responses to critical changes in the evaluation criteria of the RAEs (research assessment exercises) by Moed, "UK Research Assessment Exercises: Informed Judgments on Research Quality Or Quantity?" *Scientometrics* 74, no. 1 (2008): 153–61.

69. Moravcsik, "Assessing the Methodology for Finding a Methodology for Assessment," *Social Studies of Science* 16, no. 3 (1986): 538.

70. As representatives of the two extreme positions that can be taken, respectively, see Godin, "The Knowledge-Based Economy: Conceptual Framework or Buzzword," *Journal of Technology Transfer* 31, no. 1 (2006): 17–30, and Leydesdorff, *The Knowledge-Based Economy: Modeled, Measured, Simulated* (Boca Raton, Fla.: Universal Publishers, 2006).

71. For a survey on economic-oriented patent literature analyses, see Zvi Griliches, "Patent Statistics as Economic Indicators: A Survey," *Journal of Economic Literature* 28, no. 4 (1990): 1661–1707.

72. The 1957 test is reported in Garfield, "Breaking the Subject Index Barrier: A Citation Index for Chemical Patents," *Journal of the Patent Office Society* 39, no. 8 (1957): 583–95, www.garfield.library.upenn.edu/essays/v6p472y1983.pdf.

73. Reisner, "A Machine Stored Citation Index to Patent Literature Experimentation and Planning," in *Automation and Scientific Communication: Short Papers Contributed to the Theme Session of the 26th Annual Meeting of the American Documentation Institute,* ed. Hans Peter Luhn, 71–72 (Washington, D.C.: American Documentation Institute, 1963).

74. Philip Ellis, G. Hepburn, and Charles Oppenheim, "Studies on Patent Citation Networks," *Journal of Documentation* 34, no. 1 (1978): 12–20. See also the further extension of this technique to the fields of prostaglandins and X-ray tomography: Hilary Nunn and Charles Oppenheim, "A Patent-Journal Citation Network on Prostaglandins," *World Patent Information* 2, no. 2 (1980): 57–60; Lasley Dunlop and Charles Oppenheim, "The Effect of Recycling on a Patent Citation Network," *World Patent Information* 2, no. 3 (1980): 100–102.

75. For an overview of patent citation studies until the late 1990s, see Oppenheim, "Do Patent Citations Count?" in *The Web of Knowledge: A Festschrift in Honor of Eugene Garfield*, ed. Blaise Cronin and Helen Barsky Atkins, 405–32 (Medford, N.J.: Information Today, 2000). A collection of essays on patent citation analysis from an economic perspective is Jaffe and Trajtenberg, *Patents, Citations & Innovations: A Window on the Knowledge Economy* (Cambridge, Mass.: MIT Press, 2002).

76. Jaffe, Trajtenberg, and Henderson, "Geographic Localization of Knowledge Spillovers as Evidenced by Patent Citations," *The Quarterly Journal of Economics* 108, no. 3 (1993): 577–98; Peter Thompson, "Patent Citations and the Geography of Knowledge Spillovers: Evidence from Inventor- and Examiner-Added Citations," *The Review of Economics and Statistics* 88, no. 2 (2006): 383–88, www.fiu.edu/~economic/wp2004/04-05.pdf.

77. For a bibliometric perspective on the "scientification" of U.S. technology, see Narin, Hamilton, and Olivastro, "The Increasing Linkage Between U.S. Technology and Public Science," *Research Policy* 26, no. 3 (1997): 317–30.

78. See, for example, Diana Hicks et al., *Quantitative Methods of Research Evaluation Used by the U.S. Federal Government* (Japan: National Institute of Science and Technology Police, 2002), www.nistep.go.jp/achiev/ftx/eng/mat086e/pdf/mat086e.pdf. In particular, on the evaluation of firms' R&D knowledge stock, see Bronwyn H. Hall, Adam B. Jaffe, and Manuel Trajtenberg, "Market Value and Patent Citations," *RAND Journal of Economics* 36, no. 1 (2005): 16–38, repositories.cdlib.org/postprints/2454/.

79. Zhen Deng, Baruch Lev, and Francis Narin, "Science & Technology as Predictors of Stock Performance," *Financial Analysts Journal* 55, no. 3 (1999): 20–32. The Tech-Line model is underpinned by two U.S. patents (numbers 6.175.824 and 6.832.211).

80. Narin, "Patent Bibliometrics," *Scientometrics* 30, no. 1 (1994): 150–51.

81. A recent confirmation of the scale-free property comes from a large-scale analysis of the NBER (National Bureau of Economic Research) utility patent and patent citation data file conducted, with the tools of complex network analysis, by Thomas F. Brantle and M. Hosein Fallah, "Complex Innovation Networks, Patent Citations and Power Laws," in *PICMET 2007 Proceedings*, 540–49 (Portland, Oregon: PICMET, 2007). For a disciplinary case study, see Xin Li et al., "Patent Citation Network in Nanotechnology (1976–2004)," *Journal of Nanoparticle Research* 9, no. 3 (2007): 337–52.

82. Martin Meyer, "What Is Special About Patent Citations? Differences Between Scientific and Patent Citations," *Scientometrics* 49, no. 1 (2000): 93–123.

83. This is, for instance, the thesis held by Vinkler, "The Origin and Features of Information Referenced in Pharmaceutical Patents," *Scientometrics* 30, no. 1

(1994): 283–302. It may be contrasted with the opposite opinion on the value of the examiner's citation set forth in Trajtenberg, Henderson, and Jaffe, "University Versus Corporate Patents: A Window on the Basicness of Invention," *Economics of Innovation and New Technology* 5, no. 1 (1997): 19–50.

84. Michael B. Albert et al., "Direct Validation of Citation Counts as Indicators of Industrially Important Patents," *Research Policy* 20, no. 3 (1991): 251–59.

85. Carpenter and Narin, "Validation Study: Patent Citations as Indicators of Science and Foreign Dependence," *World Patent Information* 5, no. 3 (1983): 185.

Chapter Seven

On the Shoulders of Dwarfs: Citation as Rhetorical Device and the Criticisms to the Normative Model

Citation statistics are widely used, in a science policy perspective, to build indicators of scientific performance. Whether we like it or not, despite all precautionary caveats and fine-grained methodological warnings issued by number-makers, the end-consumer, be it the lonely digital-bookworm going through a list of retrieved hits ranked by citation frequency or the university tenure and promotion committee assessing a candidate's impact on the field of study, will inevitably attach a rude qualitative meaning to the typical final product, the unidimensional ranking (of articles, authors, universities, countries): those who rank first are qualitatively better than the rest because they contribute more to the advancement of knowledge. A further dramatic, albeit qualified, endorsement of this principle came in December 2006 from the UK Higher Education Funding Council (HEFCE) with the announcement that, upon completion of the 2008 Research Assessment Exercise (RAE), a new horizon for university research evaluation and funding will be opened up by the adoption of an entirely different, hopefully more efficient and cost-effective rating system ruled by bibliometric criteria. Two preliminary reports on the feasibility and methodological prerequisites of this shift, prepared by, respectively, Evidence Ltd. on behalf of Universities UK and the Leiden CWTS on behalf of HEFCE, don't leave much room for uncertainties: although recognizing the caveats originating from forty years of debates on citation analysis, specifically the impossibility of a uniform approach to data management coupled with the challenge issued by the choice of a precise level of normalization for citation counts across heterogeneous subject domains, the former clearly designates "citations per paper" as "a central quality index" for the new system, at least as far as science, engineering, and technology subjects are primarily concerned, while the latter places the long-running Dutch expertise in constructing, testing, and implementing advanced "source expanded" (i.e., beyond ISI coverage) bibliometric indicators at the RAE new course's service.[1]

As soon as one begins to explore the set of questions raised by the bibliometric view of the world, however, it immediately becomes clear that the concept of quality of a scientific or technical publication is just as elusive as that of influence encountered in chapter 5. Citation analysis, in turn, redoubles the elusiveness by establishing a direct genealogical relationship between quality and influence. Modern science places value on scientific work on the score of its being essential for further work to be successfully accomplished, so if citations, at least in the aggregate, indicate the impact of previous literature on current research, and if impact is a good operational measure of the influence exercised by that literature, then the more a document is influential, the higher its quality. But does citation impact capture a true manifestation of quality? Is the set of properties distinguishing a piece of scientific work that possesses quality to a certain degree from those that possess it to a lesser degree, or do not possess it at all, so closely linked to the formal act of citing? Let's take an extreme example.

Even though no one would disagree that Einstein's papers on special and general relativity represent a touchstone for quality assessment, it can hardly be denied that their theoretical commitment, their unique capability of rethinking space and time in such a way as to disclose a new world that was already there, form just but one of the many possible embodiments of good science: quality is measured, in this case, more against the ability of scientific discourse to reshape human understanding of physical reality, than against the ordinary scenario of scientists pursuing minimal research objectives within the boundaries of a shared conceptual framework (a paradigm). In a sense, one could even be satisfied with the rarity of such epochal breakthroughs because, if they were the norm, very few certainties would actually exist, and science would live in a state of permanent and radical conceptual revolution. But if exceptional scientific breakthroughs are mostly associated with the historical occurrences of genius, then it might be construed that, during a normal state of affairs, scientific quality is a matter of comparison among, as Newton would say, "dwarfs" standing on the shoulders of (and paving the way for) "giants." Hence one might rightfully wonder how short dwarfs actually are and what kind of quality, if any, is exhibited by the work of nonrevolutionary researchers ordinarily contending for tenure, promotion, or funding. Underlying these questions, however, other more fundamental questions of epistemological nature are at stake: Does science advance only on the spur of super-talented people's brainwaves, or do the micro-transformations brought about by the apparently less significant work of average researchers play a role as well? Can citation analysis detect qualitative differences among the relative contributions of researchers engaged in ordinary tasks? Bibliometrics is perhaps too narrow a perspective on science for tackling all these questions, but let's examine in more detail what kind of evidence it does produce on scientific quality, starting from the atomic level of the individual act of citing.

7.1. REASONS TO CITE: DO THEY REALLY MATTER?

Is it worthwhile counting, analyzing, and comparing citation scores? What does a bibliographic citation say? What kind of link does it establish between citing and cited papers? How does its significance vary in relation to the text surrounding it? Above all, when do authors cite, and why? Bibliometricians have constantly been aware that there are many more reasons to cite than merely settling an intellectual account.[2] Tracking them systematically within selected samples of authors and texts became, starting in the mid-1970s, a major task on the research agenda, in the conviction that encompassing the variability of citation styles and practices into a plausible framework was the first step toward a comprehensive theory of citation. Such a theory would, in turn, place citation-based bibliometric indicators on more firm theoretical grounds.[3]

Thanks to Merton, Price, Garfield, and Small, the conceptual foundations of citation analysis were laid down between the 1960s and the 1970s. Merton had linked the scientists' referencing behavior to a professional "etiquette"; Price had shown how to apply mathematical reasoning to the patterns of intersection deriving from a multitude of individual acts of citing; and Garfield and Small had turned citations into the building blocks of a specialized symbolic language decipherable through citation indexes.

Sociologists and information scientists, however, remained unsatisfied with the abstractedness of Merton's norms and the positivistic attitude of citation analysts toward the structure of scientific communications, feeling it more and more urgent to reach a deeper understanding of the citation process through live examples of scientific practice. Such an effort took three main directions, focusing, respectively, on the psychology of the citer and the self-reported identification of citing motivations by the authors themselves (citations as a private process); the interrelationship between citations and the text invoking their occurrence (citations as textual devices); and the comparison of private processes and formal communications with true samples of scientific practice or, stated alternatively, the comparison of what texts and authors say (and cite) with what they actually do at the lab bench.

7.1.1. What Scientists Say

An apparently straightforward way to investigate citation behavior is asking the authors themselves why they cite. Citations, after all, are private acts, and direct questioning can highlight strategies and motivations lurking behind the overt dialogue with past literature enacted in published reports of scientific practice. So, by way of illustration, having surveyed twenty chemists at the Central Research Institute for Chemistry of the Hungarian Academy of the Sciences, Peter Vinkler found that Mertonian rules were basically complied with, since authors cited predominantly for documentary reasons.[4] But elsewhere things went quite differently, and it would be fundamentally wrong to draw general conclusions from experiments performed in such a wide variety of disciplinary and institutional settings; empirical evidence, once and for all, is not additive.

There is little doubt that, if two authors were in a position to add citations to exactly the same paper before submitting it for publication, they would hardly cite the same sources, but how different their choices would be depends on too many variables to be estimable with a reasonable degree of accuracy. Experiments of this kind, nonetheless, were performed by Eugene Garfield in the 1960s and Blaise Cronin in 1981. The latter submitted an unpublished experimental paper, stripped of all bibliographic references, to a sample of British psychologists, asking them to supply the bibliographic notes. The outcome revealed a considerable lack of unanimity about the number and location of references they felt necessary to back up each argument.[5] It might even be construed that the set of authors cited and re-cited by a scientist throughout an oeuvre—what White has dubbed "citation identity"—is as distinctive as fingerprints in characterizing the scientist's research style, that is, the way he or she, driven by the perceived relevance of cited works, lets the bibliographic universe speak in unison with that oeuvre so as to publicize the intellectual context surrounding its growth.[6]

Interestingly enough, when scientists are asked to account for referencing behavior, they do not seem to feel the need of explaining their choices by appealing to optimal criteria or universal rules of good conduct. Terrence Brooks, for instance, recorded the reasons for citing given by twenty-six scholars of various disciplinary extraction at the University of Iowa. The answers were distributed along seven motivational scales, ranging from the Mertonian credit-giving to the rhetorical desire to persuade the audience. Surprisingly, persuasiveness ranked as a dominant motivation, a somewhat curious result if one considers that philosophers and science historians have often spent tons of erudition demonstrating that, at least since the scientific revolution of the sixteenth century, dissimulated rhetorical skills have played a crucial role in the establishment of the modern art of reasoning.[7] Does the degree of philosophical and historical awareness of current authors or their post hoc rationalizations affect the self-reporting of the reasons to cite? To what reasonable extent might we expect that, asking authors to self-assess their citing preferences, they will tell them exactly as they are? And what about their grasp of the proposed scales: Does the variability in responses reflect true differences in the behavior of the respondents or simply the differences in their interpretation of the proposed motivations?

Experiments addressing the epidermal layer of citation behaviors, that is, the nature of the judgments made by authors about the works they cite, have been carried out, since the 1980s, by several authors, including the following:[8]

1. Chandra Prabha, with members of a Department of Business Administration
2. Mengxiong Liu, with Chinese physicists
3. William Shadish and colleagues, with psychologists

4. Donald Case and Georgann Higgins, with scholars in communications studies
5. Marylin White and Peiling Wang, with agricultural economists
6. Rong Tang and Martin Safer, with biologists and psychologists

Numbers 1 and 2 showed that the number and choice of specific cited references were not as critical for the surveyed authors as they are for citation analysts; in the second study, specifically, the number of references seemed well correlated with authors' use of an institutional library, while the choice of a particular reference appeared significantly influenced by the desire to strengthen a knowledge claim or by the prestige attached to the cited paper/author. Numbers 3 and 4 pointed out one important aspect relative to the way highly cited papers are perceived, namely that they need not be exceedingly creative works, their appreciation often being associated with the status of exemplar references, concept markers, or simply review papers. White and Wang's results in number 5, although reestablishing the importance of internal, strictly cognitive motivations for citing, highlighted also the possible "noncognitive" reasons behind the decision not to cite: documents judged too old, too specific, or simply too difficult to obtain. In number 6, finally, Tang and Safer deliver evidence to the effect that, though self-citations and citees whom the citing author has a close relationship with are treated somewhat "emotionally," the choice and physical allocation of references are weighted mostly through a rational, rule-based process, so that strong predictors of author-rated citation importance can be found by a closer inspection and categorization of textual variables, including the number of times a reference is cited, the length of the cited reference, the depth of its treatment in the text, and the specific section of the article in which it is located.

When self-reports are compared with evidence from the text itself, a gap between the two occasionally emerges, as documented by Leydesdorff and Amsterdamska's 1990 study in which 239 authors, who had cited one of four most highly cited biochemistry papers published by a research group of the University of Amsterdam, were asked to qualify their citing perspective both from a theoretical (reasons to cite, scientific evaluation of the cited work) and a social (personal link with cited authors) point of view. In a significant percentage, the reported motivations for citing didn't correspond with the specific use of cited references actually made in the citing text. Self-reports, indeed, tended to underestimate the cited papers' substantive role in drawing particular conclusions or in setting forth specific knowledge claims. The authors consequently suggested that "reported motives and the subjective reasons given by authors for their citing should be distinguished from the various cognitive functions which citations play in argumentation presented in scientific articles."[9]

7.1.2. What Texts Say

Instead of pretending to step inside the author's mind, the text itself can be dissected in the hope of inferring the use and function of a cited reference from the context surrounding it and the conceptual content of the citing passage. In principle, this shortcut can reveal which aspects of the cited paper attracted the attention of the citer and why, thus allowing a skilled examiner to assess whether a particular reference was truly needed for the completion of the citing work. In a classic 1975 paper, using references from a single journal, in a single specialty of a single branch of science, and in a narrow time period, Moravcsik and Murugesan performed a context analysis on a sample of thirty high-energy physics papers published in the *Physical Review* between 1968 and 1972. Their findings cast serious doubt on the usefulness of citation counts in influence tracing and quality control: 41 percent of the references were perfunctory, not truly needed for the understanding of the citing paper, nor did the cited documents appear indispensable for the working out of its content; 31 percent of the references were redundant, making the same point as previous references and added perhaps to "make everyone happy" in view of priority disputations. Anybody—concluded the authors—could increase his or her own citation score by writing mediocre publishable papers on "hot" topics that would subsequently get perfunctory citations.[10] Chubin and Moitra integrated the Moravcsik-Murugesan's scheme with three further variables: the form of the paper (letter or full-length article), the type of research (experimental or theoretical work), and the research outlet (three American journals and one Dutch journal). Their content analysis of references in forty-three high-energy physics papers yielded some interesting insights into physicists' citation habits, such as, for example, that positive citations are much more frequent than negative ones, that irrelevant or subsidiary citations are less common in letters than in full-length articles, and that letters are cited more quickly and more extensively than the corresponding full-length version, although, in the long run, it is the latter that advance the discipline, turning into certified knowledge and continuing to live, at best, a life rich in citations. In addition, they found a lower global percentage of perfunctory citations than in Moravcsik and Murugesan's experiment and claimed that the observed regularities pointed to "the operation of certain norms" in citation behavior.[11]

Many subsequent authors added to this tradition. Their findings convey further evidence that evidence itself in citation studies is a truly complex matter. Susan Cozzens, for example, confronted the patterns of use of two papers from the fields of neuropharmacology and sociology of science. She found that references to the former, unlike those to the latter, changed markedly over time, shifting gradually from an in-depth concern with the details of the results, the experimental procedures, and even the peripheral knowledge claims, to a standardized endorsement of the source paper's main knowledge claim. A similar prevalence of standardized use emerged from Oppenheim and Renn's analysis of the reasons why twenty-three highly cited old papers in the field of physics and physical chemistry continued to be cited: in 40 percent of the cases, they were purely historical reasons. Oppenheim's team also performed a two-stage citation content analysis of the classic 1953 Watson and Crick paper disclosing the structure of DNA. In line with the previous findings, they discovered that the percentage of authors citing that landmark paper for historical reasons increased over time, while an increasingly lower number of citing papers invoked it to obtain specific information, data, and methodologies.[12]

Content and context analysis were, in the eyes of early bibliometricians, nothing more than necessary steps toward increasingly refined indicators of scientific performance. Once upon a time, simple output measures, such as the number of scientists and publications, dominated the stage; then, with the advent of the *SCI*, those primordial measures were overshadowed by citation impact; finally, the evaluative meaning of citation scores could be further enhanced by sifting out the relevant citations from the mess of perfunctory or subsidiary ones. In the long run, however, this kind of micro-analytic approach generated a paradox: if citations are the building blocks of a highly structured language for the transmission and manipulation of scientific concepts (Garfield, Small), taken individually and projected onto the background of the texts calling them in, they betray a variety of meanings hardly consistent with the trust placed in the usefulness of their counting for quality assessment purposes.

As early as the mid-1960s, the suspicion had already arisen that, contrary to Merton's belief, only occasionally do citation practices comply with the mission of awarding credits and peer recognition to colleagues and predecessors. "How often," wondered Norman Kaplan in 1965, "are the works of others cited without having been read carefully? How often are citations simply lifted from the bibliography in someone else's work without either reading or giving credit to the man who did the original search of the literature? How often are citations tacked on after the paper is completed as an afterthought and window dressing?"[13] The misuses can be of several types and encompass a variety of intermediate degrees of "dishonesty": a good scientist can generously give out credits to a colleague's publication for repaying extra-scientific debts or simply for professing institutional modesty; conversely, a mediocre scientist may wish to transfer partially or totally the responsibility for errors or omissions to the "authoritative" referenced sources, while a dishonest scientist can under-cite or cite "en passant" an unpublished work of a colleague upon which he or she has built a successful publication. In such a case, being footnoted would be too little of a reward for the defrauded colleague if compared to the lost opportunity of scoring a major point in the publication game. At any rate, inasmuch as bibliographic citations can be imprecisely, incompletely, or even deceitfully allotted, the rules of their production and circulation appear to be far more loose than Merton taught.

One might go even further and assert that if a method does exist in the apparent madness of citation behavior, it is at the level of discourse construction. Citations embed the cited document into an argument. By placing the source in the argument, they make the point that there is no need to argue again for the adequacy of the cited materials, which is assumed to be already accomplished in the original study. Thus they work, at least to a certain extent, as rhetorical devices whose final purpose is perfectly in tune with the formal organization of journal papers in the current scientific communication system: to persuade the scientific community of the validity (novelty, importance, truth) of the author's theses. Citations' meaning, accordingly, vary with the function of the texts (and contexts) calling them in: "The [cited] materials themselves, the uses to which they can be put, and the relations between an existing result and a new one," argued Jerome Ravetz in 1971, "are as complex as the history of a solved problem. The material may be crucial, or merely incidental in the argument; it may have been central to the first formulation of the problem, or merely a late addition; and it may have been used as it was published or required extensive re-working. In all these dimensions, there is a continuous and complex scale from complete dependence to near independence."[14]

7.1.3. Comparing What Scientists and Texts Say with What Scientists Do

Asking scientists about their inner mental processes or letting texts speak on their behalf led to quite similar conclusions about the weakness of the ties connecting citations to the norms of scientific behavior. When such weakness was carried to the extreme, citations lost any duty as carriers of peer recognition and were left to the mercy of the authors' will. Along this path it is not surprising that, at least since the mid-1970s, they started to show up on the research agenda of the more critical and constructivist trends in contemporary sociology of science, including Pierre Bourdieu's project to "trap *Homo Academicus*, supreme classifier, in the net of his own classifications,"[15] the "strong program" of the Edinburgh school (David Bloor, Barry Barnes), the empirical relativist program of the Bath group (Harry Collins, Trevor Pinch), ethnomethodology (Harold Garfinkel, Bruno Latour, Steve Woolgar, Karin Knorr-Cetina), and the actor-network theory (Latour, Callon). From within different social theories, these authors set out to demonstrate, through historical case studies or live recordings of laboratory life, that the "black box" of science, before which Merton had stopped without attempting to open it, contains little more than a mirror reflecting the warped picture of the scientists themselves, rather than a faithful image of the external world.

By comparing what scientists actually do and say in the laboratory with what scientists write and say they have done in official reports, constructivist sociologists have sought to demolish the myth of the laboratory as a workshop for the generation and packaging of the absolute, neutral, and disinterested truth. The main outcome of that comparison is the unbridgeable gap supposedly existing between the register of formal accountings—the politically correct journal paper—and that of everyday research practice. In the former, the physical world seems to act and speak by itself, the style is impersonal, the presentation of materials and methods is stylized in generic formulations drawn from current textbooks, and any allusion to the dependency of observations on theoretical speculation is obscured. The latter, inversely, is the realm of pure contingency, where the choice, processing, and completion of relevant tasks is made possible by the application of practical skills drawn from an entire repertoire of "tricks of the trade," escaping any form of official encoding, let alone the final report in a journal paper. The *Dictionary of Useful Research Phrases*, stumbled across by Nigel Gilbert and Michael Mulkay during their frequentation of biochemical laboratories, exemplifies the distance between the two universes in a playfully extreme fashion: Where the formal register says "It has long been known that . . . ," it was perhaps originally meant, "I haven't bothered to look up the references"; a phrase like "Typical results are shown . . ." is a respectable substitute for "The best results are shown"; likewise, to say "of great theoretical and practical importance . . ." is a paper-compliant substitute for the more utilitarian "interesting to me"; and the concessive introductory clause "While it has not been possible to provide definite answers to these questions . . ." replaces the more prosaic "The experiment didn't work out, but I figured I could at least get a publication out of it."[16]

A classical way to handle these fouls in the game of science is to keep conceptually distinct the informal stratum—the context of discovery—from its final consolidation in logically interconnected pieces of discourse to be narrated—the context of justification—and to maintain that the two strata jointly contribute to shape the scientific method ending up reconciled in the journal paper, which forms a communication device fully functional to the advancement of knowledge. "All this confusion among the way research is conducted, the way manuscripts are idealized, and the way articles are finally written," admits William Garvey, "certainly seems strange to the outsider. But the combination of these three processes has been an enduring method of reporting scientific ideas and observations because it has proven very successful in leading scientists to conduct new research which further explains the ideas and extends the observations described in the article. And, of course, the reporting of this new research follows anew the same pattern. Somewhere embedded in this process is the basis of the extraordinary forward progress of science."[17] Scientific articles, reasserts Jack Meadows, "are as carefully constructed as any archaeological artefact: they reflect the requirements of the contemporary scientific community, changing as it changes."[18]

Constructivists, on their part, cast a dark shadow on the fairness of the entire communication process. It is not only the fact that, as already pointed out by Peter Medawar in 1963, the scientific paper is "fraudulent" since it gives "a totally misleading narrative of the processes of thought that go into the making of scientific discoveries."[19] Even more deceitfully, the transition from laboratory experience to the formal account of empirical data is marked by extensive textual adjustments intended to "make things work" in one way or another. Scarcely anyone will notice the trick as long as the formal organization of the scientific paper resembles a perfect crime scene, where the gray zones of scientific discourse are brightened up by the deceitful transparency of evidence-based argumentation. What happens afterward, whether the paper's knowledge claims will be ratified or dismissed by the community of peers, and whether its authors are going to be fairly rewarded for true or alleged merits, is largely a matter of "military" strategy and informal bargaining carried out in the spirit of the clique. Laboratory life, accordingly, is a mere production and circulation of inscriptions, enunciations, and texts, materials that, to a considerable extent, embellish the reality they are supposed to mirror, narrating things not as they actually occurred, but as they should have occurred if the ideal norms of scientific activity were put into action.

A closer look in a "constructivist style" at the structure of the modern experimental report will help illustrate how rhetoric is dissembled by the well-polished architecture of the finished product. Let's take a typical research article in a biomedical journal. It is organized in five standard sections:

1. *Introduction.* Provides a rationale for the results by discussing the state of the art through a concise literature review, which culminates in the question to be answered. In John Swales's classical terms, the basic rhetorical moves of a typical research article's introduction are establishing the territory, summarizing its history, showing a gap, and introducing the solution that is going to fill the gap.[20] Many cited references congregate here, never to be summoned again significantly in the course of the argument. The relative strengths and weaknesses of the

cited works, when explicitly addressed, are discussed in a prevailingly tactful fashion.

2. *Methodology*. Illustrates the methodology employed in data collection and processing. Unlike what happens in a doctoral dissertation, the choice does not need to be defended, and its impersonal unfolding is exactly what marks the gap with laboratory life: experimental results are publicized here as the mechanical product of the experimenters' interaction with nature mediated by an unobtrusive sequence of standard procedures.
3. *Results*. Reports the salient research findings through a considerable amount of factual information arranged in a logical sequence and typically reorganized in tables and figures.
4. *Discussion and conclusion*. Delves into the interpretation of the experimental data in light of the literature review performed in the first section, so as to corroborate or contest previous findings. In one way or another, data turn out to be always significant, otherwise the paper would hardly get published, but a certain amount of rhetorical skill is lavished to temper the appreciation of the article's strong points with statements addressing the limitations of the study design and the room for further inquiry.
5. *Bibliography*. The polite list of titles cited in the various sections is finally presented; lined up in a sort of martial arrangement, they seem all equal, but the textual location where each reference has been summoned up—the most powerful being those in the introduction—and the specific task accomplished in the argument by each cited source let a hierarchy arise. Some references turn out more equal than others on account of their stronger propensity to comply with the article's rhetorical endgame.

With the advent of citation indexes, the bibliographic section, upon whose optical recognition the ISI has built its fortunes, experienced a dramatic increase in popularity along with a further push toward standardization. The price of fame, however, was hardly negligible in terms of homologation and loss of documentary power. The subjection of bibliographic references compilation to the rigorous editorial standards of international style manuals, introduced between the end of the nineteenth and the beginning of the twentieth centuries and gradually implemented by journal publishers worldwide (*Harvard Style*, *MLA Style*, *Vancouver Style*, *APA Style*), accelerated the transition from the extensive use of commented references in foot- or endnotes to a more impersonal style, where a list of full references at the end of the manuscript is combined with abbreviated pointers to the list entries scattered throughout the text (a number in parentheses for the *Vancouver Style,* an abbreviated "author-date" notation for the *Harvard Style*). Such a transition was technically congenial to the capture and automatic processing of bibliographic records, but its underground scope, in constructivists' opinion, was much more momentous than sheer machine-readability insofar as it turned into one of the most effective strategies implemented by journal publishers and editorial boards to mark papers with the hallmarks of scientific objectivity.[21]

The exclusion of discursive notes, in particular, has been extremely effective in the homologation of citations, which is a prelude to the two main bad side effects of a totalizing empiricist epistemology: first, to obscure parallel or alternative lines of discussion while sacrificing every research path irregularity to the linearity of the central claims; second, to prevent the author (and the reader) from better qualifying or contextualizing the use of cited sources, thereby flattening the text on the primary function of reporting "facts" and "information," in the naïve conviction that language is a plain instrument used to transfer objective knowledge from the author's to the reader's mind. A "neutralized" text of this kind has little or nothing to say on the sociocognitive dynamics preceding the final publication and says even less about the strategies employed by the authors to bring into line the initial research hypothesis with the empirical data. Rather, it narrates a cleverly arranged fairy tale. It is the tale of an empty and receptive mind, the scientist's, into which data flow spontaneously from laboratory sensorial experiences and experiments, growing into elementary observations. Elementary observations are eventually condensed into simple propositions, each describing an elementary state of the world, whose combination yields complex propositions useful to confirm or disprove an existing theory. The twentieth-century philosophy of science, from Bertrand Russell and Karl Popper onward, would demolish such a fictional picture of the process of scientific discovery along with the genuine inductive logic it ratifies, demonstrating that there is no place in scientific research for neutral observation, just as there is no laboratory experiment without a set of hypotheses and theoretical expectations that condition not only the methodological choices and the gathering of empirical data, but also the way final results are weighed and evaluated. The role of bibliographic citations in the foregoing scenario appears sharply at variance with that advocated by Garfield, Merton, and Small: citations do not incorporate relatively stable concept symbols but social transactions; their mission is not to acknowledge intellectual debts but to inhibit criticisms, to overcome opponents with the weight of authority, to persuade the peer audience of the validity of a thesis by listing all of the author's potential allies. "A paper that does not have references," asserts Latour, "is like a child without an escort walking at night in a big city it does not know."[22]

7.1.4. Counting the Uncountable: Bibliometric Models of Citation Practices

In a sense, the constructivist perspective on citation practices is not to be disdained, inasmuch as, by dint of digging into the hidden stratum of scientists' court etiquette, it provides an antidote against any indiscriminate, uncritical trust in the unobtrusiveness of citation data. Citation, as Cronin pointed out in 1984, "needs to be thought of as a process" whose outcomes, the unidimensional lists of bibliographic references attached to scholarly papers, "reflect authors' personalities and professional *milieux*."[23] Plunged into individual idiosyncrasies, however, citations become irremediably elusive, multidimensional objects, dwelling at the crossroads of distinct and mutually conflicting systems, basically polarized around the social reward system of institutionalized disciplines and the rhetorical-linguistic system of text dressing up.[24] To escape the traps set by the micro-level of analysis, bibliometricians and information scientists have traveled over two parallel avenues, heading for opposite attitudes toward citation behavior: on the negative side, to completely reject the utility and even feasibility of a theory of citation rooted in the phenomenology of citing behavior; on the positive side, to keep on pursuing mathematical models sustained by carefully designed empirical tests aimed at quantifying the relative weights exercised by factors of different nature on citation rates.

1. *No theory at all.* Paul Wouters has insisted on the fundamental difference between references (perspective of the citing authors) and citations (perspective of the cited authors) to argue for the impossibility of a theory of citation resting on the study of reference behavior. The invention of citation indexes made it possible to treat citations as having a life of their own in the context of indicator theories rooted in particular representations of science. The two levels of analysis pursued by citation and reference analysis have consequently drifted apart to such a degree that the only way to capture citation meaning is to work at the top level of the indicator theories themselves, to establish a general theory explaining how they can be related to one another.[25] Less diplomatically, van Raan has simply dismissed the recurrent quest for a citation theory on empirical and statistical grounds: even though individual referencing behavior is somewhat whimsical, it can't reasonably be maintained that researchers refer to earlier work in a completely arbitrary fashion. What is more, citation analysis deals with highly cited papers or with large numbers of citations to the oeuvres of research groups; hence, its focus is on the statistical regularities emerging from the aggregation of many individual citing acts. "Citation analysis," concludes van Raan echoing Price, "is at the 'thermodynamic side' . . . the individual characteristics of the citers are interesting, but the distributions-functions of these characteristics are the make-up of that part of the world which is relevant to bibliometric analysis."[26]
2. *Bits of mathematical and evidence-based explanations of the citation process.* Citations either occur or they do not, and the actual number of them eventually accruing to an article in any given counting period is the result of so many unrelated factors that stochastic modeling has been conveniently pursued, just as in any other source/items relationships, to elucidate the structure of the entire process. Schubert and Glänzel, for example, have adapted to citations the same dynamic Waring model based on a stochastic birth process previously tested on author productivity patterns, while Burrell has been developing since 2001 a stochastic model of citations in the presence of obsolescence that takes advantage of the same mixtures of nonhomogeneous Poisson processes applied to the description of the other informetric processes reviewed in chapter 4.[27] On a parallel track, many studies have been published, from the early 1980s onward, featuring variously refined mathemati-

cal models that typically rely on some sort of regression analysis in order to estimate the best predictors of papers' citation scores on the basis of "internal" (textual) and "external" (social reputation, journal outlet, etc.) criteria applied to selected samples of scientific literature. In a forerunner 1983 paper, John Stewart tried to determine whether citations accrued, three years after publication, to a set of geosciences articles mostly related to plate tectonics were better explained by article characteristics, including length, number of references, and relevance to the research area, or by author characteristics, including institutional prestige, professional age, and previous productivity. Though admitting the impossibility of calculating the unique contributions of specific variables to the overall variance in citation rates, he found that article characteristics were the strongest predictors.[28] About fifteen years later, Baldi extended Stewart's approach by adding variables for the measurable properties of the citing articles besides those of the cited articles, which he considered a "dyadic" approach more in tune with Small's theory of citation than Stewart's. In his "network-analytic" model, the characteristics of potentially citing and potentially cited papers influencing the probability of citation are embedded in three distinct sets of variables intended to formalize, respectively, the normative reasons to cite, such as content endorsement or perceived quality; the constructivist reasons to cite, such as cited authors' prestige; and an additional group of control variables addressing the factors likely to influence the occurrence of a citation between two articles, such as article's size or journal's visibility. The potentially cited article's theoretical content, for instance, is measured by a variable indicating the presence or absence of claims for having made a unique theoretical contribution, while its empirical content is operationalized by the number of quantitative tables and graphs per page; concomitantly, the potentially cited author's eminence is approximated by the author's citation score in the year prior to the article's publication year, whereas the existence of social ties between potentially citing and potentially cited authors is encoded by a variable expressing whether they ever worked at the same institution or received their Ph.D. from the same graduate department. Tested on a sample of articles on celestial masers, the model revealed that "the intellectual content of citing and cited articles is particularly important for a citation to occur."[29]

To sharpen our understanding of the influence exercised by social closeness on citation behavior, some authors have relied on combinations of bibliometric and sociological tools, such as social network analysis,[30] while others have followed a track similar to Stewart and Baldi's in the search for predictors of citation success mostly at the formal level of authors', texts', or journals' bibliometric regularities, thereby reaching sometimes similar, sometimes sharply diverse conclusions from within difficult-to-compare experimental and mathematical settings. By way of illustration, the best or at least some significant predictors of citation scores have been identified, from time to time, including the following:[31]

a. authors' productivity rates in chemical engineering, the most prolific authors appearing to be also the most likely to attract future citations;
b. citation records of first authors in crime-psychology;
c. first author eminence, journal prestige, article length, and number and timeliness of references in social and personality psychology;
d. prestige ranking of the publishing journals and the number of references per article in sociology;
e. ISI impact factor and the average citation rates of the publishing journals in the areas, respectively, of emergency medicine and management;
f. quality ratings of individual articles by journals' editors and editorial boards in marketing research;
g. textual and graphical elements, including readability of the abstract, number of uncommon words, and number of references, figures, tables, and graphs in two subsets of medical and high-energy physics papers dealing, respectively, with Burkitt's lymphoma and Heavy Quark Potential;
h. ratings of clinical relevance by practicing professionals in clinical medicine;
i. typology and geographical focus of research design as well as relative standing of the journal outlet in demography, where empirical research on developed countries published in core, English-language journals takes the lead in the chance for citations;
j. authors' professional standing in ecology;
k. study design and size of the *P-value*, i.e., the number supposedly expressing the statistical significance of a reported outcome, in psychiatry;

l. the computer-oriented nature of the content in librarianship (with regard to a set of research papers published in 1975 by 39 librarianship journals); and
m. number of claims and cited patents in biotechnology patents.

None of the studies documenting the above correlations—the reader is strongly advised to become familiar with the details of each specific data handling and modeling assumption before trusting any reported conclusion as a plus or minus of evidence in favor of or contrary to the normative theory—can purposefully claim to represent the cornerstone of the experimental design for citation studies everywhere, at all times, and for all disciplines. Moreover, for a component like journal impact factor or cited author prestige, being in the position to explain a significant portion of the variance in citation rates doesn't tell an ultimate truth about the nature of the citation itself: Did the citers choose to reference that particular paper by a socially prominent scientist or publish in a high impact journal because of the "showcase effect" and the consequent rhetorical gain granted by either condition or, on the contrary, because socially prominent scientists and highly cited journals are more likely the source of relevant, cutting-edge, and hence intrinsically more citable research findings? An answer to this and similar questions, if possible at all, would certainly require a widening of the bibliometric perspective to embrace aspects of the intellectual content of published papers that cannot be operationalized in a multiple regression equation just as easily as journal ranking, authors' academic affiliations, number of references, and so on.

What experimental reports suggest, however, is that the exasperation of constructivist theses lies outside the range of bibliometrics and perhaps of any realistic understanding of science as well. It can hardly be denied, in fact, that a set of standards does operate to orient science toward a comprehension of natural and social phenomena substantially different from that of art or literature: their being flexible and socially negotiated, neither universal nor eternal, and their being unfaithfully reflected by the formal communication system of science don't necessarily entail the absolute negation of their existence. Whatever the actual behavior of scientists, standards supply a conventional benchmark and a set of expectations against which deviations can be measured from time to time. That is one of the reasons why, before undertaking a conversation with a constructivist,

Belver Griffith used to prescribe a heavy dose of Mertonian normativism: "Take two Mertons," he recommended, "and see me in the morning."[32]

7.2. BETWEEN GIANTS AND DWARFS: CITATION ANALYSIS AND THE STRATIFICATION OF SCIENTIFIC COMMUNITIES

Were it to be taken too seriously by a totalitarian bibliometric mind, the stark inequality in productivity/citation rates among scientists uncovered by countless empirical studies and canonized by bibliometric laws would almost inevitably raise the suspicion that scientific life is marked by an astounding waste of energy: a flock of poorly talented, nearly invisible "Mr. Nothings" that absorbs vast organizational and financial resources despite its being inexorably cut out of the pasture of scientific eminence. What, if any, is its place in the economy of knowledge?

In a much-debated 1972 article, Jonathan Cole and Stephen Cole dismissed the theory, ascribed to the Spanish philosopher José Ortega y Gasset and named after him "the Ortega hypothesis," that the advancement of science rests on the humble and obscure work of a host of mediocre researchers who, by their unostentatious drudgery, prepare the ground for the eminent scientist, the man of genius. The citation practices of a sample of American academic physicists demonstrated, instead, that top scientists, namely those who produce highly cited publications, cite in their best papers mainly contributions of other top scientists, whose eminence is confirmed, in turn, by high citation scores and by additional, nonbibliometric kinds of evidence, such as being included in Cattell's *American Men of Science*, working in top-ranked departments, and winning honorific awards. Thus, in the stratified social system of science, progress is hardly the outcome of a democratic division of labor among intellectual workers. Rather, it is built upon the contributions of a relatively small scientific elite.[33]

Similar conclusions were reached, in the early 1980s, by other authors embarking on citation studies in the fields of criminology, American sociology, and Dutch physics,[34] but the subsequent debate questioned both the assumptions and the methodology of the Coles' findings. Quite predictably, the main assumption at stake was that scientists' citation practices generally abide by Mertonian rules; hence, bibliographic citations are roughly valid indicators of cognitive influences. Barbara MacRoberts and Michael MacRoberts completely rejected this premise, claiming that bibliographies are nothing but a faulty mirror of the actual connections between authors and documents, to such a degree that citation analysis affords a completely misleading picture of scientific communication dynamics. The key points of their argument are much the same as the criticisms of the normative theory of citing set forth in section 7.1, and can be summarized as follows:[35]

1. *What citations fail to say about influences.* As David Edge had already argued in two classic anti-citationist papers, bibliographic citations do not account for the tacit knowledge dimension placed underneath the outer layer of the published report.[36] They overlook both the informal seething of ideas taking place in the prepublication stage and the formal influences mentioned (or hinted at) in the finished text but not referenced in the bibliography. "Cited" is not the same as "used" or "influential," because many scientists contribute significantly to scientific progress while receiving little or no credit in whatever form, let alone bibliographic citations. Apart from the widely discussed cases of eponymic and obliterated antecedents, a typical instance of failed recognition occurs whenever techniques and instrumentation are used or mentioned in experimental papers without the minimal reference to who invented or improved them. Commenting on a joint science/technology validation study of bibliometric indicators carried out in the 1970s at the STW/FOM, Utrecht, Cornelius Le Pair referred to such unevenness as "the citation gap of applicable science."[37]
2. *What citations superficially or incorrectly say about influences.* Citation analysis overestimates the value of inclusive or review papers, often cited by scientists in lieu of the original sources of ideas and techniques. Scientists are not philologists; they often become aware of previous contributions through informal communication media or secondary literature, and both channels strongly affect their citing behavior. In the worst case, authors simply ignore previous pertinent and citable literature.
3. *What citations say about the citers' idiosyncrasies and motivations rather than about influences.* Citation practices are strongly biased by

> misuses, such as stealing other papers' references or padding reference lists with items that have not been read, and uses unrelated to the bricklaying integral to the cognitive mission of science, such as the adding of perfunctory or also-ran citations, and the rhetorical summoning of authoritative sources simply in view of their ability to catch the attention and respect of the audience. Furthermore, citation practices vary in accordance with too many variables to be of any practical use in science studies, including disciplinary habits, publication type, and nationality. A true test of the Ortega hypothesis, therefore, would require detailed sociological and historical work involving text analysis along with accurate reconstructions of the laboratory life.

It's clear that the contention here is not over whether the conclusions reached by citation analysts are wrong or correct in connection with specific case studies. It is the foundations of the building that the MacRobertses attack: citations cannot be trusted at all as proxies of intellectual influence, let alone scientific merits, their right of citizenship in the sociology of science being denied even before the border is crossed and any citation counting commences. Of course, any haunter of bibliometric literature will easily collect plenty of counter examples against such an attempt to raze the entire edifice. Still, the foregoing remarks deserve particular attention also from trained bibliometricians, as they point to the critical observation that "quality," "citation impact," "influence," and "contribution to scientific progress," are and ought to be maintained on conceptually different layers. An influential document—i.e., a document affecting the subsequent practice of science either positively or negatively—does not necessarily exert an impact at the formal level of citations: there could be plenty of reasons to avoid or fail citing it. Conversely, a high citation score doesn't necessarily tell the true story behind an influential paper: as demonstrated in the previous section, there could be plenty of reasons to cite it other than its actual merits (however those are defined). Even more, distasteful as it may sound, a truly influential document that is also heavily cited for some time does not necessarily bring about what would be easily and universally recognized as a significant advancement in the history of science. This point is well illustrated by the case of review or methodological papers as opposed to more creative, theoretical papers. In many areas, such as chemistry and medicine, methodological papers are cited on average much more than other kinds of contributions, holding sometimes the rank of citation stars, as happened to the most highly cited paper in the history of science: Oliver Lowry's 1951 method for measuring the amount of proteins in solutions. On the admission of Lowry himself, that paper contained "almost nothing original," and its citation score didn't signify to him any great scientific accomplishment.[38] Additional evidence on this point comes from the reasons to cite highly cited papers reviewed in subsection 7.1.1 as well as from an experiment on a cross-disciplinary sample of Dutch highly cited authors surveyed in 2000 by a CWTS team. Citation stars' responses confirmed that high citation scores do not automatically match the scientists' self-perception of the quality and relevance of highly cited papers' contributions to knowledge progress.[39] Of course, they might be underestimating the value of their own contributions, but if that is not the case, then it must be recognized that the only universe in which a methodological paper advances knowledge in proportion to its high citation score is exactly a universe governed by the Ortega hypothesis, where even the pedestrian affair of enhancing the sensitivity of a measurement technique adds another brick to the pile that, one day or another, will enable someone to climb up there and see farther than anyone else into the secrets of nature.

The Coles were certainly aware of this potential drawback in their elitist argument when they hastened to raise the height of the dwarfs by emphasizing that, if Kuhn's normal science is the usual habitat of the average researcher, "even the scientists who make these 'smaller' discoveries come principally from the top strata of the scientific community."[40] A greater indulgence toward the bulk of run-of-the-mill employees crowding the theater of science was displayed by Garfield, who acknowledged that, in spite of the apparently elitist behavior of the citation network disclosed by ISI *Citation Classics* commentaries, the same data "may support the perspective on science known as the Ortega hypothesis, which states that these giant papers are built on the contributions of the masses of less cited works. This can be visualized as a pyramid standing on its base."[41] Likewise, in Studer and Chubin's *The Cancer Mission*, the centrality of Lowry's paper on protein measurement to the advancement of reverse transcriptase research is argued on the ground that methodological papers, just like electronic microscopes and other techniques, are not simply practical tools or "means" to realize some research end, but "need to be interpreted in their substantive capacity as problems 'facilitators' which circumscribe and, at the same time, promote research in a domain."[42]

Based on the theses reviewed so far, it is arguable that quality is a relative concept, its appraisal depending on how one chooses to connect the opposites. If no lines of continuity are drawn between the giants and the dwarfs, so that the only true advancements are thought to be the relatively uncommon conceptual revolutions made possible by a small elite of top brains working in the top world universities and writing in the top international journals, then bibliometric indicators are basically as good as any other evaluative tool to detect excellence in the top strata, but dwarfs' performance levels are beyond the reach of their resolution; if, on the contrary, the two extremes are placed not too distant from one another, so that the advancement of science is deemed to be inconceivable without the painstaking job of many short or medium height researchers preparing the material and conceptual conditions of major discoveries, then bibliometric measures make sense also at low levels of productivity and performance, provided they are built with unbiased data and all the necessary methodological requirements set forth in chapter 6 are fulfilled.

At one extreme of the scale, the ideal situation, in which the citation impact of an author is a good approximation of actual influence and both citation impact and influence are good proxies of the author's positive role in the advancement of science, occurs when his or her contributions bring about a theoretical outbreak involving either the foundation of an entire new research area or a complete reorganization of existing conceptual framework. In both cases, the high citation rates that follow the reception of that contribution, while giving the idea that each subsequent paper in the same area is a building block in an edifice resting on fresh, solid foundations, are most likely the prelude of its authors' eponymic route to immortality.[43] At the opposite extreme, when we are faced with the myriad nonexplosive documents growing out of the daily work in scientific laboratories and currently listed in the individual curricula submitted for tenure, promotion, or granting, citation data alone are hardly in the position to offer a reliable picture of individual scientific value. Here the originality of a contribution, the peculiar set of motives that lead to its recognition as a valid building block of the edifice of science, are tested not so much against its potential in creating a new course of ideas (if so, probably very few scientists would get on), as against some general assumptions about how the daily business in science should be dealt with. In this somewhat narrower scenario, quality may be synonymous with accuracy in the application of techniques and methods to the analysis of empirical data, as in many clinical papers adding new evidence to the pros and cons of a drug or surgical technique. Otherwise, quality may also hint at something completely detached from empirical data, such as the elegance of the formalism in mathematical proofs and modeling. Or it may stand for the ability to further corroborate the explicative (and predictive) power of a firmly established theory. The occurrence and proper weight of these manifestations of quality in particular documents is detected at best by peer judgment. Bibliometric indicators, at this level, exhibit notable limitations induced by systematic as well as random sources of errors. The overestimation of methodological papers in certain areas is an example of the former type, while the occurrence of deviant citation behavior is an example of the latter. In reaction to the MacRobertses' paper, however, bibliometricians argued that, under certain conditions, these two sources of errors can be minimized, and quantitative measures can effectively support peer review in monitoring research performance. In short, they advocated two principles that would sustain the bibliometric research front in the years to come: improving the quality of data handling techniques and enhancing the quality of data themselves.

Regarding the quality of data handling techniques, more refined mathematical tools and a greater number of variables have to be taken into account than the Coles did in determining the eminence of scientists by bibliometric methods. As emphasized by Stephen Stigler and Harriet Zuckerman, the persistence of sources of variability or error in bibliographic citations doesn't involve the impossibility of accurately measuring the underlying communicative processes, if only plausible stochastic models are developed to estimate the relative weight of citation data in the presence of those sources of systematic and random error.[44] Regarding the quality of data, a basic requirement for any statistical analysis aimed at minimizing the effect of random behavior is to work with unbiased and sufficiently large samples.[45] If large collections of documents are analyzed instead of single items, such as the oeuvres of research groups, the publication output of institutes, universities, and nations, the network of citations assumes a thickness consistent with a more in-depth interpretation of its structure. That is exactly what the advanced bibliometric methodologies outlined in chapter 6 set out to do. But if citations are taken in isolation, and the focus is on individual documents and authors, many assumptions that can be relied upon at an aggregate level of analysis fail, the crucial one being the trust in the overall prevalence of a (Mertonian) normative use of citations in drafting the final reports of scientific discoveries. And if it is no longer possible to exploit the benefits of statistical reasoning, then the ambiguity of the individual, of what makes the individual unique and somewhat unclassifiable, comes to the surface and has to be dealt with in a different (sociological, psychological) conceptual framework.

7.3. LIFE AND DEATH OF THE SCIENTIFIC AUTHOR

Bibliometricians have long been conversant with the problem of measuring the formal properties of scientific collaboration. Their point of departure has been the basic index of collaborative activity traceable in published scientific literature: coauthorship of journal papers. Coauthors' names, institutions, and nationalities can be easily picked out in recent publications, where they are supposed to leave the footprints of the pre-publication activity leading to the discoveries being communicated. That is why the byline of coauthored scientific papers has become the treasure house of bibliometric information on the patterns of local, national, and international collaborative networks. Just like any other reductive assumption in the bibliometric style, however, this is not to be taken as gospel truth. In the real world, especially if the focus is on the network of interactions among members of the same institution or research group, collaboration is mostly a matter of social convention, its patterns being variable not only as time goes by but also, at any given time, across institutions and research fields, often in reaction to individual, social, and political variables of a contingent nature.[46] Nonetheless, the statistical analysis of coauthorship networks is one of the most fruitful lines of inquiry into the complex matter of collaborative research patterns at least since the 1970s, when, following in Derek Price's footsteps, Beaver and Rosen envisioned the first comprehensive bibliometric theory of scientific collaboration.[47]

During the last three decades, a growing interest of the bibliometric community in the cooperative organization of research and its bearings on science policy applications took hold and culminated, in 2000, in the creation of a global interdisciplinary research network, coordinated by Hildrun Kretschmer and virtually centered in Berlin, whose focus is explicitly placed on collaboration in science and technology: COLLNET.[48] Many studies, before and after the launching of COLLNET, have investigated the statistical patterns of coauthorship at the individual, cross-national, and multinational level. At each layer, the increase in collaboration, as manifested by the share of multiauthored papers, has been found to characterize all areas of science even in domains, like mathematics, traditionally devoted to an almost exclusive dialogue of the individual genius with the Absolute. The relation of this trend to other sensitive bibliometric targets, such as the stratification of scientific communities, publication productivity, and citation impact, have been also pinned down from a variety of perspectives.

Interestingly enough, coauthorship networks in invisible colleges of tightly defined specialties have been found to manifest the same behavioral patterns of real-life social networks in which "birds of a feather flock together."[49] The relative frequency of coauthorship, in other words, seems statistically higher between authors of the same invisible college that are "similar" with respect to the number of publications than between authors of incomparable productivity. By contrast, the networks of collaboration between authors belonging to the same institution display a marked tendency toward mingling authors of different bibliometric stature. Publication productivity, in any case, has been found to correlate rather ambiguously with coauthorship: there seem to be peaks of productivity around some thresholds of cooperation specific for each research area (for example, five to six coauthors in biomedicine), beyond which publication activity decreases.[50] Citation impact, on the other hand, has turned out at least partially correlated with the number of coauthors and, more specifically, with international cooperativeness: multiauthored papers seem to attract more citations than single-authored papers; international co-publications, in particular, receive on average a higher number of citations than domestic papers, even though notable variations do exist across disciplines and country pairs. It is not the case, however, that coauthorship can replace citation impact as a quality indicator, because the two indicators point to quite different dimensions of the communication process.[51]

If the register switches from the descriptive to the evaluative side of bibliometrics, then a whole array of new issues comes up. Now authorship has to be, in principle, clearly defined and identifiable before any statistical analysis, because the author is the point of departure and the final destination of every citation count for science policy purposes. Scientometricians tally publications and citations produced by "someone"—an individual, a group, an institution, a country—with the objective of establishing if that "someone" is comparatively more or less meritorious than "someone else," having contributed to a greater extent to the advancement of knowledge in a particular domain. The academic reward and reputational system, in fact, rests on the postulate that it is always possible to identify and assign the individual intellectual responsibility of a piece of scientific or technical work, a postulate perfectly in tune with both the classic (Mertonian) model of distributional justice in science and the liberal doctrine of copyright that, from the eighteenth century onward, has locked up individual creativity in the cage of private intellectual property rights.

At a certain point in Western intellectual history, however, the author's monarchy over literary artifacts began to collapse on various fronts. Beginning in the 1950s, a group of philosophers and literary critics set out to pull down any attempt to discover the subjective origin of literary texts. Briefly, they argued against the very possibility of tracing that origin back to the author's interiority. "A text," contended Roland Barthes in 1968, "does not consist of a line of words, releasing a single 'theological' meaning (the 'message' of the Author-God), but is a space of many dimensions, in which are wedded and contested various kinds of writings, no one of which is original: the text is a tissue of citations, resulting from the thousand sources of culture."[52] Just one year later, a similar distrust was echoed in Michel Foucault's thesis that the name of an author is nothing more than a variable useful to characterize the existence and circulation of certain discourses within a society, since "the writing of our days has freed itself from the necessity of 'expression'; it only refers to itself, yet it is not restricted to the confines of interiority."[53]

Notwithstanding that subsequent literary criticism would have questioned Barthes's and Foucault's statements, the author's identity crisis they alluded to was by no means incidental. During the same years, in a growing number of publications, especially within the realms of high-energy physics and biomedical sciences, the author, that "someone" addressed by scientometric exercises, enjoyed a new status. Far from being an individual indivisible unit responsible for the entire intellectual content of a publication, it appeared merely as a member of a class of coauthors that could number anywhere from a few units to hundreds or even thousands of individuals. By contrast, the romantic picture of the man of genius testing revolutionary ideas in a homemade laboratory, then shouting "Eureka" and disclosing the results for the benefit of both colleagues and humankind, betrayed all of its mythical character. This trend, as Derek Price insightfully argued in the 1960s, was closely linked to the emerging entrepreneurial dimension of scientific research. Since the post–World War II period, in fact, multiauthorship has become a distinctive feature of the "big science," the science fueled by government (mostly military) funding and massive research facilities that was going to plan expensive, path-breaking enterprises, such as moon landings, nuclear fission, and human genome mapping. Undertakings of this sort were conceivable only within a strictly collaborative organizational framework where the main project is subdivided into many subtasks, assigned to participating universities or research institutes, which in turn break them down into micro-objectives assigned to young, entry-level researchers or laboratory technicians and clinicians, whose understanding of the ultimate goal and theoretical concern of their work need not be complete at all.[54] Each certainly expects an amount of recognition at the day of reckoning, but his or her participation in activities such as experiment design, final report drafting, and critical revision is not necessary, nor can it be taken for granted in any evaluation exercise. Thus, even the act traditionally connoting the "aura" of intellectual paternity, that of writing, cannot recompose the lost unity as, at least in the above-mentioned disciplines, "to be an author is not necessarily to be a writer."[55]

The diluted intellectual responsibility implicit in author inflation questions the idea that authorship is per se a measure of intellectual achievement. At the same time, it hinders the use of both publications and citations for evaluative purposes insofar as it tears the respective points of reference asunder: the claim for intellectual property over a published idea in the former case (one might legitimately ask, who speaks?); the supposedly indivisible atoms of recognition awarded to the cited document in the latter (one might ask as well, who is awarded for what part of the work?). In addition, multiauthorship bears upon the very mathematical roots of evaluative bibliometrics, insofar as it introduces a pattern unfamiliar to other metrics: the situation in which one item (a paper) is linked to multiple sources (authors). Empirical evidence exists, in this regard, that a bibliography including multiauthored articles with fifty or more coauthors is not fitted by the original Lotka's distribution anymore, even though the misfit can be accommodated and adequately explained within more refined mathematical frameworks.[56]

In an ideal scenario, the algorithm for assigning credits to the coauthors of a paper in a particular field, should comprise the following steps:

1. The different types of activities required to produce the paper are adequately classified and weighted in advance with reference to a standard (as argued in chapter 6, this entails issues of classification and standardization faced at all levels of evaluative bibliometrics).
2. Each author's percentage contribution to the different activities underlining the production of the paper is estimated, for instance by means of questionnaires.
3. Each author's fractional credit is properly awarded, weighting number 3 by means of number 1.

In practice, while an approximation to such an algorithm is conceivable for small experiments confined to well-institutionalized and tightly defined research areas,[57] in most large-scale, cross-institutional evaluative exercises performed in real-science contexts, the relative contribution to multiauthored papers is assessed solely on the basis of publicly available data, such as the number and ranking of author names in the byline. So, in order to cope with the issue of counting multiple authorship for informetric purposes, different source crediting systems have been tested from time to time. None is the "true" system, but the choice of one scoring method over another has been found to dramatically affect the final outcome, thereby suggesting the simultaneous use and comparison of different methods. Let's briefly review them.[58]

1. *Full or total or standard count.* Each author receives full credit for the multiauthored paper, just as if the creative and organizational effort behind manuscript production were equally distributed and identi-

cally repeated by each head. Clearly, this solution inflates the score of scientists producing many multiauthored papers and makes hardly comparable papers with different numbers of authors.

2. *Straight count.* Full credit is given to just one author. It may be either the first named or the senior author. In Lotka's original solution, for example, the credit of 1 was assigned only to the senior author, while the remaining authors were not given any credit at all. Likewise, the Coles recommended disregarding all but the first author. It is well known, however, that being named as first in natural sciences papers does not necessarily imply a higher share of responsibility (in many papers authors are listed alphabetically). On the other hand, being the last may allude either to a prominent position (the project leader) or to an honorific mention for a ghost or guest authorship. On the methodological ground, as the sociologist Duncan Lindsey noted in 1980, handling multiauthored papers as if they were single-authored, thus giving full credit only to the first author, introduces a strong sampling bias in that it treats the papers on which a scientist's name occurs first as a representative sample of all of that scientist's papers.[59]
3. *Fractional or adjusted count.* Each author is awarded a $1/n$ fraction of the credit, where n is the number of authors (for example, in the case of four authors, each gets one-fourth). This solution, defended among others by Price and Lindsey, is conceptually similar to the standard one in that it treats all authors equably. At the same time, it regards credit as a single unit to be parceled out, thereby making the individual shares of credit dependent on the number of authors. Adjusted credit can also be defined according to a more complex rule that takes into account both the author name's rank in the byline and the number of authors, so that the first or senior author receives a higher percentage of credit than the others.[60]

Science managers and journal editors, though perfectly aware of the new status of the scientific author, simply ignore it, thereby avoiding the definition of authorship protocols more adequate to the new landscape. As a consequence, the "author" is alive and kicking in all the official communications channels, independent of what happens before and beyond the byline. "Authorship credit," state the International Committee of Medical Journal Editors (ICMJE) guidelines,

> should be based on 1) substantial contributions to conception and design, or acquisition of data, or analysis and interpretation of data; 2) drafting the article or revising it critically for important intellectual content; and 3) final approval of the version to be published. . . . When a large, multi-center group has conducted the work, the group should identify the individuals who accept direct responsibility for the manuscript. . . . All contributors who do not meet the criteria for authorship should be listed in an acknowledgments section.[61]

If these guidelines are to be valid, one cannot help but recognize their cinematographic style. As in a movie, which relegates to the closing credits the list of technicians who took part in crafting each individual scene without sharing in the overall design, a scientific article should end with a generic section of acknowledgments featuring the names of all the supposedly "minor" contributors to whom even the dignity of figuring in bibliographic indexes is usually denied. It is the group itself, or a few influential members of the group, to decide who succeeded in passing the authorship threshold. Thus being included is ultimately a social affair, the result of an act of co-optation. This is a pragmatic though by no means original solution, something quite similar to what occurred in seventeenth-century experimental philosophy, when laboratory technicians, artisans, engineers, and instrument makers, even if collaborating actively in the production and ratification of sound empirical knowledge, enjoyed no scientific credibility (and visibility) whatsoever, due to their low social status.

NOTES

1. Evidence Ltd., *The Use of Bibliometrics to Measure Research Quality in UK Higher Education Institutions* (Research Report) (2007), bookshop.universitiesuk.ac.uk/downloads/bibliometrics.pdf; van Raan, Moed, and van Leeuwen, *Scoping Study on the Use of Bibliometric Analysis to Measure the Quality of Research in UK Higher Education Institutions [Report to HEFCE]* (2007), www.hefce.ac.uk/pubs/rdreports/2007/rd18_07/rd18_07.pdf.

2. An early categorization of reasons to cite appeared in Garfield, "Can Citation Indexing Be Automated?" in *Essays of an Information Scientist 1962–1973*, 85 (Philadelphia: ISI Press, 1977), www.garfield.library.upenn.edu/essays/V1p084y1962-73.pdf (note that this is a reprint of a 1964 conference paper), while a later paper specifically devoted to the subject is Garfield, "When to

Cite?," *Library Quarterly* 66, no. 4 (1996): 449–58, www.garfield.library.upenn.edu/papers/libquart66(4)p449y1996.pdf. See also the classification proposed in Melvin Weinstock, "Citation Indexes," in *Encyclopaedia of Library and Information Science*, Vol. 5 (New York: Marcel Dekker, 1971), 19.

3. Two useful reviews of citation studies are Mengxiong Liu, "The Complexities of Citation Practice: A Review of Citation Studies," *Journal of Documentation* 49, no. 4 (1993): 370–408; and Lutz Bornmann and Hans-Dieter Daniel, "What Do Citation Counts Measure? A Review of Studies on Citing Behavior," *Journal of Documentation* 64, no. 1 (2008): 45–80, www.lutz-bornmann.de/icons/BornmannLutzCitingBehavior.pdf. The latter covers materials published up to mid-2005.

4. Vinkler, "A Quasi-Quantitative Citation Model," *Scientometrics* 12, nos. 1–2 (1987): 47–72.

5. Cronin, "Agreement and Divergence on Referencing Practice," *Journal of Information Science* 3, no. 1 (1981): 27–33. On the theoretical justification of this approach, see Cronin, "The Need for a Theory of Citing," *Journal of Documentation* 37, no. 1 (1981): 16–24. An experiment similar to Cronin's is described in Garfield, "Can Citation Indexing," 89.

6. White, "Authors as Citers Over Time," *Journal of the American Society for Information Science and Technology* 52, no. 2 (2001): 87–108.

7. Brooks, "Private Acts and Public Objects: An Investigation of Citer Motivations," *Journal of the American Society for Information Science* 36, no. 4 (1985): 223–29. A classic on the role of rhetoric in early modern scientific argumentation is Maurice A. Finocchiaro, *Galileo and the Art of Reasoning: Rhetorical Foundations of Logic and Scientific Method* (Dordrecht; Boston; London: Reidel, 1980).

8. The references corresponding to the cited studies are 1) Prabha, "Some Aspects of Citation Behavior: A Pilot Study in Business Administration," *Journal of the American Society for Information Science* 34, no. 3 (1983): 202–6; 2) Liu, "A Study of Citing Motivations of Chinese Scientists," *Journal of Information Science* 19, no. 1 (1993): 13–23; 3) Shadish et al., "Author Judgements About Works They Cite: Three Studies from Psychology Journals," *Social Studies of Science* 25, no. 3 (1995): 477–98; 4) Case and Higgins, "How Can We Investigate Citation Behavior? A Study of Reasons for Citing Literature in Communication," *Journal of the American Society for Information Science* 51, no. 7 (2000): 635–45; 5) White and Wang, "A Qualitative Study of Citing Behavior: Contributions, Criteria, and Metalevel Documentation Concern," *Library Quarterly* 67, no. 2 (1997): 122–54; 6) Tang and Safer, "Author-Rated Importance of Cited References in Biology and Psychology Publications," *Journal of Documentation* 64, no. 2 (2008): 246–72.

9. Leydesdorff and Amsterdamska, "Dimensions of Citation Analysis," *Science, Technology & Human Values* 15, no. 3 (1990): 322–23.

10. Moravcsik and Murugesan, "Some Results on the Quality and Function of Citations," *Social Studies of Science* 5, no. 1 (1975): 86–92. The classification was subsequently put to test in Moravcsik and Murugesan, "Citation Patterns in Scientific Revolutions," *Scientometrics* 1, no. 2 (1979): 161–69, and V. Cano, "Citation Behavior: Classification, Utility, and Location," *Journal of the American Society for Information Science* 40, no. 4 (1989): 284–90.

11. Chubin and Moitra, "Content Analysis of References: Adjunct or Alternative to Citation Counting?" *Social Studies of Science* 5, no. 4 (1975): 433.

12. Cozzens, "Comparing the Sciences: Citation Context Analysis of Papers from Neuropharmacology and the Sociology of Science," *Social Studies of Science* 15, no. 1 (1985): 127–53; Oppenheim and Renn, "Highly Cited Old Papers and the Reasons Why They Continue to Be Cited," *Journal of the American Society for Information Science* 29, no. 5 (1978): 225–31; Tanzila Ahmed et al., "Highly Cited Old Papers and the Reasons Why They Continue to Be Cited, Part II: The 1953 Watson and Crick Article on the Structure of DNA," *Scientometrics* 61, no. 2 (2004): 147–56.

13. Kaplan, "The Norms of Citation Behavior: Prolegomena to the Footnote," *American Documentation* 16, no. 3 (1965): 181.

14. Ravetz, *Scientific Knowledge and Its Social Problems* (Oxford: Clarendon Press, 1971), 257. See also, for a similar perspective, G. Nigel Gilbert, "Referencing as Persuasion," *Social Studies of Science* 7, no. 1 (1977): 113–22.

15. Bourdieu, *Homo Academicus* (Stanford, Calif.: Stanford University Press, 1988), preface to the English edition, xi.

16. Gilbert and Mulkay, *Opening Pandora's Box: A Sociological Analysis of Scientists' Discourse* (Cambridge: Cambridge University Press, 1984), 176–77. See also Karin D. Knorr-Cetina and Michael Mulkay, eds., *Science Observed: Perspectives in the Social Study of Science* (London: Sage, 1983).

17. Garvey, *Communication: The Essence of Science* (Oxford: Pergamon Press, 1979), 78–79.

18. Meadows, "The Scientific Paper as an Archaeological Artefact," *Journal of Information Science* 11, no. 1 (1985): 30.

19. Medawar, "Is the Scientific Paper Fraudulent? Yes; It Misrepresents Scientific Thought," *Saturday Review*, 1 (August 1964): 43. The paper is a slightly modified version of a talk on BBC delivered by the author in September 1963 (transcript available at contanatura-hemeroteca.weblog.com.pt/arquivo/medawar_paper_fraud.pdf).

20. Swales, *Genre Analysis: English in Academic and Research Settings* (Cambridge: Cambridge University Press, 1990).

21. A series of contributions on the hidden empiricist epistemology of style manuals come from psychologists: Derek Gregory, "Editorial," *Environment and Planning. D: Society and Space* 8, no. 1 (1990): 1–6; Robert Madigan, Susan Johnson, and Patricia Linton, "The Language of Psychology: APA Style as Epistemology," *American Psychologist* 50, no. 6 (1995): 428–36; James Hartley, "What Do We Know About Footnotes? Opinions and Data," *Journal of Information Science* 25, no. 3 (1999): 205–12. For a similar perspective in the field of geography, see Michael R. Curry, "On the Possibility of Ethics in Geography: Writing, Citing and the Construction of Intellectual Property," *Progress in Human Geography* 15, no. 2 (1991): 125–47.

22. Latour, *Science in Action: How to Follow Scientists and Engineers Through Society* (Cambridge, Mass.: Harvard University Press, 1987), 33. Latour's theory of citation is reappraised in Terttu Luukkonen, "Why Has Latour's Theory of Citations Been Ignored by the Bibliometric Community? Discussion of Sociological Interpretations of Citation Analysis," *Scientometrics* 38, no. 1 (1997): 27–37.

23. Cronin, *The Citation Process: The Role and Significance of Citations in Scientific Communications* (London: Taylor Graham, 1984), 83, garfield.library .upenn.edu/cronin/citationprocess.pdf.

24. On the multidimensionality of citations from a theoretical perspective, see Cozzens, "What Do Citations Count? The Rhetoric-First Model," *Scientometrics* 15, nos. 5–6 (1989): 437–47, and Leydesdorff, "Theories of Citation?," *Scientometrics* 43, no. 1 (1998): 5–25, and the recent review by Jeppe Nicolaisen, "Citation Analysis," *Annual Review of Information Science and Technology* 41 (2007): 609-41. An attempt to tease out the concrete (joint) operation of both the reward and rhetoric components of citation practices in a small sample of research articles by two groups of chaos theorists is reported in Danette Paul, "In Citing Chaos: A Study of the Rhetorical Use of Citations," *Journal of Business and Technical Communication* 14, no. 2 (2000): 185–222. Small has recently built on citation context analysis to develop a qualitative framework purporting to account for all the possible combinations of two elementary dimensions of the citation process, namely "literalness" versus "consensus." The former measures the convergence between cited work and citing context; the second gauges the degree of consensus on the content of the cited work among the citers. At one extremity, citation patterns driven by high literalness and high consensus are best modeled by Mertonian normativism, while, at the opposite extremity, the low consensus and low literalness that often characterize negative and paradigm-breaking citation behaviors are best captured by constructivist theories. See Small, "On the Shoulders of Robert Merton: Towards a Normative Theory of Citation," *Scientometrics* 60, no. 1 (2004): 76–78.

25. Wouters, "The Citation Culture" (Ph.D. thesis, University of Amsterdam, 1999), www.garfield.library.upenn.edu/wouters/wouters.pdf.

26. van Raan, "In Matters of Quantitative Studies of Science the Fault of Theorists Is Offering Too Little and Asking Too Much," *Scientometrics* 43, no. 1 (1998): 136.

27. Glänzel and Schubert, "Predictive Aspects of a Stochastic Model for Citation Processes," *Information Processing & Management* 31, no. 1 (1995): 69–80; Burrell, "Stochastic Modelling of the First-Citation Distribution," *Scientometrics* 52, no. 1 (2001): 3–12.

28 Stewart, "Achievement and Ascriptive Processes in the Recognition of Scientific Articles," *Social Forces* 62, no. 1 (1983): 166–89.

29. Baldi, "Normative Versus Social Constructivist Processes in the Allocation of Citations: A Network Analytic Model," *American Sociological Review* 63, no. 6 (1998): 843.

30. Such a combination of citation and social network analysis in the study of academic networking, formerly advocated by Evelien Otte and Ronald Rousseau, "Social Network Analysis: A Powerful Strategy, Also for the Information Sciences," *Journal of Information Science* 28, no. 6 (2002): 441–53, is applied to the citation network of three information scientists in Ben Johnson and Charles Oppenheim, "How Socially Connected Are Citers to Those That They Cite?" *Journal of Documentation* 63, no. 5 (2007): 609–37.

31. References for a limited number of studies exemplifying the above-mentioned results on the predictors of citation rates are given here cumulatively in the same order as they are listed in the text:

a. Peters and van Raan, "On Determinants of Citation Scores: A Case Study in Chemical Engineering," *Journal of the American Society for Information Science* 45, no. 1 (1994): 39–49.
b. Glenn D. Walters, "Predicting Subsequent Citations to Articles Published in Twelve Crime-Psychology Journals: Author Impact Versus Journal Impact," *Scientometrics* 69, no. 3 (2006): 499–510.
c. Nick Haslam et al., "What Makes an Article Influential? Predicting Impact in Social and Personality Psychology," *Scientometrics* 76, no. 1 (2008): 169–85.
d. Michael J. Lovaglia, "Predicting Citations to Journal Articles: The Ideal Number of References," *American Sociologist* 22, no. 1 (1991): 49–64.
e. Michael Callaham, Robert L. Wears, and Ellen Weber, "Journal Prestige, Publication Bias, and Other Characteristics Associated with Citation of Published Studies in Peer-Reviewed Journals," *JAMA* 287, no. 21 (2002): 2847–50, jama.ama-assn.org/cgi/content/full/287/21/2847; Timothy A. Judge et al., "What Causes a Management Article to Be Cited: Article, Author, or Jour-

nal?" *Academy of Management Journal* 50, no. 3 (2007): 491–506, journals .aomonline.org/amj/editorials/Judge.etal.Citations.2007.pdf.

f. Stefan Stremersch, Isabel Verniers, and Peter C. Verhoef, "The Quest for Citations: Drivers of Article Impact," *Journal of Marketing* 71, no. 3 (2007): 171–93.

g. William E. Snizek, Kay Oehler, and Nicholas C. Mullins, "Textual and Non-textual Characteristics of Scientific Papers: Neglected Science Indicators," *Scientometrics* 20, no. 1 (1991): 25–35.

h. Cynthia Lokker et al., "Prediction of Citation Counts for Clinical Articles at Two Years Using Data Available within Three Weeks of Publication: Retrospective Cohort Study," *BMJ* 336, no. 7645 (2008): 655–57, bmj.com/cgi/content/full/336/7645/655.

i. Hendrik P. van Dalen and Kène Henkens, "What Makes a Scientific Article Influential? The Case of Demographers," *Scientometrics* 50, no. 3 (2001): 455–82, ssrn.com/abstract=250885.

j. Roosa Leimu and Julia Koricheva, "What Determines the Citation Frequency of Ecological Papers?" *Trends in Ecology and Evolution* 20, no. 1 (2005): 28–32.

k. Pentti Nieminen et al., "Statistically Significant Papers in Psychiatry Were Cited More Often Than Others," *Journal of Clinical Epidemiology* 60, no. 9 (2007): 939–46.

l. Bluma C. Peritz, "Predictors of Citation Impact: An Exploratory Study," *Scripta Hierosolymitana* 29 (1989): 205–16.

m. Bou-Wen Lin, Chung-Jen Chen, and Hsueh-Liang Wu, "Predicting Citations to Biotechnology Patents Based on the Information from the Patent Documents," *International Journal of Technology Management* 40, nos. 1–3 (2007): 87–100.

32. Quoted in Small, "On the Shoulders of Robert Merton," 72.

33. Cole and Cole, "The Ortega Hypothesis," *Science* 178, no. 4059 (1972): 368–75. Ortega y Gasset was a truly elitist philosopher who held that societies advance when the creative minority is allowed to govern. His view of the role played by twentieth-century scientists, however, was more complex than that filtered in the scientometric and sociological debate via the "Ortega hypothesis." On the one hand, he regarded a specialist who knows very well only his own tiny corner of the universe while being radically ignorant of all the rest as the prototype of the mass-men; at the same time, he praised the intellectual breadth of "the man Einstein" for being "capable of greater exactitude and liberty of spirit than the man Newton." Gasset, *The Revolt of the Masses* (New York & London: W.W. Norton, 1999), 42, www.19.5degs.com/ebook/revolt-of-the-masses/1419/

read (note that the original Spanish was published in 1930). On the alleged misinterpretations of Ortega's theory and the sequence of errors and miscitations triggered by the Coles' article, see Heidi Lee Hoerman and Carole Elizabeth Nowicke, "Secondary and Tertiary Citing: A Study of Referencing Behavior in the Literature of Citation Analysis Deriving from the Ortega Hypothesis of Cole and Cole," *The Library Quarterly* 65, no. 4 (1995): 415–34, and Endre Száva-Kováts, "The False 'Ortega Hypothesis': A Literature Science Study," *Journal of Information Science* 30, no. 6 (2004): 496–508. Yet such a misinterpretation does not amount to a "philological crime" as maintained by Száva-Kováts; the Coles and many others who referred to the "Ortega hypothesis" often without even reading Ortega's book were not interested in the philological reconstruction of the philosopher's claims, but in using an eponymic shortcut for the theory of the social stratification of science they were coping with. That's all that matters from the point of view of the history of bibliometrics.

34. Gary S. Green, "A Test of the Ortega Hypothesis in Criminology," *Criminology* 19, no. 1 (1981): 45–52; Mark Oromaner, "The Ortega Hypothesis and Influential Articles in American Sociology," *Scientometrics* 7, nos. 1–2 (1985): 3–10; William E. Snizek, "A Re-Examination of the Ortega Hypothesis: The Dutch Case," *Scientometrics* 9, no. 1 (1986): 3–11.

35. MacRoberts and MacRoberts, "Testing the Ortega Hypothesis: Facts and Artifacts," *Scientometrics* 12, no. 5 (1987): 293–95.

36. Edge, "Quantitative Measures of Communication in Science: A Critical Review," *History of Science* 17 (1979): 102–34; Edge, "Why I Am Not a Cocitationist," *Society for the Social Studies of Science Newsletter* 2 (1977): 13–19, www.garfield.library.upenn.edu/essays/v3p240y1977–78.pdf.

37. Le Pair, "The Citation Gap of Applicable Science," in *Handbook of Quantitative Studies of Science and Technology*, ed. Anthony F.J. van Raan, 537–53 (Amsterdam: Elsevier, 1988).

38. Lowry's statements are reported in Garfield, "Citation Frequency as a Measure of Research Activity and Performance," in *Essays of an Information Scientist 1962–1973*, 406 (Philadelphia: ISI Press, 1977), www.garfield.library.upenn.edu/essays/V1p406y1962-73.pdf.

39. Tijssen, Visser, and van Leeuwen, "Benchmarking International Scientific Excellence: Are Highly Cited Research Papers an Appropriate Frame of Reference?" *Scientometrics* 54, no. 3 (2002): 381–97.

40. Cole and Cole, "The Ortega Hypothesis," 369.

41. Garfield, "The Articles Most Cited in the SCI, 1961–1982. 9. More Contemporary Classics of Science," in *Essays of an Information Scientist 1986*, 59 (Philadelphia: ISI Press, 1988), www.garfield.library.upenn.edu/essays/v9p055y1986.pdf.

42. Studer and Chubin, *The Cancer Mission*, 201.

43. Garfield, "What's in a Name? The Eponymic Route to Immortality," in *Essays of an Information Scientist 1983*, 384–95 (Philadelphia: ISI Press, 1984), www.garfield.library.upenn.edu/essays/v6p384y1983.pdf.

44. Stigler, "Precise Measurement in the Face of Error: A Comment on MacRoberts and MacRoberts," *Social Studies of Science* 17, no. 2 (1987): 332–34; Zuckerman, "Citation Analysis and the Complex Problem of Intellectual Influence," *Scientometrics* 12, no. 5 (1987): 329–38.

45. Nederhof and van Raan, "Citation Theory and the Ortega Hypothesis," *Scientometrics* 12, no. 5–6 (1987): 325–28.

46. Katz and Martin, "What Is Research Collaboration?" *Research Policy* 26, no. 1 (1997): 1–18, www.sussex.ac.uk/Users/sylvank/pubs/Res_col9.pdf. In the same vein are the conclusions reached, on the basis of interviews with scientists, by Grit Laudel, "What Do We Measure by Co-Authorships?" *Research Evaluation* 11, no. 1 (2002): 3–15, www.ingentaconnect.com/content/beech/rev/2002/00000011/00000001/art00001.

47. For a quick review of Beaver and Rosen's 1978–1979 studies and their meaning in view of subsequent developments in the structure of collaboration networks, see the retrospective account in Beaver, "Reflections on Scientific Collaboration (and Its Study): Past, Present, and Future," *Scientometrics* 52, no. 3 (2001): 365–77. An early review of studies on research collaboration is K. Subramanyam, "Bibliometric Studies of Research Collaboration: A Review," *Journal of Information Science* 6, no. 1 (1983): 33–38.

48. An update on COLLNET's activity and planned meetings can be found at www.collnet.de/. The first meeting was held in 2000 on the occasion of the Second Berlin Workshop on Scientometrics and Informetrics. Its proceedings are available online: Frank Havemann, Roland Wagner-Döbler, and Hildrun Kretschmer, eds., *Collaboration in Science and in Technology: Proceedings of the Second Berlin Workshop on Scientometrics and Informetrics, September 1–3, 2000* (Berlin: Gesellschaft für Wissenschaftsforschung, 2001), www.collnet.de/workshop/Berlin2000.pdf.

49. The above-reported findings on the stratification of coauthorship networks derive from Kretschmer, "Coauthorship Networks of Invisible Colleges and Institutionalized Communities," *Scientometrics* 30, no. 1 (1994): 363–69, and Kretschmer, "Patterns of Behaviour in Coauthorship Networks of Invisible Colleges," *Scientometrics* 40, no. 3 (1997): 579–91.

50. Glänzel, "Coauthorship Patterns and Trends in the Sciences: A Bibliometric Study with Implications for Database Indexing and Search Strategies—1980–1998," *Library Trends* 50, no. 3 (2002): 461–73, findarticles.com/p/articles/mi_m1387/is_3_50/ai_88582625. A similar ambiguity has been reported

for authors in the neurosciences by Braun, Glänzel, and Schubert, "Publication and Cooperation Patterns of the Authors of Neuroscience Journals," *Scientometrics* 51, no. 3 (2001): 499–510. For an overview of the development and application of indicators based on coauthorship, see Glänzel and Schubert, "Analysing Scientific Networks Through Co-Authorships," in *Handbook of Quantitative Science and Technology Research: The Use of Publication and Patent Statistics in Studies of S&T Systems*, ed. Henk F. Moed, Wolfgang Glänzel, and Ulrich Schmoch, 257–76 (Dordrecht: Kluwer Academic, 2004).

51. On the positive correlation between coauthorship and high citation rates see, among others, John C. Smart and Alan E. Bayer, "Author Collaboration and Impact: A Note on Citation Rates of Single and Multiple Authored Articles," *Scientometrics* 10, no. 5 (1986): 297–305. The beneficial effects of international cooperation on citation rates are extensively documented in Narin, Stevens, and Whitlow, "Scientific Co-Operation in Europe and the Citation of Multinationally Authored Papers," *Scientometrics* 21, no. 3 (1990): 313–23, and Glänzel, "National Characteristics in International Scientific Co-Authorship Relations," *Scientometrics* 51, no. 1 (2001): 69–115. The limitations inherent in the statistical correlation between coauthorship and quality are discussed, among others, in Martin Bridgstock, "The Quality of Single and Multiple Authored Papers: An Unresolved Problem," *Scientometrics* 21, no. 1 (1991): 37–48, and more recently in Ulrich Schmoch and Torben Schubert, "Are International Co-Publications an Indicator for Quality of Scientific Research?" *Scientometrics* 74, no. 3 (2008): 361–77.

52. Barthes, "The Death of the Author," in *Image, Music, Text*, 146 (London: Fontana Press, 1977).

53. Foucault, "What Is an Author?" in *Language, Counter-Memory, Practice: Selected Essays and Interviews by Michel Foucault*, ed. Donald F. Bouchard, 116 (Ithaca, New York: Cornell University Press, 1979). The essay had originally appeared in the *Bulletin de la Societé Française de Philosophie* 63, no. 3 (1969): 73–104.

54. The epistemological debate surrounding the appearance of massive multiauthorship in modern physics and biomedicine is reviewed in Mario Biagioli, "The Instability of Authorship: Credibility and Responsibility in Contemporary Biomedicine," *The FASEB Journal* 12, no. 1 (1998): 3–16, www.fasebj.org/cgi/content/full/12/1/3.

55. Cronin, "Hyperauthorship: A Postmodern Perversion or Evidence of a Structural Shift in Scholarly Communication Practices?" *Journal of the American Society for Information Science and Technology* 52, no. 7 (2001): 567. For a more detailed insight into the author's perspective on the relationships between the reward system of science and scholarly communication (and publication) practices

across different "epistemic cultures," see the recent Cronin, *The Hand of Science. Academic Writing and Its Rewards* (Lanham, Md.: The Scarecrow Press, 2005).

56. Rousseau, "Breakdown of the Robustness Property of Lotka's Law: The Case of Adjusted Counts for Multiauthorship Attribution," *Journal of the American Society for Information Science* 43, no. 10 (1992): 645–47; Kretschmer and Rousseau, "Author Inflation Leads to a Breakdown of Lotka's Law," *Journal of the American Society for Information Science and Technology* 52, no. 8 (2001): 610–14. The mathematical issues inherent in multiple authorship have been tackled, for example, in the framework of Lotkaian informetrics, by Egghe, *Power Laws in the Information Production Process: Lotkaian Informetrics* (Amsterdam: Elsevier Academic Press, 2005), chapter 6.

57. See, for example, Vinkler, "Research Contribution, Authorship and Team Cooperativeness," *Scientometrics* 26, no. 1 (1993): 213–30.

58. The implications of different accrediting methods are reviewed in Egghe, Rousseau, and Van Hooydonk, "Methods for Accrediting Publications to Authors or Countries: Consequence for Evaluation Studies," *Journal of the American Society for Information Science* 51, no. 2 (2000): 145–57, hdl.handle.net/1942/792. For an early review on multiple authorship from a library and information science perspective, see Martha A. Harsanyi, "Multiple Authors, Multiple Problems: Bibliometrics and the Study of Scholarly Collaboration; A Literature Review," *Library & Information Science Research* 15, no. 4 (1993): 325–54.

59. Lindsey, "Production and Citation Measures in the Sociology of Science: The Problem of Multiple Authorship," *Social Studies of Science* 10, no. 2 (1980): 145–62.

60. See, for example, the proportional counting system propounded by Van Hooydonk, "Fractional Counting of Multiauthored Publications: Consequences for the Impact of Authors," *Journal of the American Society for Information Science* 48, no. 10 (1997): 944–45.

61. International Committee of Medical Journal Editors, *Uniform Requirements for Manuscripts Submitted to Biomedical Journals: Writing and Editing for Biomedical Publication* (2007), www.icmje.org/.

Chapter Eight

Measuring Scientific Communication in the Twentieth Century: From Bibliometrics to Cybermetrics

The onset of the World Wide Web in the mid-1990s made Garfield's citationist dream more likely to come true. In the world network of hypertexts, not only is the bibliographic reference one of the possible forms taken by a hyperlink inside the electronic version of a scientific article, but the Web itself also exhibits a citation structure, links between web pages being formally similar to bibliographic citations ("sitations").[1] The dual nature of the hypertext citation, which on the one hand consolidates and enriches with new strongholds its traditional location in the bibliographic section of academic papers, and on the other is freed from all discursive and stylistic constrictions to involve the organizational unit of the entire web space, accounts for the two avenues covered by citation analysis in the new milieu: first, the design of automatic citation indexing systems aimed at capturing, storing, and making readily searchable the bibliographic network woven with citations from papers stored in e-journals and open archives of scholarly material; and second, the application of statistical analysis to the hyperlink structure of the Web itself.

The first avenue exploits the full potential of hypertext technologies to breathe fresh life into the classic investigation of communication patterns embedded in scholarly papers and bibliographic citations. Nowadays hypertext is a dominant information retrieval technique, and moving seamlessly from a bibliographic reference to the full text of the cited work is a widespread facility among the electronic journal services of commercial publishers and database aggregators. The standards for the definition of markup languages, such as SGML and above all XML, have enormously favored this trend since, by virtue of such codifications, the information units composing a text, specifically the items listed in a reference list, can be marked and made recognizable by a label that facilitates their automatic connection with the full text of the cited document as much as their localization by an ad hoc crawler's algorithm for metadata harvesting. Encouraged by the amazing flexibility of the new environment, several new tools for trawling the citation network of scholarly literature have been introduced with either a disciplinary or multidisciplinary vocation. At the disciplinary level, outstanding databases supplying practitioners of the respective specialties with autonomous citation crunching machines are, among others, those provided by the *Chemical Abstract Service* (*CAS*) for chemistry and related sciences; the *SAO/NASA Astrophysics Data System (ADS)* for astronomy; the *SPIRES HEP* database for high-energy physics; American Mathematical Society's *MathSciNet* for mathematics; *CiteSeer* and *IEEE Xplore* for computer science, electrical engineering. and electronics; *Citebase* for subsets of physical and life sciences; and the *Citations in Economics* service (processing data from the *RePec Digital Library*) for economics. In the multidisciplinary arena, instead, the two large-scale emerging competitors to the *Web of Science* are *Google Scholar*, sponsored by the world's most famous search engine, and *Scopus*, maintained by Elsevier on a commercial basis. Both data sources are currently being examined and benchmarked in view of their possible use, as alternatives or complements to ISI databases, for the construction of scientometric indicators capable of correcting the limitations in coverage of the *SCI* and sister indexes. Their recent history, however, does not permit any conclusive appraisal for the time being.

In the second direction, alongside the land populated with objects familiar to bibliometricians of the old school, the medium itself, that is the Web-based communication network for the online dissemination and retrieval of scholarly literature, is setting up the bibliometric research agenda. Web search engines are expected to accomplish now a task comparable to the one accomplished (also) by citation indexes in the 1960s: to counteract the information overdose and assist scientists in the daily struggle to pick the most relevant information out of a huge quantity of irrelevant stuff. The design of effective search engine crawling and indexing algorithms profits by two main sources of information: text words and hyperlinks. The exploitation of word statistics has a long history rooted in similarity metrics developed within the classic tradition of text information retrieval: the more two pages are similar in terms of their lexical content, the more the vectors representing them in the vector-space model are close to each other. In the case of hyperlinks, instead, the chief underlying assumption stems directly from Garfield's use of citations as building blocks of an information retrieval system. Just as in scholarly literature a cited document is more likely related to the content of the citing one than any noncited document, a linked web page is more likely related to the content of the linking one than any nonlinked page. Therefore, the relevance of a page to a user query can be estimated by looking at the link rates and topology of the other pages pointing to it: the more a page gets linked, the more it is potentially useful; similarly, the fewer the links that separate a page from another deemed relevant by the user, the greater the probability that the first will turn out to be relevant too. The best known application of this scheme is Sergey Brin and Lawrence Page's *PageRank,* Google's ranking algorithm. Instead of simply counting link rates, *PageRank* assigns different "prestige" scores to individual pages according to their position in the overall network: more weight is assigned to the pages receiving more links and, among them, to those linked by the most frequently linked pages, just as if—bibliometrically speaking—top documents were automatically singled out in a citation index being more heavily cited by other top documents.[2] Before *PageRank* and partly inspiring it, IBM's *Clever Project*, an extension of Jon Kleinberg's *HITS* (*Hypertext-Induced Topic Search*), employed link analysis to quickly and efficiently locate the most relevant web pages on a given subject. Starting from the Mertonian assumption that hyperlinks, like bibliographic citations, confer some sort of authority to the linked document, the *Clever Project* staff devised an algorithm for the automatic capture of Web "authorities" and "hubs." An "authority" is a page that receives many links from quality "hubs" (the cybertech equivalent of a citation classic); a quality "hub" is a page providing many links to "authorities" (the rough equivalent of a good review paper). Not accidentally, in spotting the forerunners to their algorithm, the authors mentioned, next to Garfield's citation indexes, Pinski and Narin's 1976 work on the computation of influence weights for scientific journals.[3]

Along the two avenues outlined above, the digital life of the bibliographic citation in the new universe can be paralleled, almost point-by-point, by the steps marking the coming-of-age of scientometrics outlined in the foregoing chapters:

1. design and development of the tool of the trade (there, the *SCI*; here, search engines and in-house software for citation and hyperlink mining)
2. construction and testing of citation-based metrics (there, the impact factor of journals and similar ISI-dependent measures; here, the Web impact factor and the new usage and impact metrics made possible by digital repositories)
3. exploration of the cognitive and social structure of the network (there, co-citation analysis; here, hyperlink network analysis)
4. advanced mathematical modeling of the network structure and dynamics (there, the mathematical offspring of classic bibliometric laws; here, complex network analysis)

8.1. CITATIONS IN E-JOURNALS AND OPEN ARCHIVES

The discovery that most, if not all, of the information relevant to a scholarly field is simply a few clicks away from one's workstation, just in time for enriching the live research and discovery process with new, potentially decisive stimuli, is perhaps the most practical revolution that has occurred in the working style of scientists since the time of Galileo and Newton. Over the past decade, profiting considerably by the dramatic upheaval in the modes of circulation of scientific literature triggered by the Internet and the World Wide Web, major international publishers and database vendors have uploaded in their servers a huge mass of scholarly material in electronic form. Most of this stuff is actually no more than the digital counterpart of what previously had existed in the paper world: peer-reviewed scientific journals accessible on a subscription-toll basis. Online availability has been quite a godsend to journal papers and to all those (formerly) print-only materials dwelling in the lethargic limbo from where only serendipitous local catalog queries could have shaken them out. Journals jumping into cyberspace in full dress significantly augmented the chance to increase citation impact, and their local online usage became one of the best predictors of future citations.[4] On the other side, because of the structural shift in the way online articles versus print-only ones are delivered to the reader—so easy to be singled out through user-friendly web interfaces and so deeply immersed in a hyperlinked environment that prearranges the whole set of possibilities for extending search and retrieval operations beyond the initial focus—it is no coincidence that some recent experiments point to an apparently paradoxical situation: as more and more journal backfiles are uploaded on publishers' servers—at least according to an extensive citation analysis of ISI articles recently performed by James Evans—surprisingly fewer distinct articles from those journals are cited more, and citations tend to concentrate on more recent publications.[5] Aside from obvious "internal" explanations in terms of Price's immediacy effect and Kuhn's sociocognitive consensus around paradigmatic (disciplinary) cores, this narrowing effect tells something about the new medium's potential in building prefabricated roads that enable the road builders to drive users' attention toward specific spots.

Almost simultaneously with the digital burgeoning of scholarly stuff, by virtue of its ability to make timely and universally accessible any type of digital content, the Web started to fill up with a new class of inhabitants, previously confined to the semiprivate circles of scholars sharing the same invisible college: all the freely available preprint or postprint versions of scientific articles posted directly by the authors to personal or institutional websites and disciplinary repositories. These materials for the most part naturally evolve into (or result from) standard journal papers, thereby finding their way to existing abstracting and indexing services. Frequently, however, they live a fruitful life on their own, just waiting for someone to become aware of their nonjournal existence and use them before or after any official investiture by a traditional gatekeeper.

To the extent that a scholarly work, whatever its material composition (print or digital), publication status (preprint, full article, or postprint), or access restrictions (open or toll-access), is cited by ISI-covered journals, its citation life doesn't escape the close-mesh network of traditional citation indexes. Consequently, a reasonable estimate of its impact is secured by long-established metrics. But if one admits that, in spite of Bradford's and Garfield's laws, this is not enough to capture the full effects of its exposition to the enlarged community of potential readers (and citers) set up by the Web, then the issue of finding a reliable system to quantify the Web-wide cognitive and social life of scientific literature must be tackled on a different ground.

In a digitally interconnected environment, the impact of a set of documents outside the ISI circuit can be estimated using one of the following methods: 1) counting, by means of usage mining techniques, the number of document views or downloads over a certain period of time; 2) interviewing or submitting a self-report questionnaire to a significant sample of readers, for instance researchers of a university department; 3) counting, by means of search engines' facilities, the number of links to the website hosting the documents; or 4) identifying and counting, as ISI indexes do, the bibliographic citations to those documents from non-ISI sources.

Solution 1 exploits the ability of modern computers to store and process huge amounts of rough data on the network traffic. As long as the full-text download of a document is considered a valid proxy of its actual usage, web-mining techniques disclose a potential treasure house of information on the impact of any type of digital resource in a stage prior to its joining the citation game. The monitoring of Internet sessions log files generated by the web servers of digital libraries and web information services is, from this point of view, the direct heir to the long-established tradition of usage studies based on circulation and reshelving of physical libraries' materials. Standards and protocols have been developed, in the context of national and international projects such as COUNTER (Counting Online Usage of Networked Electronic Resources), SUSHI (Standardized Usage Harvesting Initiative), and MESUR (MEtrics from Scholarly Usage of Resources), to make uniform the recording and reporting of online usage statistics; to facilitate their automatic harvesting from different vendors or content providers; and to define, validate, and cross-validate journal usage factor metrics, which mimic and will hopefully complement the ISI IF as measures of journal importance. Usage statistics, on the other hand, are even less reliable and unobtrusive data than formal citations because their provision is influenced by a variety of external factors, including publishers' electronic interfaces and the risk for spurious inflation by manual or automatic methods. Taken as indicators of actual use, moreover, what they purport to indicate is hardly generalizable beyond the sample community for which usage data have been originally recorded.[6]

Solution 2 is well-established in the field of social sciences. Unlike the previous solution, it profits by explicit qualitative judgments of expert users, but relies on a narrower empirical basis and entails additional biases because of the preliminary job of manually defining, collecting, and classifying survey data. Both methodologies have advantages, but neglect one important aspect, namely that the hyperlink/citation analogy points unambiguously to a completely different solution of the quality assessment problem; a solution where, in accordance with Price and Garfield's project, it is not the judgment formulated by an individual or the individual usage session, but rather the "spontaneous" network of links among documents that accounts for the bibliometric value of the specific item under examination (document, author, institution). So, if the Web can be roughly assimilated to a big, anarchic library of hyperlinked documents, then the extraction of quantitative information from hyperlinks is expected to provide a unique insight into their relative value and place on the digital shelves.

Solution 3, counting links to the websites of scholarly journals, is quite in the spirit of Price's network model, but the equivalence it suggests between link rates and quality of the hosted journals is of limited value, inasmuch as it strikes short of the target: it is the links to the container that get counted here, not directly the links to the scholarly information units. Applied to the websites of journals indexed by the ISI and equipped with standard IF scores, this methodology has yielded rather divergent outcomes depending on the composition and subject homogeneity of the samples, thereby suggesting that caution should be taken in the design of web-based journal impact measures allegedly similar to ISI-derived metrics.[7]

Solution 4 is the *Citebase* and *CiteSeer* one: automatic indexing systems specifically designed to mine the citation network out of the huge mass of scholarly documentation scattered throughout the universe of open access digital archives. Before introducing them, a cursory look at the role played by bibliographic citations in the new landscape is necessary.

8.1.1. Citations and Open Access

> Imagine a universal bibliographic and citation database linking every scholarly work ever written—no matter how published—to every work that it

> cites and every work that cites it. Imagine that such a citation database was freely available over the Internet and was updated every day with all the new works published that day, including papers in traditional and electronic journals, conference papers, theses, technical reports, working papers, and preprints. Such a database would fundamentally change how scholars locate and keep current with the works of others. In turn, this would also affect how scholars publish their own works, in light of the increased visibility of research regardless of publication venue and the increased potential to demonstrate the value of works through citation analysis. In short, a universal citation database would serve as an important catalyst for reform in scholarly communication.[8]

When, in February 1997, Robert Cameron envisioned an openly accessible universal network of scholarly literature interconnected through bibliographic citations, the scientific community was on the verge of experiencing the full advantages of the free circulation of scientific information. Early peer-reviewed open access journals had already appeared nearly a decade earlier, including *New Horizons in Adult Education*, *Psycoloquy*, *Postmodern Culture*, and *Surfaces*. In May of the same year, the *RePEc—Research Papers in Economics* information network got started; on June 26, the National Library of Medicine in Bethesda announced free online access to the *Medline/PubMed* bibliographic database, thus setting the stage for the forthcoming *PubMed Central* open archive of biomedical journal literature. On August 19, Stevan Harnad, who three years before had come up with the systematic self-archiving of publications by scientific authors, launched the *CogPrints* server for cognitive sciences. Meanwhile, computer scientists and high-energy physicists had further strengthened the online scaffold of their long-established informal communication networks for article preprint sharing. In 1991, Paul Ginsparg had saved time and effort by setting up *arXiv*, a milestone preprint (and postprint) central repository initially devoted solely to high-energy physics, but later expanded to include computer science, astronomy, mathematics, nonlinear science, quantitative biology, and statistics.

The revolution sparked by the Internet and the World Wide Web in the structure of scientific communication grafted on to this rich soil, resulting in the multiplication of literature distribution channels alternative or complementary to the commercial peer-reviewed journal. That was when, exasperated by the inflationary surge in journal subscription rates over the last decades, a transversal party of scientists, librarians, and science users worldwide started to chart a new course for scholarly publishing beyond the traditional paradigm of authors-giving-journals-their-articles-for-free and journal-publishers-selling-the-same-articles-for-a-fee. Under the slogan "public access to publicly funded research," and hammering away at the strong, almost self-evident claim that research results must be widely disseminated and read to be useful, the open access (OA) movement, which took off between the late 1990s and the early 2000s, has since publicized theoretical and business models, along with an OA-compliant technical infrastructure, to support the free online dissemination of peer-reviewed scientific literature. Two options have ultimately come into focus for authors following this way of publication: a gold and a green road. The gold road consists of submitting a paper directly to a true OA journal, that is, a peer-reviewed journal that makes freely available all of its content for all users webwide while shifting editorial management's costs onto subjects other than the publisher (typically the author or the funding institution). So far, over 3,200 such journals have been listed in the *Directory of Open Access Journals* (www.doaj.org), still too few to foresee a balanced competition of gold OA soldiers against the ironclad of subscription-based journals. The green road, by contrast, is faster and far less financially demanding than the gold; it lets authors keep publishing in traditional, toll-access journals, but requires the self-archiving of a peer-reviewed version of the same content into an openly accessible repository, be it a fully accessorized institutional or disciplinary open archive or the humble author's personal website. The affordability of green self-archiving, its ability to isolate the access problem from the all-or-nothing equation established by publishers between scholarly communication and (gold or nongold) journals, and its perfectly fitting the decentralized network of OAI-compliant institutional repositories promoted by universities over the last years have resulted in some current OA advocates to make the point that institutional self-archiving is actually the shortest and most effective path to pursue OA's endgame. Yet, even if scientific authors would scarcely deny the willingness to secure universal free access to their work for the sake of progress (noblesse oblige), the scarce interest in the color of the road covered by their manuscripts upon acceptance for publication is testified by the fact that, more often than not, they miss the chance currently offered by about 70 percent of journal publishers to self-archive an e-print version of the final paper. One reasonable explanation is that scientists are primarily committed to career advancement through publication in highly esteemed journals. Promoting access to research papers beyond the clique of whoever takes an active part in the career move doesn't, therefore, seem to be their chief concern. That's why, in the OA political project, green self-archiving is not, at best, a spontaneous gift bestowed by enlightened benefactors of the republic of letters, but the completion of an official mandate of the funding body or employer. Two landmark mandates of this kind, precursors to several other similar provisions, were issued in December 2007 by the European Research Council (ERC) and the National Institutes of Health (NIH), both requiring that peer-reviewed research articles sprung from their respective funding policies be self-archived and made publicly available by investigators within a fixed time span (twelve months for the United States, six months for Europe) after the official date of publication.[9]

The avenues of bibliometrics and OA intersect at the crossroads where research dissemination and impact/quality assessments take place. In the absolute, if the absolute were within the reach of humans, quality and access would be kept distinct. A quality paper is a quality paper, whether toll access, open access, or not accessible at all, even though in the last case its cognitive potential would lie dormant indefinitely. In the real world, though, quality judgments are social artifacts just as are the objects they are referred to, and the concrete manifestations of quality are irremediably conditioned by scientists' ability to effectively communicate their practical and theoretical accomplishments to the right people in the right places and via the most effective channels: scientific achievements are in fact, to a greater extent, communication achievements. So, in light of the Mertonian view of citations as rudimentary acts of peer recognition, the question of whether the channel controlling the dissemination of research products affects their chances of peer recognition is not trivial at all.

In recent years, a strategic goal of the OA movement has been to demonstrate that, aside from all obvious benefits, open access also substantially increases research impact as it automatically increases the probability of being cited (OA citation advantage). A seminal paper by Steve Lawrence in 2001 set the course, providing evidence to the effect that citation rates in a sample of computer science conference articles appeared significantly correlated with their level of accessibility. "There is a clear correlation," contended the author, "between the number of times an article is cited, and the probability that the article is online. More highly cited articles, and more recent articles, are significantly more likely to be online."[10] A correlation between two variables, as statisticians constantly remind us and Lawrence promptly recognized, doesn't entail a direct causal relationship, nor does it rule out the possible existence of a third explanatory factor behind it. Nevertheless, to the extent that the correlation discloses a regular pattern in the events under analysis, it suggests hypotheses intended for further testing in search of independent evidence. The simplest hypothesis here is that free online articles are more frequently cited because of their easier availability (OA postulate or Lawrence postulate). To test this, many experiments similar to Lawrence's have been carried out that supply a good deal of evidence in favor of the OA citation advantage.[11] Most notably, Harnad and Brody's team has been detecting OA citation advantage across all disciplines in a twelve-year sample of ISI articles (1992–2003), seeking also to clarify the relationship between citation patterns and online usage measures generated by OA databases. Their methodology is straightforward: compare citation scores of individual OA and non-OA articles issued by the same non-OA journal. The results obtained for physics, engineering, biology, business, health, political science, education, law, business, management, psychology, and sociology confirm, if not accentuate, the trend disclosed by Lawrence's findings, pointing to a citation impact that is 25 to 250 percent higher for OA papers.[12] Similar conclusions were reached by Kristin Antelman with a sample of ISI journal articles published in 2002 across the domains of political science, philosophy, mathematics, and electrical and electronic engineering; by Gunther Eysenbach with a sample of articles issued in 2004 by the *Proceedings of the National Academy of Sciences (PNAS)*; and by Michael Norris and colleagues with a sample of articles published in 2003 in high-impact mathematics, ecology, economics, and sociology journals.[13] Eysenbach's approach, remarkably, takes into account the potential confounding role of many variables, including number of authors, individual lifetime publication counts, geographical provenance, and funding organizations, and features a longitudinal analysis that deepens the perspective by comparing citation data at three different moments in time.

Against the OA postulate stream, new, carefully "controlled" evidence backed up by alternative interpretations of citation statistics has been delivered by information scientists. Counter-arguments typically focus on a narrower and more easily manageable empirical base than Harnad and Brody's, such as *arXiv*'s subsections or selected groups of journals from publishers granting different access status to their products, in order to compare the citation history of journal articles posted to e-print archives or otherwise openly accessible with that of unposted/pay-per-access items. Posted materials still appear to enjoy considerably higher citation rates in many cases, but the time dimension and a subjectivity factor in the selection of postable items have been called in, next to increased visibility and readership, as potentially confounding variables behind OA inflated scores. In the first place, because of journal article publication delays, the longer "shelf-exposition" might cause citation counts to start earlier for manuscripts previously posted as preprints than for unposted ones (early view postulate). Second, it might as well be hypothesized that best authors, those who are naturally inclined to produce high-impact papers, tend to be overrepresented in open archives, and even less prominent authors do not post whatever spouts out of their word processors, but preferably papers of special significance or time-critical value, themselves predestined to citation paradise (self-selection bias postulate).

The effects of both confounding factors have been investigated via citation analysis in astronomy, physics, mathematics, biomedicine, and economics. Although limited support has been found for the former, the thesis that authors tend to self-archive high-quality, naturally citation-catching stuff has gained a higher degree of consensus, and consequent attempts to control for this variable in empirical case studies have resulted in a more complex picture of the impact/access mechanism that partially scales down the demands of OA standard-bearers.[14]

Not surprisingly, the OA citation affair has provided the community of professional bibliometricians and the ISI with a timely entry point in the debate on the merits of competing publication models, and the former's positions, exemplified by a 2007 Moed paper, are predictably in tune with the latter's, set forth in three "official" reports published over the last four years. Moed's case study builds on the work of Harnad and Brody but is replete with methodological caveats of noble bibliometric lineage hinting at the complexity of citation practices and the skewness of citation distributions.[15] The study performs a thorough citation analysis of papers posted to the *arXiv*'s condensed matter section before being published in scientific journals and compares the results with those of a parallel citation analysis for unposted articles published in the same journals. The approach is longitudinal, as in Eysenbach's study, with citations counted over a seven-year period using both fixed and variable citation windows. The two main outcomes of Moed's experiment cast serious doubts on Lawrence postulate:

1. Articles posted to the preprint server are actually more cited than unposted ones, but the effect varies with the papers' age. The difference between citation age distributions of posted and unposted items can be partially explained by the publication delays of the latter, thereby lending positive evidence to an early view bias of OA materials. The preprint server accelerates citation, and this is undoubtedly an overall positive effect on scientific communications, but free availability does not ensure, per se, an increased impact.
2. If one estimates authors' prominence independently of *arXiv*'s effect, then the citation advantage of many OA papers fades into the individual performances of the authors themselves, whose ability to gain higher visibility appears largely unaffected by the publishing strategy. In other terms, top authors publish (and deposit) more papers than less prominent authors, and those papers are also likely to be top papers, hence to attract more citations.

ISI's interest in open access has a close relationship with the journal selection process set up by Garfield in the 1960s. The selection, which places much emphasis on citation statistics, is meant to pick out the highest quality sources from the pool of extant journals, but is independent of the business or distribution model. High-quality open access journals, to the extent that scientists use and cite them in a consistent way, don't escape the citation network, so monitoring their bibliometric performance has always been a primary concern of the company. Two studies on the citation impact of OA journals indexed in the *Web of Science* appeared between April and October 2004, while a third report, sponsored by the Publishing Research Consortium and coauthored by members of ISI-Thomson, Elsevier, and Wiley-Blackwell, has been circulating since May 2007.

The first report, focused on 148 natural science journals included in the 2002 edition of the *JCR*, supplied evidence to the effect that, with few notable exceptions, the impact factor of ISI OA journals is variable according to the subject category, but generally lower than that of non-OA titles, and "though there is some suggestion in aggregate of a slightly more rapid accumulation of citations, this effect is, so far, minimal."[16] In short, scientists did not seem to use OA articles more substantially and more quickly than non-OA articles, despite their immediate availability.

The second study, conducted on 219 OA journals included in the 2003 edition of the *JCR,* partially rectified the previous conclusions. Although, as to the impact factor, many OA journals still ranked in the lower half of their subject category, in physics, engineering, and medicine there appeared a clear citation advantage for more recent OA articles, which received a higher percentage of total citations than non-OA articles of the same age.[17] In some ways, this pattern resembles the above "early view" effect: citations accrue faster to freely available journals, but the same does not necessarily hold true in the long run.

The third study is a review that draws together the threads of nearly all the previous literature on the subject and, in line with Moed's call for a more sound methodology in OA citation studies, marshals the counter-evidence gathered so far against Lawrence's postulate.[18]

It would be erroneous, or at least fruitless, to reject these arguments as if they were merely self-serving by-products of the multinational companies' vested interests in conspiring against free, democratic access to information for the sake of profit. In historical perspective, they are anything but the formal expression of some basic principles firmly rooted in the bibliometric paradigm outlined in the foregoing chapters. If the bibliometric game is played according to its rules, namely if one admits that the core of truly important literature continues to thicken around a small set of highly esteemed and heavily cited journals; if nothing, in a scientific career, succeeds in competing with the importance of a publication in top journals; and if, in spite of the warnings set forth in chapter 7 against any uncritical assimilation between citation impact, influence, quality, and scientific progress, citation statistics are trusted as proxies of the rate at which good ideas affect other researchers' good ideas, then the access problem should be kept methodologically distinct from the impact problem. In fact, for the circle to be squared in the bibliometric style, one has to recognize that it is ultimately up to the invisible colleges of relevant researchers in each specialty to judge the relevance and utility (and, in some cases, citability) of a document, and such a judgment is largely independent of its access mode.

Open access to research, it might reasonably be argued, does not need the evidence-building machine of citation metrics to prove its desirability. Aside from ethical arguments, which line up the right to know implicit in the OA principle with other basic human rights,[19] there are other substantive reasons to advocate OA. In the first place, shortening the distance between the invisible colleges' members disseminated in the webwide scientific communication network and turning it into a live, "real time" collaboration network, OA makes scientific communication faster and more effective. Second, at least for the "non-Big Science" research areas, it increases the chance that, beyond economic, social, and geographical constraints, a higher number of talented people enter scientific enterprises and pursue research goals successfully. Third, and maybe most important, prior to the formal level of uptaking and citing other researchers' work, OA finds its way to a place quite separate from the lively social club where citations dwell, specifically to the semiprivate cave where a scientist gathers all the necessary evidence relative to the research project underway, thereby shaping new ideas in constant interplay with those of other researchers picking apples from the same tree. Although the most creative minds do not need to be avid consumers of someone else's contributions, any obstacle to this essential confrontation bears upon the context of discovery and is de facto a true obstacle to scientific progress.

On the other hand, a true challenge for OA, if it is to assert itself as worthy of any continuing role in a bibliometric contention, would be to adjust or enhance the resolution of the observation instrument, ISI indexes, in the hope of perceiving fresh details within the new, decentralized landscape of scientific communication unveiled by the World Wide Web: Does a piece of research have a pre-ISI (or non-ISI) life that sheds a brighter light on its level of acceptance? To what degree and at which rate are the documents easily pulled from open digital shelves actually used by scientists? Do alternative, nonjournal publication and citation cultures stand a chance of winning a better place in bibliometric investigations if citation statistics are gathered outside the perimeter of ISI databases? Is a non-Bradfordian pattern of communications conceivable, at least for some research areas in particular phases of their development?

An intriguing perspective, in the foregoing debate, comes from Harnad's recent quest for a multidimensional, field-sensitive, and carefully validated (against external criteria) "open access scientometrics," taking definitive advantage of the wealth of usage and impact metrics enabled by the multiplication of online, full-text, open access digital archives.[20] To be feasible, however, such an enterprise must preliminarily gain better control over the growing volume of free online scholarly documentation, whose bibliographic management and bibliometric exploitation depend ultimately on the design and implementation of software capable of doing, more quickly and on a larger scale, what the ISI has been doing for nearly half a century with paper journals: locating and storing relevant documents, indexing the bibliographic reference sections, keeping data clean and consistent, and turning the page upside down to analyze and count citations. The magic word here is "metadata," that is, the set of encoded data attached to the information units processed by the automatic indexing system to help identify, retrieve, and manage them in a timely and effective fashion. At best, all existing or sprouting digital repositories should adhere to the same metadata standard for keeping citations of stored documents consistent and up-to-date through the removal of duplicate records; the tracking and proper labeling of different versions of the same document (preprint or published); and the identification and disambiguation of sensitive scientometric data, such as personal author names, institutional affiliations, and titles of journals or publication venues.[21] To date, faced also with the dramatic increase in the number of available digital libraries and the growth rate of their content, automatic indexing algorithms have been unable to perform these operations in a bibliometrically satisfactory way. The two facilities described in the next section, *Citebase* and *CiteSeer*, nevertheless provide instructive case studies that shed light on the potentialities of citation indexing in the openly accessible portion of the cyberworld.

8.1.2. *Citebase, CiteSeer*: The Road Toward an Open Access Citation Index

Citebase (www.citebase.org), issued in December 2001, is an automatic indexing system of a limited number of open access repositories, including *arXiv, PubMed Central, BioMed Central,* and *CogPrints*, developed by Tim Brody's team at the University of Southampton. In a sense, it represents the culmination of the OAI—Open Archives Initiative (1999) and the OpCit—Open Citation Project (1999–2002). The former, focused on open archives' interoperability, led to a widely shared technical solution, the *OAI-Protocol for Metadata Harvesting* (*OAI-PMH*), for the gathering together (or harvesting) of metadata from a number of distributed repositories into a combined, centrally searchable data store; the latter pointed to the automatic generation, in the *arXiv* digital repository, of hyperlinks between bibliographic references and the full text of the cited documents.[22]

The *Citebase* software parses the bibliographic references of the full-text papers hosted by the aforementioned servers and, every time a reference matches the full text of another paper deposited in the same repositories, it creates a link with it. A series of operations familiar to users of traditional citation indexes can be carried out thereafter:

1. Retrieve all the (freely available) papers referenced in a given paper or all the papers that cite a given paper.
2. Rank the retrieved records by number of citations received and display the *Top 5 Articles* that cite a given article.
3. Retrieve all the articles co-cited with a given article and the *Top 5 Articles Cocited.*
4. Retrieve articles that share one or more references with a given article.
5. Display a trend-graph of citation scores and download hits per article over time.

In addition, a *Usage/Citation Impact Correlator* produces a live correlation table comparing the number of times an article has been cited with the approximate number of times it has been downloaded. The resulting correlation index is intended to furnish a rudimental test of the extent to which the use (or download) of an article predicts its future citation, but with a couple of cautionary warnings: log entries are limited to the English mirror of *arXiv* since 1999 and, like other types of information derived from web logs, download data can be altered by various disturbing factors.[23]

CiteSeer, formerly *ResearchIndex* (citeseer.ist.psu.edu), is an automatic digital library search and management system developed by Steve Lawrence, Lee Giles, and Kurt Bollacker at the Princeton NEC Research Institute (now NEC Labs) during the late 1990s, and currently hosted, under a limited open source license, at the Pennsylvania State University's College of Information Sciences and Technology.[24] Unlike *Citebase*, it doesn't parse publications self-archived centrally by their authors in a fixed number of e-print servers, but gathers together research article preprints and postprints from several distributed nodes of the open access Web. The relevant documents, for the most part computer science conference papers in postscript and pdf format, are located mainly through web crawling techniques, even though alternative methods, such as indexing journals at publishers' website and direct posting by the authors, are also within the system's capabilities. Once a document is downloaded, *CiteSeer* parses its text to pick out relevant data, including authors' names, title, abstract, word frequencies, and the reference list. Then it indexes these pieces of information and places them into a local database, thereby allowing, beyond conventional keyword searches, citation searches and citation link browsing as well as searches for similar or related documents on the basis of similarity criteria drawn from word co-occurrences, co-citation, and bibliographic coupling (here dubbed "Active Bibliography").

As a bibliometric facility, *CiteSeer*'s autonomous citation indexing software produces rough citation and co-citation counts "on the fly" for each paper cited in the database; identifies self-citations; and performs automatically what early bibliometricians considered a potential remedy against the ambiguity of citation analysis, namely the extraction of the context surrounding the citation in the body of the paper. In addition, the *Next Generation CiteSeer* (*CiteSeer^X*), available in beta release at citeseerx.ist.psu.edu/, features *Estimated Venue Impact Factors*—a rudimentary counterpart of ISI journal impact factor ranking for computer science journals and proceedings indexed by *The DBLP Computer Science Bibliography*—and plans the extension of automatic indexing to acknowledgments and funding information from research publications. When perfected, this function should allow the retrieval of all the papers acknowledging a given grant and the visualization of an interactive graph for authors' collaboration networks.[25]

CiteSeer took a solitary path for some time, gaining approval and prestigious sponsors (National Science Foundation, Microsoft Research, NASA) until, in February 2004, Thomson ISI and Princeton NEC Laboratories announced a commercial agreement for the creation of a multidisciplinary citation index to the scholarly literature stored in institutional and subject-based repositories scattered throughout the Web. The new *Web Citation Index*, based on *CiteSeer* technology, was launched officially in 2005 as part of the *ISI Web of Knowledge Platform*. In line with the company's source selection policies, it covers materials from OA repositories that meet a dcfault set of quality criteria, such as *arXiv*, the *Caltech Collection of Open Digital Archives*, the *Australian National University Eprints Repository*, the *NASA Langley Technical Library Digital Repository*, and the open access content available through Proquest's *Digital Commons*, an OAI-compliant institutional repository service.[26]

Citebase and *CiteSeer*, at least for now, are not advisable tools to be used for bibliometric evaluations because they are still pilot projects or tests. All the same, they unambiguously portend two alternative courses of citation indexing in the Web era. On the commercial front, the ISI project points to the gradual absorption, into a tried and tested citation crunching machine, of OA slices that will eventually find their way to a web portal for the centralized search of all academic literature "of value" according to ISI quality criteria, no matter its distribution and commercialization methods. On a side, still unsurfaced road, open access literature calls for open access search and analysis tools whose technical feasibility, as *CiteSeer* and *Citebase* demonstrate, has been dramatically enhanced by the spread of green OA practices and by the multiplication of interoperable, OAI-compliant repositories. A warning signal for this achievable scenario materialized when, in February 2004, an international task force formed on the occasion of the third OAI Workshop on Innovations in Scholarly Communication held at CERN, in Geneva, envisioned an OACI (Open Access Citation Index) Project for the construction of an open access citation indexing framework stretching beyond the material and conceptual limits of ISI databases. Needless to say, for the foreseeable future sudden changes are not likely to occur, but signals do exist that, thanks to the current level of technological development, researchers heading in the direction indicated by *CiteSeer* and *Citebase* are on the right track.

8.2. THE CITATION AS HYPERLINK AND THE CURRENT TRENDS IN QUANTITATIVE WEB STUDIES

A website or web page linking to another website or web page is closely akin, from a structural point of view, to a scholarly document citing another document: a hyperlink represents, as it were, the cybertech version of a bibliographic reference.[27] Upon further reflection, the analogy is even unbalanced in favor of the former; a hyperlink digs "deeper" than a scholarly reference because it is not just a typographically and (for many authors) conceptually marginal component of the document, but the basic structure regulating text construction and organization in a distributed hypertext environment. Even more, the probability that a web page be included into a search engine database, hence the probability that it will ever come out of obscurity and be brought to the user's attention, increases as the web crawler fetches other pages linking to it. In some sense, scientific writing, too, is an act of hyperlinking texts, authors, concepts, and theories, with bibliographic citations partially bringing to the surface this hidden texture, but obvious differences between citations and hyperlinks immediately stand out. Links are rarely Mertonian; they do not acknowledge intellectual debts, at least not as a primary function; and they lack any form of peer review. Their provision, moreover, is not unidirectional (a paper usually cites a previously published paper, whereas websites can hyperlink to each other regardless of their publication date). Finally, unlike citations, links are not indelible footprints in the landscape of recorded scholarly activity, but often vary unpredictably within short time periods.

In the second half of the 1990s, the World Wide Web was recognized as a complex system and a tool for modeling the behavior of other natural and social complex systems. Its hyperlink structure, concomitantly, came to the forefront along three main directions of inquiry, whose domains intersect at different points:[28]

1. complex network analysis, a subfield of statistical physics, which investigates the topological properties of the Internet and the Web as particular cases of an evolving complex network
2. hyperlink network analysis, a direct descendant of social network analysis, which interprets the connections between websites as tech-

nological symbols of social ties among individuals, groups, organizations, and nations
3. webometrics, grown in the riverbed of information science, which extends to the web space concepts and methods originally developed in the field of bibliometrics

Only the last area has a direct, genealogical connection with the bibliometric tradition of citation analysis, but the overlap with the other two is made explicit by the common concern with the network of links/citations, whose structural properties set boundary conditions for the individual behaviors without mechanically determining them. In the wake of Price's thermodynamics of science, the study of these properties is performed, at a higher level of abstraction than that of the individual units (and often at variance with it), by means of statistical methods.

8.2.1. Bibliometric Laws in the Cyberworld: Complex Network Analysis

The reader of this book should already be accustomed to the concept of network. Networks of scientific papers and authors held together by bibliographic citations are the leading actors of the story told hitherto. Garfield invented the tool for mining them out of the formal publishing practices of scientists, while Price laid the conceptual foundations for the statistical study of their global behavior. Early on in the history of the Internet, scientists of diverse origins realized that objects similar to a citation network are ubiquitous; nature, society, language, and history teem with networks whose structural and dynamic properties can hopefully be tackled within a common mathematical frame.[29] The job, nonetheless, is bristling with difficulties. It is apparently safe, for example, to characterize the Internet as a collection of countable units: computers, routers, and cable connections in between. But when the machines are plugged in and running, an accurate overall estimate of the network's size, link texture, and growth dynamics has proved extremely difficult to achieve. A similar difficulty is experienced, at a different organizational layer, with the World Wide Web, where the units are web pages and the connections are URL links between them. Both the Internet and the World Wide Web, in fact, are heterogeneous, self-organizing networks, comprising many diverse and differently behaving nodes, sensitive to nonlocal interactions and often in competition with each other for connectivity, which grow exponentially at different rates. Altogether, they obey an evolutionary principle irreducible to any kind of preconceived algorithm or blueprint. In such a system, local properties are not necessarily representative of global patterns, so the latter cannot simply be inferred by the former using conventional statistical tools.

An antidote against the puzzles of complexity came by theoretical physicists. In their hands, the Internet and the Web gradually lost any reference to individual behavior and turned into particular cases of the general construct of complex network, through which an impressive array of natural, technological, and social configurations—from U.S. electrical power grids to bacterial metabolic and neuronal networks—are currently being deciphered. At the same time, the ease with which web servers' logfiles and connectivity data can be harvested for mining purposes led to placing a special experimental value on the Web itself, which was entrusted with the task of disclosing important details in the architecture of complexity. The web topological structure, i.e., the number and distribution of links between the nodes, initially played the crucial role of the guinea pig, affording a good opportunity to gain a deeper insight into a wide range of critical issues, including the way users surf the Web and the ease with which they gather information; the formation of Web communities as clusters of highly interacting nodes; and the spread of ideas, innovations, hacking attacks, and computer viruses. In a step further, theoretical physicists have recently shifted the attention from the static, structural aspects of networks to the dynamics of their evolution responsible for the emergence of a given structure by progressive addition or removal of nodes and links. A key role in this modeling exercise is played by the construct of "graph," a mathematical structure made up, at simplest, of a collection of vertices or nodes connected by links called edges or arcs. Building on graph theory, scientists assimilated the Web to an oriented graph, with web pages as nodes and hyperlinks as edges standing for the interactions between the nodes. To develop an accurate and reliable stochastic model of the Web graph, to measure its width, to reduce it into manageable chunks, and to survey the behavior of the users passing through, have since become critical tasks in scientific and commercial settings alike. The most basic questions, then, turned out to be as follows: What kind of graph is the Web? What pattern, if any, is revealed by the hyperlink distribution among the nodes? Do the links tend to be evenly distributed? If not, why not?

In the late 1950s, when the Hungarian mathematicians Paul Erd□s and Alfréd Rényi supplied graph theory with a coherent probabilistic foundation, the conviction gained ground that complex social and natural systems could be represented, in mathematical terms, by random graphs. Unlike regular graphs (or lattices), where all nodes have the same number of links (or degree), each node of a random graph has an equal probability of acquiring a link, and the frequency distribution of links among nodes is conveniently described by an ad hoc probability distribution (the Poisson distribution). There exists, in random graphs, a dominant average number of links per node called the network's "scale." It is a sort of upper threshold that prevents the system from having nodes with a disproportionately higher number of links. In random graphs, moreover, nodes are not clustered and display statistically short distances between each other, whereas regular graphs tend to be clustered but with comparatively long distances between nodes.

As computerized data on technological, natural, and social networks became more and more available, the random-graph hypothesis underwent a process of testing and critical scrutiny. Empirical evidence seemed to contradict the random model, in that it suggested the structure of complex networks was somewhere between a totally regular graph and a random graph. A landmark finding, in this regard, was made by Duncan Watts and Steven Strogatz who, in a 1998 *Nature* article, set forth a model of complex networks brushing up a construct until then confined to the neighborhoods of sociology: the small world.[30] A small world is said to exist, in a social milieu, whenever members of any large group are connected to each other through short chains of intermediate acquaintances. The strange story of this counterintuitive phenomenon bears witness to the winding course followed by some ideas in percolating through commonsense and literary or anecdotal evidence toward rigorous stages of scientific investigation in both social and physico-mathematical sciences. Let's briefly review the basic steps.

1. *Common sense.* It is an ordinary experience when two strangers meeting by chance discover a common acquaintance, prompting them to

make the well-known comment, "It's a small world, isn't it?" It is likewise predictable that such an event will be more frequent among individuals belonging to "comparable" clusters, i.e., not divided by social barriers, such as race and class.

2. *Poetical and playful transposition of common sense.* As early as 1929, the shrinking of the world on account of ever-increasing interpersonal connectedness was described by the Hungarian poet Frigyes Karinthy in a short story entitled "Chain-links." Karinthy put in the mouth of a character the statement that at most five chain links, one being a personal acquaintance, would have sufficed to connect any two individuals in any part of the world. A similar belief, reinforced by subsequent experimental findings, generated the "six degrees of separation" myth, later made popular by theater, cinema, and parlor games. Scientists, for instance, invented the Erd□s number to express the collaborative distance of any author from the Hungarian mathematician Paul Erd□s (a scientist publishing an article in collaboration with Erd□s has Erd□s number = 1; a scientist publishing an article in collaboration with another scientist having Erd□s number = 1 has, in turn, Erd□s number = 2, and so on). Similar indexes of proximity exist for the physicist Wolfgang Pauli, the ufologist Leonard Stringfield, and the Hollywood actor Kevin Bacon: in the game *Six Degrees of Kevin Bacon*, the focus is on actors costarring with Bacon and the goal is to find the shortest path from any actor to him.
3. *From common sense to social theory.* Sociological research on small worlds dates back to the 1950s, when Manfred Kochen and Ithiel de Sola Pool transformed a topic of party conversation and recreational mathematics into a matter of rigorous scholarly inquiry. The authors recognized the incidence of the small world effect in shaping political influence and everyday life behaviors, as exemplified by the widespread practice of using friends (and friends of friends of friends . . .) in high places to gain favors or to get a job. But they went far beyond a simple qualitative inspection and put forward mathematical descriptions of social contact based on statistical mechanics methods, encompassing graph-theoretic models and Monte Carlo simulations. In a prophetic passage, they even hinted at "the tantalizing possibility that the small world phenomenon could shed light on the secret of how networks more generally give rise to emergent properties, such as the

higher mental functions of neural nets and their analogs in social nets or computer nets."[31] Draft manuscripts containing Pool's and Kochen's ideas circulated for over two decades before being published in 1978 and likely contributed to creating the interest in small worlds of the American psychologist Stanley Milgram.

4. *Empirical foundations of social theory.* In 1967, Milgram initiated a series of experiments to test the small world conjecture in real social networks.[32] Although the validity of his experimental design and of subsequent attempts to replicate it has recently been questioned, his finding that, on average, the acquaintance chain required to connect two random individuals is composed of about six links, provided a first rough estimate of how small the world actually is or is supposed to be. It also generated a great deal of interest in the structural properties of social networks that had caused the world to shrink the way it did. A clue to the puzzle resolution came from the American sociologist Mark Granovetter, who levered on the observation that the social ties of an individual are not all equal: some are strong, comprising only a person's close friends, most of whom are in touch with one another and form a densely knit set of social relationships; others are weak, pointing to a loose network of acquaintances, few of whom know one another. But each acquaintance is, again, enmeshed in a network of close friends, so that the weak tie between any two individuals is also a bridge between the respective densely knit clumps of close friends. The bridge would not exist if not for the presence of the weak ties operating as critical paths for the flow of information between otherwise densely connected but substantially isolated cliques. Granovetter's work went unnoticed for many years, but the path-shortening effect of weak ties he uncovered turned out to be crucial in ensuring the unity and cohesion of the entire social system.[33]
5. *Transposition of social theory in mathematical modeling and complex network theory.* Watts and Strogatz showed that a complex network is a small world displaying both the highly clustered sets of nodes typical of regular graphs and the small path lengths between any two nodes typical of random graphs. They computed a characteristic path length and clustering coefficient for the small world network and recognized the importance of "short cuts," that is, the few long-range edges topologically equivalent to Granovetter's weak ties, which reduce

> the distance between two nodes as well as between their immediate neighborhoods, neighborhoods of neighborhoods, and so on. Further experiments confirmed the small world features of the Web, where two randomly chosen documents were estimated to lie on average nineteen clicks away from each other.[34] At a lower scale, a similar behavior has been observed within definite World Wide Web subsets, such as the web spaces of the 109 UK universities checked by Lennart Björneborn for the emergence of world-shrinking phenomena.[35]

Just one year after Watts and Strogatz's paper, Réka Albert and Albert László Barabási issued an alternative class of models for the large-scale properties of complex networks.[36] In some respects, their attempt to describe the mechanism behind network topology marks a turning point in the study of complex systems. It might be thought of as a paradigmatic change, if it were not for the fact that, at least from Derek Price onward, the bibliometric tradition had worked out a set of mathematical models inspired by quite similar principles in addressing the structural and dynamic aspects of bibliographic networks (of papers, citations, and authors).

Supported by empirical evidence drawn from databases spanning as diverse fields as the World Wide Web, the bibliographic citation network, and the author collaboration network in science, Albert and Barabási did not assume, as Erdös/Rényi and Watts/Strogatz had done, that the number of nodes in the graph model is fixed, nor that the probability of connection between two nodes is random and uniform. Networks, instead, grow by the addition of new nodes linking to already existing ones, and the addition is not random but follows a mechanism of preferential attachment that replicates, on a larger scale, the Matthew Effect well known to bibliometricians: new nodes have a higher probability to link with highly connected nodes than with poorly connected or isolated ones. In more precise terms, the probability $P(n)$ that a node has to establish a link with n other nodes is expressed by a power law of the form

$$p(n) = \frac{1}{n^a}$$

The reader can easily notice the similarity of this formula to the algebraic expression of Lotka's Law presented in section 4.2. Not surprisingly, the degree exponent *a* is as important in characterizing the network's behavior as Lotka's exponent was in the classic bibliometric distribution, and a good deal of work is currently being done in estimating its value through empirical data on real-world networks. Meanwhile, the picture continued to enrich with further details. The connectivity scheme of more than 200 million web pages, examined in 1999 by a group of researchers from AltaVista, Compaq, and IBM, returned the snapshot of a Web clustered into qualitatively different domains displaying a curious bow-tie geometry.[37] The experiment indicated also that, after all, the World Wide Web is not that small because the probability for any path between two randomly chosen documents to exist is only 24 percent. What appeared fully confirmed, on the other hand, is the fundamental large-scale validity of power law distributions also in the cyber-counterpart of classic bibliometric distributions: a few nodes seem to attract an unusually high number of links, compared to a much higher number of nodes that are less frequently connected, and a significant number of totally disconnected ones. The occurrence of power laws involves both the absence of a dominant average value (the scale) for the system's critical quantities and the stability of its mathematical description at different scales, thus resulting in what is commonly dubbed a "scale-free network."

Not all complex networks, it is worth stressing, are scale-free, nor does a complex network evolve uniquely by addition of new nodes and links because removals and decline are likewise possible. Preferential attachment, furthermore, generates a scale-free structure only when particular conditions are met. Yet the emergence of scale-free stationary states and preferential attachment processes in real-world networks highlights one of the crossroads at which science studies and new communication channels have met during the past few years, namely the quantitative analysis of the World Wide Web from a scientometric perspective. Like scientific authors' productivity patterns, citation distribution among papers, word distribution in texts, and many other natural and social phenomena involving a source–item relationship, the Web hyperlink structure exhibits a power law distribution: most nodes are expected to be poorly connected, while a small but significant set of nodes are likely to get a disproportionately higher number of links, thus working as hubs or connectivity providers for the entire network. The mechanism behind this structural collapse is not yet clear, but its analogy with Price and Merton's success-breeds-success or cumulative advantage process is evident in the preferential accumulation of links controlling the formation of hubs. Skewed distributions of connectivity are the norm within the science communication system, so it is reasonable to expect that the models of the web graph worked out by complex network theories accomplish in the cyberworld a task similar to that of the Lotka/Bradford/Zipf laws in the classic bibliometric tradition, leading to the mathematical foundation of web indicators of impact, visibility, and excellence. Stated differently, if the Web is an evolving, complex network, with the characteristics of both a small world and a scale-free system, and if hyperlinks are assumed, at least in the aggregate, to indicate some kind of endorsement of the linked nodes just as citations do for scientific papers, then it should be possible to single out reference values for connectivity patterns against which the relative performance of World Wide Web subspaces could be evaluated.[38] Needless to say, much more caution would be necessary here than in classic citation analysis: hyperlinks, as noted above, are not equivalent to bibliographic citations; the Web is not a carefully selected group of scholarly sources, nor does an instrument comparable to ISI citation index actually exist outside commercial search engines, with all their limitations; furthermore, the above preferential attachment mechanism provides each node with an unequal chance of being selected in any sampling procedure, so here, too, the first step of inferential statistics, the generation of an unbiased sample of web pages, is a fairly complicated task.[39]

8.2.2. Citation Analysis in the Cyberworld: Hyperlink Network Analysis, Webometrics, and the Promise of Web Scientometric Indicators

Hyperlinks, just like bibliographic citations, are human artifacts that someone, somewhere, fabricates for supposedly definite motives. Thanks to hyperlinks, the web graph is something more than a mathematical abstraction: hundreds of millions of users worldwide move every day to and fro along the graph's edges, and the type, quality, and effects of their encounters are necessarily conditioned by this labyrinth of prefabricated roads. Hyperlinks are technological constructs, but each time somebody creates and arranges them in a particular fashion, thereby lending support to a vast array of transfers of tangible (e.g., in commercial transactions) or immaterial (information exchanges) goods, they turn into harbingers of social ties among the potential passengers: a computer network connecting people, institutions, and knowledge is thus also, unavoidably, a social network.

Sociologists have long dealt with social networks among people, organizations, and nation-states. Within the bibliometric tradition, Derek Price's conceptualization of invisible colleges and the substantial empirical work on coauthorship and scientific collaboration patterns are true examples of quantitative insight into the structure of social networks latent in scientific communities. As digitally interconnected microcomputers started to ascend the throne of scholarly desktops and lab benches, making available a new live publication platform, new sources of data (and metadata), and new web-based research tools, the same interest spilled over into computer-mediated communication systems. The network, then, came increasingly to represent not simply a communication facility, but a tool for building online collaboration platforms where new knowledge can be created, modified, and negotiated, in a sort of virtual laboratory without walls. The World Wide Web is a particular type of computer-mediated communication network; hence, it can be considered, in its own right, a social network of unprecedented magnitude and complexity, providing instant and decentralized support to a wide variety of loosely bounded and sparsely knit communities, which leave vague and provisional footprints of their operation in the hyperlink layout. The texture of hyperlinks constructed, maintained, and modified by website owners, whether individuals, organizations, or nation-states, is supposed to reflect their interests, objectives, and communicative commitments. So, whereas sociologists, at least since the 1970s, have been using social network analysis (SNA) to uncover the structural properties of the networks of relationships between actors in social spaces, the extension of SNA methods to the World Wide Web hyperlink texture—suggested by Michele Jackson as early as 1997—has been quite seamless.[40]

The cyber-counterpart of SNA is called hyperlink network analysis (HNA). One of its typical objectives, comparable to the traditional search for leaders and brokers in social groups, is to check whether the hyperlink network is organized around central nodes, or websites that, often by virtue of their owner's prestige or visibility, play the role of authorities or hubs for other nodes. Centrality measures are carried out, as in classic citation analysis, by counting the number of ingoing (indegree centrality) and outgoing (outdegree centrality) links for a given website. In addition, centrality has an aspect of "closeness," intended to single out the website with the shortest path to all others, and another of "betweeness," which is meant to estimate a website's "brokering potential," that is, the frequency with which it falls between the paths connecting other sites. The reader will not be surprised to find, as the basic HNA dataset, a matrix similar to the citation matrix used, from Price onward, to dig out the structural properties of a bibliographic citation network, save that now the matrix cells are occupied by numbers representing the frequency of hyperlinks between two web nodes instead of citations between papers, journals, or authors. One of the possible ways to process the data matrix is cluster analysis, the same tool used by co-citationists to build science war maps, which enables the analyst to classify websites according to hyperlink connectivity patterns, so that central and periphery groups with different degrees of link density do eventually emerge.

HNA techniques have been promisingly applied in a series of case studies dealing with topics as diverse as e-commerce; social movements; and interpersonal, interorganizational, and international communication, altogether confirming the feasibility of web sociometric mining.[41] A further step, then, seemed quite natural in a scientometric perspective: even though it can safely be assumed that only a small percentage of hyperlinks are dictated by scholarly interests, the question arises of whether and to what degree links can be used, just like citations in quantitative sociology, as proxies for scientific communication flows and, from an evaluative standpoint, as building blocks of new, web-inclusive scientometric indicators of research prominence. In other words, is it possible to derive from hyperlink analysis a meaningful insight into information exchanges among scientists, research groups, institutions, and countries? Can the quality or relative standing of a web unit (page, site, domain) be reasonably and unobtrusively estimated using quantitative link analysis? To what extent can the bibliometric notion of impact, traditionally dependent on the assimilation of a scientific paper by a clear-cut community of specialists, be extended to such an anarchical milieu as the Internet, and to dynamic objects with such indefinite boundaries as web documents? And to what extent is the cybermetric impact correlated with research quality of the website owner as measured by independent, less unconventional performance indicators? These questions, while sounding familiar to anyone attending bibliometric literature, press bibliometricians to find new solutions to the old problem of research quality control. The label currently attached to this ongoing effort is "webometrics," a deceptively familiar neologism.

Webometrics was born, as it were, on the shoulders of a giant—the nearly fifty-year tradition of journal-based citation analysis—with the ambition of looking beyond the giant's view to see if, and under what conditions, fresh insights into scientific communication can be obtained, carrying to the extreme the citation/hyperlink analogy. In the French tradition of Callon and Latour's epistemology, thc Brazilian scicntist Marcia Bossy struck out in this direction as early as 1995, suggesting that the digital network layer offered an unprecedented source of information on the scholarly sociocognitive activities that predate the formal publication output. The standard framework for tracking information flows among scientists could then be supplemented with new, sensitive measures of their symbolic status on the Internet.[42] "The hypertextual nature of the web" as Cronin puts it, "means that the principles of citation indexing can be applied much more widely than at present. Web-based retrieval systems will allow us to go beyond traditional citations and track acknowledgments, diffuse contributions, and other input measures."[43] The trail, after all, was already blazed: it could have sufficed, at the outset, to replace the old information units—articles, books, citations—with web pages, websites, and links from carefully "controlled" sources, namely universities', departments', research institutes', and individual scientists' web spaces. Accordingly, the terminology had to be slightly adapted: an outgoing link from a web page becomes an outlink (bibliographic reference), an ingoing link, instead, is an inlink (citation), both terms obviously being perspective-dependent for a link between two given pages; a self-link is a link from a web page to itself (self-citation); two pages are co-linked when both have inlinks from a third page (co-citation), while they are co-linking if both have outlinks to a third page (bibliographic coupling). Commercial search engines, in particular AltaVista, with its ability to retrieve pages containing outlinks to a given page, supplied a tool of the trade roughly comparable to ISI citation indexes. A similar use, theorized in the mid-1990s by Rodríguez i Gairín and Don Turnbull,[44] was tested by Ray Larson in his exploratory attempt to map, with co-citation analysis techniques, the intellectual structure of a World Wide Web subset across the fields of earth sciences and geographical information systems, and by Peter Ingwersen in his seminal calculation of a web impact factor.[45] AltaVista searching facilities (and limitations), moreover, were instrumental in Judit Bar-Ilan's attempt in 1997 to investigate the applicability of classic bibliometric laws to news items appearing in Usenet newsgroups on mad cow disease immediately after its outbreak.[46]

The extension of well-established informetric methods to the World Wide Web eventually placed webometrics on a solid launchpad because, as stated by Almind and Ingwersen, it allowed "for analyses to be carried out almost in the same way as is traditional in the citation databases."[47] That is why, from the very beginning, link analysis made its way toward the same research posts attended by bibliometrics and classic citation analysis, further emphasizing the elective affinities between the evaluative use of citation/link statistics and their practical application to the enhancement of web information retrieval systems. Two analytical tools bearing witness to this kinship are co-links and co-words statistics.

After Larson's seminal clustering experiment, the statistical analysis of co-linked web units has found applications as diverse as[48]

a. determining the patterns of relationships between multiple web spaces "closed" with respect to a particular theme
b. mapping the network of connections among European academic institutions (Xavier Polanco calls this "co-site analysis")
c. comparing, in a business intelligence perspective, the relative positions of competing companies
d. identifying the graph "signatures" of emerging web communities gathering virtually around a specific issue
e. investigating the patterns of relationships among a certain number of institutions participating in the self-organization of the European Information Society
f. enhancing the ranking algorithms and the algorithms for retrieving pages thematically related to a starting page
g. experimenting with semiautomatic metadata propagation methods for improving the search for specific information on the Web
h. classifying academic blogs on a given subject (having previously wiped out nonsubstantive links)

Likewise, Web co-word analysis has served both sociological and information retrieval goals. An advanced algorithm of co-word mapping, for example, has been introduced by Prabowo and Thelwall for automatically tracking subjects within broader topics in ongoing debates, and the co-occurrence of key words in national and top level domains has been used by Leydesdorff and Curran to identify the connectivity patterns of the "Triple-Helix" (government/university/industry).[49] In the information retrieval arena, instead, a multivariate analysis of word pairs occurring in web search engines' user queries has been tested to enhance ranking algorithms,[50] while Polanco's team has suggested outstepping classic web page co-word techniques by co-usage analysis (i.e., the number of times two documents are referred to in a user query), so as to facilitate the identification and visualization of usage and searching behavior patterns in free online databases.[51] Notwithstanding their walking side by side, however, Web information retrieval and webometrics address different tasks and are expected to comply with quite distinct sets of theoretical obligations. Whereas the former is primarily concerned with enhancing the probability that a user will extract the most relevant information out of a digital morass, the latter points to socially relevant measures (of influence, impact, quality, and visibility) and makes one wonder what the computed quantities signify and whether they actually measure what they were supposed to measure.

At its simplest, a rough estimate of the impact of a website or the ranking of subject-related websites according to the number of inlinks is obtained, like the ISI IF of journals, by drawing a core list of top source sites, indexing their outlinks, and classifying the results by "sitation" frequency.[52] At a higher level of comprehensiveness, relying on the ability of AltaVista-type search engines, such as Yahoo! Search and Windows Live Search (formerly MSN Search) to count the number of web pages linking to at least one page of a given website, bibliometricians have been speculating on the properties of a numerical index, a sort of cybermetric equivalent of Garfield's IF, useful for the comparative evaluation of websites' performance. The web impact factor (WIF) of a site or area of the Web, introduced in 1998 by Ingwersen, may be defined, in perfect symmetry with the journal IF, as a measure of the frequency with which the "average" web page of the site has been linked at a certain time. In the most accredited formulation, it is the ratio between the number of web pages linking to that site or area (self-links excluded) and the number of pages it contains (as indexed by the search engine) at a given point in time.[53] If, for instance, at time t, the total number of link pages (self-links excluded) to the site S is $I = 100$, while the number of web pages published in S that are indexed by the search engine is $P = 50$, then

$$WIF(S) = \frac{I}{P} = \frac{100}{50} = 2$$

This apparently straightforward measure, upon closer inspection, is affected by methodological shortcomings that reach far beyond the technicalities of its calculation to involve the crucial issue of data collection methods on the Web for informetric purposes. Where do link data come from? How reliable and valid are the tools for gathering them? That is to say, to what extent are the results reproducible under different conditions by different collectors? And to what extent do those results actually address the phenomenon they are meant to describe?

A major obstacle to answering these questions is the inability of commercial search engines to restore a reliable and consistent picture of global and local connectivity rates over time. This is because of a number of structural limitations, three of which appear especially severe.[54]

1. In spite of the exponential growth of the body of indexable documents, search engines crawl and index only a small portion of the World Wide Web. A striking demonstration of this coverage deficiency is, for instance, their technical inability to account for the "invisible Web," a virtually infinite no-man's land where provisional hospitality is given to a wide array of hardly classifiable materials, such as web pages subjected to crawler exclusion criteria or screened by access restriction policies, and the hundreds of millions of dynamic web pages returned daily by web database servers in response to user queries.
2. Different search engines use distinct crawling algorithms, notoriously opaque to the user, which strongly affect their effectiveness in trawling sites, pages, and links across World Wide Web subdomains.
3. The overlapping between competing search engines' databases is small, and thus results obtained by collecting data from multiple sources are difficult to compare. Even within the same database, more-

over, average daily fluctuations exist, causing a marked instability in the results.

Coverage and timeliness, it is worth remembering, are not as crucial issues for search engine programmers as they are for informetricians because of the prevailing market-driven interest of the former. After all, a search engine is primarily expected to supply the general user with the most relevant information in response to a query rather than to span the universal knowledge network. Hyperlink counts, therefore, are to some extent an artifact of proprietary (and secret) crawling and indexing algorithms, whose biased and time-dependent output results in an ever-changing snapshot of an ever-changing landscape. As we shall see below, commercial search engines are not the sole viable data-gathering tool, nor is their use mandatory in academic web studies, but if web-scale research is to be conducted, they still represent a fast track for experimenting on very large datasets.

Even if the above shortcomings were minimal and search engines had no coverage biases, the WIF would still leave much to be desired as a webometric measure. Both the numerator and the denominator of the WIF formula are difficult to define unambiguously, and the ambiguity is closely related to the assumption that web pages are the best counting units. This assumption is patently weak. It is not immediately clear, for instance, what to count as an inlink in the case of web pages belonging to sites of cognate institutions (a link from the site of a university to that of an affiliate department is an inlink or a self-link?). Similarly, the number of web pages at the denominator is not a reliable estimate of the share of linkable web resources; it can be spuriously inflated by a huge number of unlinkable files, and the page formats can influence it in that a single document, for example an online HTML book, may be hosted as a single page or split into 100 pages for increased readability, weighing into the calculation either as 1 or 100. Over and above that, the choice of web pages as counting units causes link frequency to manifest quite different properties than conventional citation frequency of scholarly journal papers, as is clear from the following example. If a paper in journal *A* cites two papers issued by journal *B*, the latter's IF increases by two, whereas if a web page in the site *C* links two different pages of the site *D*, the latter's WIF increases by just one unit. Conversely, if a single paper in journal

A cites two times, at different pages, a paper published in journal *B*, the latter's IF increases by just one, whereas if two web pages in the site *C*, even though belonging to the same institutional or conceptual unit, link the same page of the site *D*, the latter's WIF increases by two units.

Not only do web pages not exhibit the same bibliometric status as journal papers, but they also display a completely different pedigree in terms of structural properties and stability. Web pages often lack a definite authorship and, given the low level of standardization, the use of codes in their creation, for example HTML tags, is subjective and variable from page to page. Web pages' half-life, too, is extremely variable; they come and go; change name, address, structure, and text content; and the variation rates of all these elements are different for different categories of materials.[55] As a result, quantitative analysis can be performed only asynchronously, on a sample that is temporally and qualitatively shifted with respect to the original, while the construction of time series for longitudinal studies has severe limitations, being restricted to data collected within the scope of gigantic online archival projects, such as *The Internet Archive Wayback Machine* (www.archive.org).

Content variability and structural instability imply that the initial stage of any webometric analysis, namely the collection of raw link data, is inherently problematic, and the scenario is much worse here than in classic bibliometrics. For a citation analyst, in fact, the data—the totality of documents referenced in the articles of ISI-covered journals—are packaged and ready for use. Even though one might question the bias inherent in the data being derived from ISI journal selection policies, they secure a sound and relatively stable starting point for any bibliometric analysis, letting the differences emerge at the level of data usage and hypothesis testing. Webometrics, by contrast, works on network data "constructed," filtered, and reengineered each time a specific study is underway.

Amid this perplexing array of issues, link analysts are trying to calibrate the observation instrument and define more reliably the unit of analysis in order to exert a better control over the experimental setting. On the one hand, they set aside the ready-made ratings supplied by commercial search engines and manage to build their own web crawlers for trawling carefully selected areas of the Web in a "domesticated" environment. Since 2000, by way of illustration, the Academic Web Link Database Project set up by the Statistical Cybermetrics Research Group at the

University of Wolverhampton has been collecting, through a homemade distributed web crawler, university link data relative to the academic web spaces of New Zealand, Australia, the United Kingdom, Spain, China, and Taiwan.[56] At the same time, webometricians discard web pages as the sole analytical target and resort to heuristic methods to identify that indissoluble piece of coherent material that properly constitutes a "web document." Mike Thelwall's "Alternative Document Models (ADMs)," for example, allow modulating link analysis by truncating the linking URLs at a higher level than that of the web page: the directory (all pages in the same directory are counted as one link target), the domain (all pages with the same domain name in their URL are counted as one link target), and the site (all pages belonging to subsites with a specified domain name ending are counted as one link target).[57] Here, also, as in the advanced bibliometric methodologies reviewed in chapter 6, the appropriate unit of investigation has been located mostly at the level of the scholarly institution, above all the university web space.

A recent experiment in large-scale webometric analysis is the *Webometrics Ranking of World Universities* (www.webometrics.info), launched in 2004 by the Cybermetrics Lab, a research group belonging to the Centro de Información y Documentación (CINDOC) of the Spanish National Research Council. The site ranks web domains of academic and research organizations worldwide according to volume, visibility, and impact of their content. The WIF is employed to capture the ratio between visibility, measured by inlink rates returned by commercial search engines, and size, measured by number of hosted web pages. Two additional measures, dubbed "Rich File" and "Scholar" indexes, are introduced to capture, respectively, the volume of potentially relevant academic output in standard formats (Adobe Acrobat .pdf, Adobe PostScript .ps, Microsoft Word .doc, Microsoft Powerpoint .ppt) and the number of papers and citations for each academic domain in Google Scholar.[58]

On a smaller scale, Thelwall and colleagues' methodology of link analysis zooms in on the patterns of connections between selected groups of academic sites at the national level. "Analyzing the interlinking between universities within a single country—they assert—offers the perfect scale for a study. The number of objects to analyze (one site per university in a country) is manageable and counting all links to a whole university site seems to give a sufficiently high level of aggregation to produce reliable results."[59] University websites have been found to be relatively more stable than other cyber-traces in longitudinal studies, and, despite a large variability at the level of the individual institution, a relative stability of their inlink and outlink rates is also documented.[60] Thus, if the research-oriented nature of academic institutions is also taken into account, it is quite reasonable to expect their linking policy to be in some measure driven by their intellectual mission. But if this holds true, then it might also be asked whether the counting of links between different academic institutions correlates in any way with independent indicators of research performance, notably peer-reviewed research ratings, publication productivity, and citation rates. Such a correlation has been actually observed in some cases, and the closer the web pages are to research-related interests, the stronger appears the correlation,[61] but no causal relationship has ever been established thereafter. Although the reputation and visibility of an academic institution are partially reflected in the "sitation impact" of its website, no evidence exists, so far, that link rates might be determined by (or used as an indicator of) research performance. Web visibility and academic performance are, once and for all, different affairs.

It is nothing but a déjà vu in the manifold lives of the citation/link that, faced with the difficulty of interpreting the quantitative patterns disclosed by statistical analysis, bibliometricians resort to more qualitative tools of inquiry, such as direct surveys of webmasters' "reasons to link" or hyperlink context and content analysis, to investigate the psychological side of the link generation process. This kind of analysis has been extended to a variety of areas, both academic and commercial, yielding the quite expectable outcome that, even in academic websites, links serving a strict scholarly purpose are rare. For the most part, indeed, they are meant to facilitate navigation toward quarters of loosely structured and generically useful information, or to suggest related resources for reasons broadly connected to research and educational interests. Sometimes they even do not perform any conceivably sociocognitive function, a circumstance all the more true if one thinks of many academicians' personal web pages.[62] Links alone, then, just like bibliographic citations alone, do not seem sufficient to pin down critical communication patterns on the Web, and their statistical analysis will probably follow, in the years to come, the same path of citation analysis, establishing fruitful alliances with other emerging qualitative and quantitative outlooks over the web landscape.

NOTES

1. The term "sitation," formerly used by Gerry McKiernan with reference to early applications of citation indexing to web resources during the mid-1990s (www.public.iastate.edu/~CYBERSTACKS/Cited.htm), has been canonized, in one of the first studies of World Wide Web hyperlink distributions, by Rousseau, "Sitations: An Exploratory Study," *Cybermetrics* 1, no. 1 (1997), www.cindoc.csic.es/cybermetrics/vol1iss1.html.

2. Brin and Page, "The Anatomy of a Large-Scale Hypertextual Web Search Engine," *Computer Networks and ISDN Systems* 30, nos. 1–7 (1998): 107–17, www-db.stanford.edu/pub/papers/google.pdf.

3. On the *Clever Project* see Soumen Chakrabarti et al., "Hypersearching the Web," *Scientific American* 280, no. 6 (1999): 54–60, www.cs.cornell.edu/home/kleinber/sciam99.html.

4. Eun-Ja Shin, "Do Impact Factors Change with a Change of Medium? A Comparison of Impact Factors When Publication Is by Paper and Through Parallel Publishing," *Journal of Information Science* 29, no. 6 (2003): 527–33; John D. McDonald, "Understanding Online Journal Usage: A Statistical Analysis of Citation and Use," *Journal of the American Society for Information Science and Technology* 58, no. 1 (2007): 39–50, resolver.caltech.edu/CaltechLIB:2006.001.

5. Evans, "Electronic Publication and the Narrowing of Science and Scholarship," *Science* 321, no. 5887 (2008): 395–99. Evans's conclusions are apparently at odds both with those of Larivière and colleagues' study reviewed in section 4.6 and with those reached, on the ground of extensive surveys of scholarly reading habits, by Carol Tenopir and Donald W. King, *Towards Electronic Journals: Realities for Scientists, Librarians, and Publishers* (Washington, D.C.: Special Libraries Association, 2000). The thesis that easy online access leads to much wider usage of older materials is also defended, on the basis of statistics of full article downloads, by Andrew Odlyzko, "The Rapid Evolution of Scholarly Communication," *Learned Publishing* 15, no. 1 (2002): 7–19, www.si.umich.edu/PEAK-2000/odlyzko.pdf. It should be noted, however, that the kind of usage measured by citations is not homologous to usage measured by other methods that resort to less formal units of analysis (direct survey, download counts).

6. On the three mentioned projects COUNTER, SUSHI, and MESUR, see the respective websites: www.projectcounter.org/, www.niso.org/committees/SUSHI/SUSHI_comm.html, and www.mesur.org. An overview of the challenging process of assembling, processing, and analyzing usage data is given, in connection with the preliminary results from the MESUR experience, by Johan Bollen, Herbert Van de Sompel, and Marco A. Rodriguez, "Towards Usage-Based Impact Metrics" (paper presented at the Joint Conference on Digital Libraries 2008

(JCDL08), Pittsburg, Pennsylvania, June 16 2008), arxiv.org/abs/0804.3791. On the issues raised by the definition of a journal usage impact factor complementary to the ISI IF, see Bollen and Van de Sompel, "Usage Impact Factor: The Effects of Sample Characteristics on Usage-Based Impact Metrics," *Journal of the American Society for Information Science and Technology* 59, no. 1 (2008): 136–49, arxiv.org/abs/cs.DL/0610154. On the technical factors affecting the validity and reliability of usage data, see Philip M. Davis and Jason S. Price, "Ejournal Interface Can Influence Usage Statistics: Implications for Libraries, Publishers, and Project COUNTER," *Journal of the American Society for Information Science and Technology* 57, no. 9 (2006): 1243–48, arxiv.org/abs/cs.IR/0602060.

7. A poor correlation has been found, in a sample of thirty-nine e-journals, in Stephen P. Harter and Charlotte E. Ford, "Web-Based Analysis of E-Journal Impact: Approaches, Problems, and Issues," *Journal of the American Society for Information Science* 51, no. 13 (2000): 1159–76, while a good correlation for library and information science journals is documented in Liwen Vaughan and Kathy Hysen, "Relationship Between Links to Journal Web Sites and Impact Factors," *ASLIB Proceedings* 54, no. 6 (2002): 356–61. It is likely that traditional citation measures and hyperlink measures account for two different aspects of a journal's performance, as suggested by Vaughan and Thelwall, "Scholarly Use on the Web: What Are the Key Inducers of Links to Journal Web Sites?" *Journal of the American Society for Information Science and Technology* 54, no. 1 (2003): 29–38, citeseerx.ist.psu.edu/viewdoc/summary?doi=10.1.1.20.3055.

8. Cameron, "A Universal Citation Database as a Catalyst for Reform in Scholarly Communication," *First Monday* 2, no. 4 (1997), www.firstmonday.dk/issues/issue2_4/cameron/index.html.

9. A clear statement of the two OA roads and their implications for authors and publishers can be found in Harnad et al., "The Access/Impact Problem and the Green and Gold Roads to Open Access," *Serials Review* 30, no. 4 (2004): 310–14, eprints.ecs.soton.ac.uk/10209/. All the contributions to that issue of *Serials Review* deal with OA policies and practices and provide an illuminating snapshot of the main trends and motifs animating the debate on the future of scholarly communication as of September 2004. For a 2008 update, in which the shift toward the increasingly central role played by institutional repositories in OA strategies is evident, see vol. 34, no. 1 of the same journal (Special Focus: Open Access Revisited), in particular Harnad et al., "The Access/Impact Problem and the Green and Gold Roads to Open Access: An Update," *Serials Review* 34, no. 1 (2008): 36–40, eprints.ecs.soton.ac.uk/15852/.

10. Lawrence, "Free Online Availability Substantially Increases a Paper's Impact," *Nature* 411, no. 6837 (2001): 521. A preprint version of the same article

has enjoyed wide circulation with the title "Online or Invisible?" citeseer.ist.psu.edu/online-nature01/.

11. The rapid growth of literature on the OA citation advantage has inspired the compilation of a dedicated online bibliography: OpCit Project, *The Effect of Open Access and Downloads ("Hits") on Citation Impact: A Bibliography of Studies*, opcit.eprints.org/oacitation-biblio.html.

12. Harnad and Brody, "Comparing the Impact of Open Access (OA) vs. Non-OA Articles in the Same Journals," *D-Lib Magazine* 10, no. 6 (2004), www.dlib.org/dlib/june04/harnad/06harnad.html; Chawki Hajjem et al., *Open Access to Research Increases Citation Impact [Technical Report]* (Montréal: Université du Québec, 2005), eprints.ecs.soton.ac.uk/11687/; Hajjem, Harnad, and Gingras, "Ten-Year Cross-Disciplinary Comparison of the Growth of Open Access and How It Increases Research Citation Impact," *Bulletin of the IEEE Computer Society Technical Committee on Data Engineering* 28, no. 4 (2005): 39–47, eprints.ecs.soton.ac.uk/11688/.

13. Antelman, "Do Open Access Articles Have a Greater Research Impact?" *College & Research Libraries* 65, no. 5 (2004): 372–82, eprints.rclis.org/archive/00002309/; Eysenbach, "Citation Advantage of Open Access Articles," *PLoS Biology* 4, no. 5 (2006): e157, biology.plosjournals.org/archive/1545-7885/4/5/pdf/10.1371_journal.pbio.0040157-L.pdf; Norris, Oppenheim, and Rowland, "The Citation Advantage of Open-Access Articles," *Journal of the American Society for Information Science and Technology* 59, no. 12 (2008): 1963–72. Contrary evidence came, on the other hand, from the biomedical field in 2001: a combined assessment of citation scores, Web usage statistics, and author perceptions on a sample of free online articles supplementing the non-OA section of the journal *Pediatrics* over the years 1997–1999 revealed that, in spite of heavier download rates, the former were significantly less cited and, at least in the survey year (2001), were perceived by scholars as somewhat "second-tier" to print papers. See Kent Anderson et al., "Publishing Online-Only Peer-Reviewed Biomedical Literature: Three Years of Citation, Author Perception, and Usage Experience," *Journal of Electronic Publishing* 6, no. 3 (2001), hdl.handle.net/2027/spo.3336451.0006.303.

14. The early view effect has been examined in relation to a sample of papers issued by *The Astrophysical Journal* in Greg J. Schwarz and Robert C. Kennicutt Jr., "Demographic and Citation Trends in Astrophysical Journal Papers and Preprints," *Bulletin of the American Astronomical Society* 36, no. 5 (2004): 1654–63, arxiv.org/abs/astro-ph/0411275. This paper's results have been subsequently extended by Travis S. Metcalfe, "The Rise and Citation Impact of Astro-Ph in Major Journals," *Bulletin of the American Astronomical Society* 37, no. 2 (2005):

555–57, arxiv.org/pdf/astro-ph/0503519. An attempt to systematically account for the effects of both early view and self-selection bias in astronomical literature has been made, within the framework of a probabilistic citation model, by Michael J. Kurtz et al., "The Effect of Use and Access on Citations," *Information Processing & Management* 41, no. 6 (2005): 1395–1402, arxiv.org/abs/cs/0503029. A substantial confirmation of Kurtz's hypothesis relative to the influence of self-promotion on the increased citation impact of articles listed at the top of the daily arXiv:astro-ph mailings is provided by Jörg P. Dietrich, "Disentangling Visibility and Self-Promotion Bias in the Arxiv: Astro-Ph Positional Citation Effect," *Publications of the Astronomical Society of the Pacific* 120, no. 869 (2008): 801–4, arxiv.org/abs/0805.0307v2. A self-selection bias for OA articles issued by four journals of the London Mathematical Society is reported in Philip M. Davis and Michael J. Fromerth, "Does the ArXiv Lead to Higher Citations and Reduced Publisher Downloads for Mathematics Articles?" *Scientometrics* 71, no. 2 (2007): 203–15, arxiv.org/abs/cs.DL/0603056. Another study, performed by Davis and colleagues on a sample of eleven journals issued by the American Psychological Society, for which the experimenters were granted the permission to manipulate the access status of individual articles, led the authors to conclude that "open access increases the readership of articles but has no effect on the number of citations in the first year after publication." Davis et al., "Open Access Publishing, Article Downloads, and Citations: Randomised Controlled Trial," *BMJ* [online first] 337 (2008): 568, www.bmj.com/cgi/content/full/337/jul31_1/a568. Similar conclusions have been reached, in the field of biological and biomedical sciences, by Davis, "Author-Choice Open Access Publishing in the Biological and Medical Literature: A Citation Analysis," *Journal of the American Society for Information Science and Technology* (2008), arxiv.org/abs/0808.2428, and Jonathan D. Wren, "Open Access and Openly Accessible: A Study of Scientific Publications Shared Via the Internet," *BMJ* 330, no. 7500 (2005): 1128, www.bmj.com/cgi/reprint/330/7500/1128. This study shows that more recent articles from higher impact journals are more likely to be available online at nonjournal websites. In the field of economics, also, authors publishing in high-impact journals have been found much more likely to self-archive than those publishing in low-impact journals: Ted C. Bergstrom and Rosemarie Lavaty, *How Often Do Economists Self-Archive?* (Santa Barbara, University of California, Department of Economics), 2007, repositories.cdlib.org/ucsbecon/bergstrom/2007a. A thorough critical analysis of Davis's findings, yet with methodological caveats that may also be extended to similar studies arguing against any OA advantage, can be found in Harnad's posts to the Open Access Archivangelism blog: Harnad, *Davis et Al.'s 1–Year Study of Self-Selection Bias: No Self-Archiving Control, No OA Effect, No Conclusion* (31 July 2008), openaccess.eprints.org/index

.php?/archives/441-Davis-et-als-1-year-Study-of-Self-Selection-Bias-No-Self-Archiving-Control,-No-OA-Effect,-No-Conclusion.html; Harnad, *Confirmation Bias and the Open Access Advantage: Some Methodological Suggestions for Davis's Citation Study* (25 August 2008), openaccess.eprints.org/index.php?/archives/451-Confirmation-Bias-and-the-Open-Access-Advantage-Some-Methodological-Suggestions-for-Daviss-Citation-Study.html.

15. Moed, "The Effect of 'Open Access' upon Citation Impact: An Analysis of ArXiv's Condensed Matter Section," *Journal of the American Society for Information Science and Technology* 58, no. 13 (2007): 2047–54, arxiv.org/abs/cs/0611060.

16. James Testa and Marie E. McVeigh, *The Impact of Open Access Journals. A Citation Study from Thomson ISI* (Stamford, Conn.: The Thomson Corporation, 2004), 10, www.isinet.com/media/presentrep/acropdf/impact-oa-journals.pdf.

17. McVeigh, *Open Access Journals in the ISI Citation Databases: Analysis of Impact Factors and Citation Patterns, A Citation Study from Thomson Scientific* (Stamford, Conn.: The Thomson Corporation, 2004), www.isinet.com/media/presentrep/essayspdf/openaccesscitations2.pdf.

18. Iain D. Craig et al., "Do Open Access Articles Have Greater Citation Impact? A Critical Review of the Literature," *Journal of Informetrics* 1, no. 3 (2007): 239–48, www.publishingresearch.net/Citations-SummaryPaper3_000.pdf.pdf.

19. See, for example, John Willinsky, *The Access Principle: The Case for Open Access to Research and Scholarship* (Cambridge, Mass.: The MIT Press, 2006), chapter 10, mitpress.mit.edu/catalog/item/default.asp?ttype=2&tid=10611.

20. Harnad, "Open Access Scientometrics and the UK Research Assessment Exercise" (paper presented at the 11th Annual Meeting of the International Society for Scientometrics and Informetrics, Madrid, 25–27 June 2007), arxiv.org/abs/cs/0703131.

21. On the "citation matching" problem faced by current digital libraries, see Dongwon Lee et al., "Are Your Citations Clean? New Scenarios and Challenges in Maintaining Digital Libraries," *Communications of the ACM* 50, no. 12 (2007): 33–38, www.cse.psu.edu/~on/cacm06.pdf. The technical and conceptual obstacles to an open access scientometrics with respect to the role of digital libraries and repositories in the collection of relevant (meta)data are discussed by Chris Ambruster, *Access, Usage and Citation Metrics: What Function for Digital Libraries and Repositories in Research Evaluation?* (29 January 2008), ssrn.com/abstract=1088453.

22. On the rationale and technical specifications of the *Citebase* project, see Brody, "Citebase Search: Autonomous Citation Database for E-Print Archives" (paper presented at the Worldwide Coherent Workforce, Satisfied Users. New

Services for Scientific Information, Oldenburg, Germany, 17–19 September 2003), physnet.physik.uni-oldenburg.de/projects/SINN/sinn03/proceedings/brody/brody_sinn03_paper.pdf; Brody, *Citation Analysis in the Open Access World* (2004), eprints.ecs.soton.ac.uk/10000/; Steve Hitchcock et al., "Open Citation Linking: The Way Forward," *D-Lib Magazine* 8, no. 10 (2002), eprints.ecs.soton.ac.uk/7717/.

23. A positive test of the correlation generator on the UK *arXiv*'s mirror is reported in Brody, Harnad, and Carr, "Earlier Web Usage Statistics as Predictors of Later Citation Impact," *Journal of the American Society for Information Science and Technology* 57, no. 8 (2006): 1060–72, eprints.ecs.soton.ac.uk/10713/. A good correlation between hit-counts and subsequent citations has been found also, for a cohort of papers published in the *British Medical Journal* in 1999, by Thomas V. Perneger, "Relations Between Online 'Hit Counts' and Subsequent Citations: Prospective Study of Research Papers in the BMJ," *BMJ* 329, no. 7465 (2004): 546–47, www.bmj.com/cgi/content/full/329/7465/546.

24. For a presentation of *CiteSeer*'s technical features and capabilities, see Giles, Bollacker, and Lawrence, "CiteSeer: An Automatic Citation Indexing System," in *Digital Libraries 98—Third ACM Conference on Digital Libraries*, ed. I. Witten, R. Akscyn, and F. Shipman III, 89–98 (New York: ACM Press, 1998), clgiles.ist.psu.edu/papers/DL-1998-citeseer.pdf; Lawrence, Giles, and Bollacker, "Digital Libraries and Autonomous Citation Indexing," *IEEE Computer* 32, no. 6 (1999): 67–71, citeseer.ist.psu.edu/lawrence99digital.html.

25. A beta test of the acknowledgment index can be accessed at cs1.ist.psu.edu:8080/acksearch/. For a full-paper presentation of this new service, see Isaac G. Councill et al., "Automatic Acknowledgment Indexing: Expanding the Semantics of Contribution in the CiteSeer Digital Library," in *Proceedings of the Third International Conference on Knowledge Capture (K-CAP 2005)*, 19–26 (New York: ACM Press, 2005), citeseer.ist.psu.edu/councill05automatic.html, and Giles and Councill, "Who Gets Acknowledged: Measuring Scientific Contributions Through Automatic Acknowledgment Indexing," *Proceedings of the National Academy of Sciences* 101, no. 51 (2004): 17599–604, www.pnas.org/cgi/content/abstract/0407743101v1.

26. For a presentation of the *Web Citation Index,* see Angela Martello, *Selection of Content for the Web Citation Index (TM): Institutional Repositories and Subject-Specific Archives* (Thomson Scientific, 2005), www.thomsonreuters.com/business_units/scientific/free/essays/wci-selection/.

27. On the analogy link/citation and the extension of bibliometric methods to the World Wide Web, see Cronin, "Bibliometrics and Beyond: Some Thoughts on Web-Based Citation Analysis," *Journal of Information Science* 27, no. 1 (2001): 1–7, and Christine L. Borgman and Jonathan Furner, "Scholarly Communication

and Bibliometrics," *Annual Review of Information Science and Technology* 36 (2002): 3–72, citeseerx.ist.psu.edu/viewdoc/summary?doi=10.1.1.88.4764.

28. A review of the three disciplinary perspectives on link analysis can be found in Han Woo Park and Mike Thelwall, "Hyperlink Analyses of the World Wide Web: A Review," *Journal of Computer-Mediated Communication* 8, no. 4 (2003), jcmc.indiana.edu/vol8/issue4/park.html.

29. There are already several good reviews of the current trends in complex network analysis, including Andrea Scharnhorst, "Complex Networks and the Web: Insights from Nonlinear Physics," *Journal of Computer-Mediated Communication* 8, no. 4 (2003), jcmc.indiana.edu/vol8/issue4/scharnhorst.html; Albert and Barabási, "Statistical Mechanics of Complex Networks," *Reviews of Modern Physics* 74, no. 1 (2002): 47–97, www.socialnetwork.hu/cikkek/Bar-AR0106096.pdf; and Mark E. J. Newman, "The Structure and Function of Complex Networks," *SIAM Review* 45, no. 2 (2003): 167–256, citeseer.ist.psu.edu/newman03structure.html. A popular science book on the subject is Barabási, *Link: The New Science of Networks* (Cambridge, Mass.: Perseus Publishing, 2002).

30. Watts and Strogatz, "Collective Dynamics of "Small World" Networks," *Nature* 393, no. 6684 (1998): 440–42, www.tam.cornell.edu/tam/cms/manage/upload/SS_nature_smallworld.pdf.

31. Kochen, ed., *The Small World* (Norwood, N.J.: Ablex, 1989), vii. The book's introductory essay is a reprint of Pool and Kochen, "Contacts and Influence," *Social Networks* 1, no. 1 (1978): 5–51, a paper taking advantage of reflections and experiments on the small world problem initiated by the authors in the 1950s.

32. Milgram, "The Small World Problem," *Psychology Today* 1, no. 1 (1967): 61–67.

33. Granovetter, "The Strength of Weak Ties," *American Journal of Sociology* 78, no. 6 (1973): 1360–80.

34. Albert, Jeong, and Barabási, "Diameter of the World Wide Web," *Nature* 401, no. 6749 (1999): 130–31, arxiv.org/abs/cond-mat/9907038.

35. Björneborn, "Small-World Link Structures Across an Academic Web Space: A Library and Information Science Approach" (Ph.D. dissertation, Department of Information Studies, Royal School of Library and Information Science, Copenhagen, 2004), vip.db.dk/lb/phd/phd-thesis.pdf.

36. Barabási and Albert, "Emergence of Scaling in Random Networks," *Science* 286, no. 5439 (1999): 509-12, arxiv.org/abs/cond-mat/9910332.

37. Andrei Broder et al., "Graph Structure in the Web," in *Proceedings of the 9th International World Wide Web Conference on Computer Networks: The International Journal of Computer and Telecommunications Networking* (Amsterdam: North-Holland Publishing, 2000), www9.org/w9cdrom/160/160.html.

38. The explicit statement that complex network theory ought to function as a heuristic tool for the methodological foundation of Web indicators can be found in Andrea Scharnhorst and Paul Wouters, "Web Indicators—a New Generation of S&T Indicators?" *Cybermetrics* 10, no. 1, paper 6 (2006), www.cindoc.csic.es/cybermetrics/articles/v10i1p6.html.

39. Besides sampling methods based on random testing of IP addresses or generation of random queries, other methodologies have been experimented that explicitly take into account the Web graph structure, for example, the random-walk models. For a survey, see Paat Rusmevichientong et al., "Methods for Sampling Pages Uniformly from the World Wide Web" (paper presented at the AAAI Fall Symposium on Using Uncertainty Within Computation, 2001), citeseer.ist.psu.edu/455785.html.

40. Jackson, "Assessing the Structure of Communication on the World Wide Web," *Journal of Computer-Mediated Communication* 3, no. 1 (1997), www.nicoladoering.de/Hogrefe/jackson.htm.

41. An overview of HNA is in Park, "Hyperlink Network Analysis: A New Method for the Study of Social Structure on the Web," *Connections* 25, no. 1 (2003): 49–61.

42. Bossy, "The Last of the Litter: 'Netometrics'," *Solaris* 2 (1995), biblio-fr.info.unicaen.fr/bnum/jelec/Solaris/d02/2bossy.html.

43. Cronin, *The Hand of Science*, 184.

44. Josep Manuel Rodríguez i Gairín, "Valorando El Impacto De La Información en Internet: Altavista, El 'Citation Index' De La Red," *Revista Española de Documentación Científica* 20, no. 2 (1997): 175–81, www.cindoc.csic.es/cybermetrics/pdf/120.pdf; Turnbull, *Bibliometrics and the World Wide Web* (Technical Report FIS-12-19-1996-1) (Toronto: Faculty of Information Studies, University of Toronto, 1996), www.ischool.utexas.edu/~donturn/research/bibweb.html.

45. Larson, "Bibliometrics of the World Wide Web: An Exploratory Analysis of the Intellectual Structure of Cyberspace," in *Global Complexity: Information, Chaos and Control: Proceedings of the ASIS 1996 Annual Meeting* (1996), sherlock.berkeley.edu/asis96/asis96.html; Ingwersen, "The Calculation of Web Impact Factors," *Journal of Documentation* 54, no. 2 (1998): 236–43.

46. Bar-Ilan, "The 'Mad Cow Disease': Usenet Newsgroups and Bibliometric Laws," *Scientometrics* 39, no. 1 (1997): 29–55.

47. Almind and Ingwersen, "Informetric Analysis of the World Wide Web: Methodological Approaches to 'Webometrics'," *Journal of Documentation* 53, no. 4 (1997): 404.

48. The references for a limited number of studies exemplifying the above-mentioned research lines are given here cumulatively in the same order as they are listed in the text:

a Cristina Faba-Pérez, Vicente P. Guerrero-Bote, and Felix Moya-Anegon, "Data Mining in a Closed Web Environment," *Scientometrics* 58, no. 3 (2003): 623–40.

b Polanco et al., "Clustering and Mapping Web Sites for Displaying Implicit Associations and Visualizing Networks" (paper presented at the NTTS&ETK 2001 Conference, Hersonissos, 2001), www.math.upatras.gr/~mboudour/articles/web_clustering&mapping.pdf.

c Vaughan and You, "Comparing Business Competition Positions Based on Web Co-Link Data: The Global Market vs. the Chinese Market," *Scientometrics* 68, no. 3 (2006): 611–28.

d Ravi Kumar et al., "Trawling the Web for Emerging Cyber-Communities," *Computer Networks* 31, no. 11 (1999): 1481–93, www8.org/w8-papers/4a-search-mining/trawling/trawling.html.

e Moses A. Boudourides, Beatrice Sigrist, and Philippos D. Alevizos, "Webometrics and the Self-Organization of the European Information Society" (paper presented at the Task 2.1 of the SOEIS Project, Rome Meeting, 17–19 June 1999), hyperion.math.upatras.gr/webometrics/.

f Jeffrey Dean and Monika R. Henzinger, "Finding Related Pages in the World Wide Web" (paper presented at the 8th International World Wide Web Conference, May 1999, Toronto, Canada, 1999), citeseerx.ist.psu.edu/viewdoc/summary?doi=10.1.1.50.1638. A method to improve the rankings based on co-citation links is reported in Jacques Savoy, "Citation Schemes in Hypertext Information Retrieval," in *Information Retrieval and Hypertext*, ed. Maristella Agosti and Alan F. Smeaton, 99–120 (Boston: Kluwer, 1996). In combination with links and bibliographic couplings, co-links' ability to identify similar academic websites is tested against human classification schemes in Thelwall and Wilkinson, "Finding Similar Academic Web Sites with Links, Bibliometric Couplings and Colinks," *Information Processing & Management* 40, no. 3 (2004): 515–26, citeseerx.ist.psu.edu/viewdoc/summary?doi=10.1.1.65.95.

g. Camille Prime-Claverie, Michael Beigbeder, and Thierry Lafouge, "Transposition of the Cocitation Method with a View to Classifying Web Pages," *Journal of the American Society for Information Science and Technology* 55, no. 14 (2004): 1282–89, citeseerx.ist.psu.edu/viewdoc/summary?doi=10.1.1.76.336.

h Junping Qiu et al., "An Exploratory Study on Substantive Co-Link Analysis: A Modification to Total Co-Link Analysis," *Scientometrics* 76, no. 2 (2008): 327–41.

49. Prabowo and Thelwall, "Finding and Tracking Subjects Within an Ongoing Debate," *Journal of Informetrics* 2, no. 2 (2008): 107–27; Leydesdorff and

Curran, "Mapping University-Industry-Government Relations on the Internet: The Construction of Indicators for a Knowledge-Based Economy," *Cybermetrics* 4, no. 1, paper 2 (2000), www.cindoc.csic.es/cybermetrics/articles/v4i1p2.html.

50. Nancy C. M. Ross and Dietmar Wolfram, "End User Searching on the Internet: An Analysis of Term Pair Topics Submitted to the Excite Search Engine," *Journal of the American Society for Information Science* 51, no. 10 (2000): 949–58.

51. Polanco, Roche, and Besagni, "User Science Indicators in the Web Context and Co-Usage Analysis," *Scientometrics* 66, no. 1 (2006): 171–82.

52. Such a method is applied to the rating of health care related websites by Lei Cui, "Rating Health Web Sites Using the Principles of Citation Analysis: A Bibliometric Approach," *Journal of Medical Internet Research* 1, no. 1 (1999): e4, www.jmir.org/1999/1/e4/.

53. The WIF's original definition is set forth in Ingwersen, "The Calculation of Web Impact Factors," 237. The subsequent adjustments to the original formula have been reviewed in Xuemei Li, "A Review of the Development and Application of the Web Impact Factor," *Online Information Review* 27, no. 6 (2003): 407–17; Alireza Noruzi, "The Web Impact Factor: A Critical Review," *The Electronic Library* 24, no. 4 (2006): 490–500, eprints.rclis.org/archive/00005543/.

54. A review of Web data collection methods and relative limitations from an informetric perspective is given by Bar-Ilan, "Data Collection Methods on the Web for Informetric Purposes: A Review and Analysis," *Scientometrics* 50, no. 1 (2001): 7–32. Early critical views on search engines' coverage and reliability are also found in Herbert W. Snyder and Howard Rosenbaum, "Can Search Engines Be Used as Tools for Web-Link Analysis? A Critical View," *Journal of Documentation* 55, no. 4 (1999): 375–84; Thelwall, "Web Impact Factors and Search Engine Coverage," *Journal of Documentation* 56, no. 2 (2000): 185–89, wlv.openrepository.com/wlv/handle/2436/3141; Viv Cothey, "Web-Crawling Reliability," *Journal of the American Society for Information Science and Technology* 55, no. 14 (2004): 1228–38, www.scit.wlv.ac.uk/~in7803/publications/cothey_2004.pdf.

55. An in-depth analysis of web pages' persistence over time has been conducted, on a single set of URLs monitored continually since 1996, by Wallace Koehler. A résumé of his findings and a mini-review of previous literature on the subject can be found in Koehler, "A Longitudinal Study of Web Pages Continued: A Consideration of Document Persistence," *Information Research* 9, no. 2 (2004), informationr.net/ir/9–2/paper174.html.

56. The database is available at cybermetrics.wlv.ac.uk/database and is described in Thelwall, "A Free Database of University Web Links: Data Collection Issues," *Cybermetrics* 6/7, no. 1, paper 2 (2002), www.cindoc.csic.es/

cybermetrics/articles/v6i1p2.html. On the Wolverhampton crawler, see Thelwall, "A Web Crawler Design for Data Mining," *Journal of Information Science* 27, no. 5 (2001): 319–25.

57. The rationale behind the four ADMs is explained in Thelwall, "Conceptualising Documentation on the Web: An Evaluation of Different Heuristic-Based Models for Counting Links Between University Web Sites," *Journal of the American Society for Information Science and Technology* 53, no. 12 (2002): 995–1005, citeseerx.ist.psu.edu/viewdoc/summary?doi=10.1.1.65.1637. A further refinement, with three additional ADMs, is presented in Thelwall and Wilkinson, "Three Target Document Range Metrics for University Web Sites," *Journal of the American Society for Information Science and Technology* 54, no. 6 (2003): 489–96, citeseerx.ist.psu.edu/viewdoc/summary?doi=10.1.1.64.2423.

58. For a presentation of the CINDOC ranking see, besides the methodological and "Best Practices" sections on the website, Isidro F. Aguillo et al., "Scientific Research Activity and Communication Measured with Cybermetrics Indicators," *Journal of the American Society for Information Science and Technology* 57, no. 10 (2006): 1296–1302.

59. Thelwall, Vaughan, and Björneborn, "Webometrics," *Annual Review of Information Science and Technology* 39 (2005): 103.

60. Payne and Thelwall, "Longitudinal Trends in Academic Web Links," *Journal of Information Science* 34, no. 1 (2008): 3–14, citeseerx.ist.psu.edu/viewdoc/summary?doi=10.1.1.83.3769.

61. Good correlations between hyperlink rates and research performance on samples of UK universities are reported, for instance, in the above-cited Thelwall, "Conceptualising Documentation," and in the following: Thelwall, "Extracting Macroscopic Information from Web Links," *Journal of the American Society for Information Science and Technology* 52, no. 13 (2001): 1157–68, citeseerx.ist.psu.edu/viewdoc/summary?doi=10.1.1.64.7031; Thelwall and Harries, "The Connection Between the Research of a University and Counts of Links to Its Web Pages: An Investigation Based upon a Classification of the Relationships of Pages to the Research of the Host University," *Journal of the American Society for Information Science and Technology* 54, no. 7 (2003): 594–602, citeseerx.ist.psu.edu/viewdoc/summary?doi=10.1.1.20.4172. Any causal interpretation of such figures, however, is prevented by the author himself on account of the peculiar functions accomplished by hyperlinks in the Web environment. A series of caveats on the analogy citation/sitation with reference to mapping techniques are set forth also in Camille Prime, Elise Bassecoulard, and Michel Zitt, "Co-Citations and Co-Sitations: A Cautionary View on an Analogy," *Scientometrics* 54, no. 2 (2002): 291–308.

62. On academic hyperlink motivations, see Thelwall, "What Is This Link Doing Here? Beginning a Fine-Grained Process of Identifying Reasons for Academic Hyperlink Creation," *Information Research* 8, no. 3 (2003), informationr.net/ir/8-3/paper151.html; Wilkinson et al., "Motivations for Academic Web Site Interlinking: Evidence for the Web as a Novel Source of Information on Informal Scholarly Communication," *Journal of Information Science* 29, no. 1 (2003): 49–56.

Conclusions

At the terminus of this long journey across bibliometric literature, a strange aftertaste arises from the impression that, at least as far as its flair for the use of citation analysis in research evaluation is concerned, the space of biblio/sciento/informetrics and its immediate vicinity seem perpetually placed on the edge of a theoretical precipice, a sort of Manichean attitude that drives any would-be inhabitant toward an exclusive choice: either you do believe—that citations are Mertonian, that the skewness of their distribution is not a problem, and so on and on—or you don't. If you do, then you join the club of those allowed to gather the streams of evidence gushing from citation databases and to pack them into the polished dress of a paper showing off *Scientometrics* (or similar) brand identity. If you don't, then everything falls apart, you're not allowed to judge a researcher by the times he or she gets cited, you cannot trace schools or research fronts by patterns of co-citations, and those who believe become a tribe of number-crunchers who count the uncountable for a living. Even the *SCI* itself, then, turns into a Yankee database that perpetuates "the myth of Garfield and citation indexing."[1]

Manicheism vanishes if things are placed in historical perspective. The availability, beginning in the 1960s, of an interdisciplinary citation index for the sciences was a unique occasion for various categories of professionals: for information scientists willing to experiment with automatic indexing techniques alternative or complementary to keyword indexing; for historians and sociologists seeking factual evidence about the structure and dynamics of disciplinary scientific communities, as well as on the winding pathways followed by ideas in their historical evolution; for mathematicians eager to face the challenge of modeling, within the frame of the fractal theory of complexity, a kind of information process in which one item (article) can have many sources (articles referencing it); and for science policymakers interested in supporting (or justifying) critical decisions with sound quantitative arguments on research productivity and performance. Though the original motivation—to improve information retrieval—is historically at the origin of citation indexing, early on the sociological and political value of citation statistics occupied a prominent place on the research agenda.

The philosophical inspiration of the pioneers in pursuing the above lines of inquiry, however, faded gradually into the background. Bernal's radical project to overturn the journal-based publication system, Price's vision of a detailed war map of science at the service of government bureaucrats planning their maneuvers, Garfield and Small's encyclopedic view of scientific literature as an integrated whole joined together by cross-disciplinary bibliographic links, were superseded by the concrete deployment of the scientometric arsenal they had made possible. Whereas Bernal's input would eventually find an ideal continuation in the open access movement, the citation machine set into motion by Garfield and Small led to the proliferation of sectorial studies of a fundamentally empirical nature. Despite the lack of a general agreement on a theory of science providing clear guidelines for the construction of scientometric indicators, subsequent bibliometricians simply took for granted the cognitive value of bibliographic citations and, without worrying too much about epistemological dilemmas, handled them in compliance with the same blueprint followed by other social scientists in cognate areas. First, a clearly delimited problem is defined in such a way that its solution is affordable without resorting to divination; then, data are collected by means of the existing tools (online databases), without imposing on them any a priori interpretative scheme, yet with a series of tacit assumptions about "how things work"; and finally, provided the twisting effect of the observation instrument is made explicit, data are processed by means of statistical techniques and a plausible, more or less refined, mathematical model for the explanation of the results is searched for.[2]

Taken individually, bibliographic citations do not necessarily imply impact or influence or any other kind of quality-driven relationship between the cited and citing authors. Unless one is dealing with self-evident occurrences of breakthrough science, their resolution is too low to ensure the singling out of individual excellence across the crowds of dwarfs thronging the neighborhood of science. If, on the other hand, their aggregate behavior is investigated in a tightly defined experimental setting, and with all the caution necessary to keep the mathematical oddities of a skewed universe under control, then their ability to capture one of the many conceivable facets of scientific quality shows up. And even so, a conviction is widely shared that the best results are still obtained by the application of different evaluation techniques. As passionately advocated by Moravcsik almost twenty-five years ago, to get the best out of their job, scientometricians, no less than other categories of social scientists, have to learn to live in a multidimensional world.[3]

NOTES

1. George A. Matter and Henri Broms, "The Myth of Garfield and Citation Indexing," *Tidskrift for Dokumentation* 39, no. 1 (1983): 1–8.
2. van Raan, "Measuring Science: Capita Selecta of Current Main Issues," in *Handbook of Quantitative Science and Technology Research: The Use of Publication and Patent Statistics in Studies of S&T Systems*, ed. Henk F. Moed, Wolfgang Glänzel, and Ulrich Schmoch, 19–50 (Dordrecht: Kluwer, 2004), www.cwts.nl/TvR/documents/AvR-HandbChKluw.pdf.
3. Moravcsik, "Life in a Multidimensional World," *Scientometrics* 6, no. 2 (1984): 75–86.

Selected Bibliography

Abt, Helmut A. "Long-Term Citation Histories of Astronomical Papers." *Publications of the Astronomical Society of the Pacific* 93, no. 552 (1981): 207–10. adsabs.harvard.edu/abs/1981PASP...93..207A.

Adams, Jonathan, Karen Gurney, and Louise Jackson. "Calibrating the Zoom: A Test of Zitt's Hypothesis." *Scientometrics* 75, no. 1 (2008): 81–95.

Adler, Robert, John Ewing, and Peter Taylor. *Citation Statistics: A Report from the International Mathematical Union (IMU) in Cooperation with the International Council of Industrial and Applied Mathematics (ICIAM) and the Institute of Mathematical Statistics (IMS)*. 2008. www.mathunion.org/fileadmin/IMU/Report/CitationStatistics.pdf.

Aguillo, Isidro F., Begoña Granadino, Jose L. Ortega, and José A. Prieto. "Scientific Research Activity and Communication Measured with Cybermetrics Indicators." *Journal of the American Society for Information Science and Technology* 57, no. 10 (2006): 1296–1302.

Ahlgren, Per, Bo Jarneving, and Ronald Rousseau. "Requirements for a Cocitation Similarity Measure, with Special Reference to Pearson's Correlation Coefficient." *Journal of the American Society for Information Science and Technology* 54, no. 6 (2003): 550–60.

Ahmed, Tanzila, Ben Johnson, Charles Oppenheim, and Catherine Peck. "Highly Cited Old Papers and the Reasons Why They Continue to Be Cited, Part II: The 1953 Watson and Crick Article on the Structure of DNA." *Scientometrics* 61, no. 2 (2004): 147–56.

Aksnes, Dag W., and Gunnar Siversten. "The Effect of Highly Cited Papers on National Citation Indicators." *Scientometrics* 59, no. 2 (2004): 213–24. english.nifustep.no/content/download/2864/28408/file/Scientometrics-paper.pdf.

Albert, Michael B., Daniel Avery, Francis Narin, and Paul McAllister. "Direct Validation of Citation Counts as Indicators of Industrially Important Patents." *Research Policy* 20, no. 3 (1991): 251–59.

Albert, Réka, and Albert László Barabási. "Statistical Mechanics of Complex Networks." *Reviews of Modern Physics* 74, no. 1 (2002): 47–97. www.socialnetwork.hu/cikkek/Bar-AR0106096.pdf.

Albert, Réka, Hawoong Jeong, and Albert László Barabási. "Diameter of the World Wide Web." *Nature* 401, no. 6749 (1999): 130–31. www.nd.edu/~networks/Papers/401130A0.pdf.

Allison, David, Derek John de Solla Price, Belver C. Griffith, Michael J. Moravcsik, and John A. Stewart. "Lotka's Law: A Problem in Its Interpretation and Application." *Social Studies of Science* 6, no. 2 (1976): 269–76.

Allison, Paul D. "Inequality and Scientific Productivity." *Social Studies of Science* 10, no. 2 (1980): 163–79.

Allison, Paul D., Tad K. Krauze, and J. Scott Long. "Cumulative Advantage and Inequality in Science." *American Sociological Review* 47, no. 5 (1982): 615–25.

Almind, Tomas C., and Peter Ingwersen. "Informetric Analysis of the World Wide Web: Methodological Approaches to 'Webometrics'." *Journal of Documentation* 53, no. 4 (1997: 404–26.

Ambruster, Chris. *Access, Usage and Citation Metrics: What Function for Digital Libraries and Repositories in Research Evaluation?* January 29, 2008. ssrn.com/abstract=1088453.

Anderson, Kent, John Sack, Lisa Krauss, and Lori O'Keefe. "Publishing Online-Only Peer-Reviewed Biomedical Literature: Three Years of Citation, Author Perception, and Usage Experience." *Journal of Electronic Publishing* 6, no. 3 (2001). hdl.handle.net/2027/spo.3336451.0006.303.

Anderson, Thomas R., Robin K. S. Hankin, and Peter D. Killworth. "Beyond the Durfee Square: Enhancing the *h*-Index to Score Total Publication Output." *Scientometrics* 76, no. 3 (2008): 577–78. www.noc.soton.ac.uk/GDD/people/tra/durfee.pdf.

Antelman, Kristin. "Do Open Access Articles Have a Greater Research Impact?" *College & Research Libraries* 65, no. 5 (2004): 372–82. eprints.rclis.org/archive/00002309/.

Antonakis, John, and Rafael Lalive. "Quantifying Scholarly Impact: IQp Versus the Hirsch *h*." *Journal of the American Society for Information Science and Technology* 59, no. 6 (2008): 956–69.

Arunachalam, S., and K. Manorama. "Are Citation-Based Quantitative Techniques Adequate for Measuring Science on the Periphery?" *Scientometrics* 15, nos. 5–6 (1989): 393–408.

Avramescu, Aurel. "Actuality and Obsolescence of Scientific Literature." *Journal of the American Society for Information Science* 30, no. 5 (1979): 296–303.

———. "Theoretical Foundation of Bradford's Law." *International Forum on Information and Documentation* 5, no. 1 (1980): 15–22.

Bailón-Moreno, Rafael, Encarnación Jurado-Alameda, Rosario Ruiz- Baños, and Jean Pierre Courtial. "The Unified Scientometric Model: Fractality and Transfractality." *Scientometrics* 63, no. 2 (2005): 231–57.

Baldi, Stéphane. "Normative versus Social Constructivist Processes in the Allocation of Citations: A Network Analytic Model." *American Sociological Review* 63, no. 6 (1998): 829–46.

Baldi, Stephane, and Lowell L. Hargens. "Re-Examining Price's Conjectures on the Structure of Reference Networks: Results from the Special Relativity, Spatial Diffusion Modeling and Role Analysis Literatures." *Social Studies of Science* 27, no. 4 (1997): 669–87.

Banks, Michael G. "An Extension of the Hirsch Index: Indexing Scientific Topics and Compounds." *Scientometrics* 69, no. 1 (2006): 161–68. arxiv.org/abs/physics/0604216.

Barabási, Albert László. *Link: The New Science of Networks*. Cambridge, Mass.: Perseus Publishing, 2002.

Barabási, Albert László, and Réka Albert. "Emergence of Scaling in Random Networks." *Science* 286, no. 5439 (1999): 509–12. arxiv.org/abs/cond-mat/9910332.

Bar-Hillel, Yehoshua. "The Present State of Research on Mechanical Translation." *American Documentation* 2, no. 4 (1951): 229–37.

Bar-Ilan, Judit. "Data Collection Methods on the Web for Informetric Purposes: A Review and Analysis." *Scientometrics* 50, no. 1 (2001): 7–32.

———. "Informetrics at the Beginning of the 21st Century—A Review." *Journal of Informetrics* 2, no. 1 (2008): 1–52.

———. "The 'Mad Cow Disease', Usenet Newsgroups and Bibliometric Laws." *Scientometrics* 39, no. 1 (1997): 29–55.

Barnett, George A., and Edward L. Fink. "Impact of the Internet and Scholar Age Distribution on Academic Citation Age." *Journal of the American Society for Information Science and Technology* 59, no. 4 (2008): 526–34.

Barnett, George A., Edward L. Fink, and Mary Beth Debus. "A Mathematical Model of Academic Citation Age." *Communication Research* 16, no. 4 (1989): 510–31.

Barthes, Roland. "The Death of the Author." In *Image, Music, Text,* 142–48. London: Fontana Press, 1977.

Basulto Santos, Jesús, and Francisco Javier Ortega Irizo. "Modelling Citation Data with Right Censoring." *Scientometrics* 62, no. 3 (2005): 329–42.

Bates, Marcia J. "Document Familiarity, Relevance, and Bradford's Law: The Getty Online Searching Project Report." *Information Processing & Management* 32, no. 6 (1996): 697–707.

Batista, Pablo D., Mônica G. Campiteli, Osame Kinouchi, and Alexandre S. Martinez. "Is It Possible to Compare Researchers with Different Scientific Interests?" *Scientometrics* 68, no. 1 (2006): 179–89. arxiv.org/pdf/physics/0509048.

Bazerman, Charles. *Shaping Written Knowledge: The Genre and Activity of the Experimental Article in Science*. Madison: University of Wisconsin Press, 1988. wac.colostate.edu/books/bazerman_shaping/.

Beaver, Donald DeB. "Reflections on Scientific Collaboration (and Its Study): Past, Present, and Future." *Scientometrics* 52, no. 3 (2001): 365–77.

Bensman, Stephen J. "Bradford's Law and Fuzzy Sets: Statistical Implications for Library Analyses." *IFLA Journal* 27, no. 4 (2001): 238–46. www.garfield.library.upenn.edu/bensman/iflabradfordslawfuzzysets.pdf.

———. "Distributional Differences of the Impact Factor in the Sciences Versus the Social Sciences: An Analysis of the Probabilistic Structure of the 2005 Journal Citation Reports." *Journal of the American Society for Information Science and Technology* 59, no. 9 (2008): 1366–82. garfield.library.upenn.edu/bensman/bensman072008.pdf.

———. "Garfield and the Impact Factor." *Annual Review of Information Science and Technology* 41 (2007): 93–155. garfield.library.upenn.edu/bensman/bensmanegif2007.pdf.

———. *Garfield and the Impact Factor: The Creation, Utilization, and Validation of a Citation Measure. Part 2: The Probabilistic, Statistical, and Sociological Bases of the Measure* (2007). garfield.library.upenn.edu/bensman/bensmanegif22007.pdf.

———. "Journal Collection Management as a Cumulative Advantage Process." *College & Research Libraries* 46, no. 1 (1985): 13–29. www.garfield.library.upenn.edu/bensman/crl85cum.pdf.

———. "Probability Distributions in Library and Information Science: A Historical and Practitioner Viewpoint." *Journal of the American Society for Information Science* 51, no. 9 (2000): 816–33. www.garfield.library.upenn.edu/bensman/jasistprobart.pdf.

———. "Urquhart's and Garfield's Laws: The British Controversy over Their Validity." *Journal of the American Society for Information Science and Technology* 52, no. 9 (2001): 714–24. www.garfield.library.upenn.edu/bensman/jasisturqgar.pdf.

Bergstrom, Carl T. "Eigenfactor: Measuring the Value and Prestige of Scholarly Journals." *C&RL News* 68, no. 5 (2007). www.ala.org/ala/acrl/acrlpubs/crlnews/backissues2007/may07/eigenfactor.cfm.

Bergstrom, Ted C., and Rosemarie Lavaty. *How Often Do Economists Self-Archive?* Santa Barbara: University of California, Department of Economics, 2007. repositories.cdlib.org/ucsbecon/bergstrom/2007a.

Bernal, John Desmond. "Review of the Science Citation Index." *Science Progress* 53, no. 211 (1965): 455–59. www.garfield.library.upenn.edu/essays/v5p511y1981-82.pdf.

———. *The Social Function of Science*. Cambridge, Mass.: MIT Press, 1967.

Bettencourt, Luís M. A., David I. Kaiser, Jasleen Kaur, Carlos Castillo-Chávez, and David E. Wojick. "Population Modeling of the Emergence and Development of Scientific Fields." *Scientometrics* 75, no. 3 (2008): 495–518. web.mit.edu/dikaiser/www/BKKCW.PopModels.pdf.

Biagioli, Mario. "The Instability of Authorship: Credibility and Responsibility in Contemporary Biomedicine." *The FASEB Journal* 12, no. 1 (1998): 3–16. www.fasebj.org/cgi/content/full/12/1/3.

Biglu, Mohammad Hossein. "The Influence of References per Paper in the SCI to Impact Factors and the Matthew Effect." *Scientometrics* 74, no. 3 (2008): 453–70. eprints.rclis.org/archive/00011935/.

Björneborn, Lennart. "Small-World Link Structures Across an Academic Web Space: A Library and Information Science Approach." Ph.D. dissertation, Royal School of Library and Information Science, Copenhagen, 2004. vip.db.dk/lb/phd/phd-thesis.pdf.

Björneborn, Lennart, and Peter Ingwersen. "Toward a Basic Framework for Webometrics." *Journal of the American Society for Information Science and Technology* 55, no. 14 (2004): 1216–27. www.db.dk/binaries/Perspectives Webometrics-Jasist.pdf.

Bloom, Harold. *The Anxiety of Influence*. 2nd. ed. Oxford: Oxford University Press, 1997.

Bollen, Johan, and Herbert Van de Sompel. "Usage Impact Factor: The Effects of Sample Characteristics on Usage-Based Impact Metrics." *Journal of the American Society for Information Science and Technology* 59, no. 1 (2008): 136–49. arxiv.org/abs/cs.DL/0610154.

Bollen, Johan, Herbert Van de Sompel, and Marco A. Rodriguez. "Towards Usage-Based Impact Metrics." Paper presented at the Joint Conference on Digital Libraries 2008 (JCDL08), Pittsburg, Pennsylvania, June 16, 2008. arxiv.org/abs/0804.3791.

Bonckaert, Patrick, and Leo Egghe. "Rational Normalization of Concentration Measures." *Journal of the American Society for Information Science* 42, no. 10 (1991): 715–22.

Bonitz, Manfred. "Ten Years Matthew Effect for Countries." *Scientometrics* 64, no. 3 (2005): 375–79. www.viniti.ru/icsti_papers/english/Bonitz.pdf.

Bonitz, Manfred, Eberhard Bruckner, and Andrea Scharnhorst. "Characteristics and Impact of the Matthew Effect for Countries." *Scientometrics* 40, no. 3 (1997): 407–22.

Bookstein, Abraham. "Explanation of the Bibliometric Laws." *Collection Management* 3, nos. 2/3 (1979): 151–62.

———. "Implications of Ambiguity for Scientometric Measurement." *Journal of the American Society for Information Science and Technology* 52, no. 1 (2001): 74–79.

———. "Informetric Distributions, Part I: Unified Overview." *Journal of the American Society for Information Science* 41, no. 5 (1990): 368–75. topology .eecs.umich.edu/archive/bookstein_I.pdf.

———. "Informetric Distributions, Part II: Resilience to Ambiguity." *Journal of the American Society for Information Science* 41, no. 5 (1990): 376–86. topology.eecs.umich.edu/archive/bookstein_II.pdf.

———. "Informetric Distributions, Part III: Ambiguity and Randomness." *Journal of the American Society for Information Science* 48, no. 1 (1997): 2–10.

———. "Patterns of Scientific Productivity and Social Change: A Discussion of Lotka's Law and Bibliometric Symmetry." *Journal of the American Society for Information Science* 28, no. 4 (1977): 206–10.

Bookstein, Abraham, and Don R. Swanson. "Probabilistic Models for Automatic Indexing." *Journal of the American Society for Information Science* 25, no. 5 (1974): 312–18.

Borgman, Christine L., and Jonathan Furner. "Scholarly Communication and Bibliometrics." *Annual Review of Information Science and Technology* 36 (2002): 3–72. citeseerx.ist.psu.edu/viewdoc/summary?doi=10.1.1.88.4764.

Börner, Katy, Chaomei Chen, and Kevin W. Boyack. "Visualizing Knowledge Domains." *Annual Review of Information Science and Technology* 37 (2003): 179–255. www.cs.sandia.gov/projects/VxInsight/pubs/arist03.pdf.

Bornmann, Lutz, and Hans-Dieter Daniel. "Convergent Validation of Peer Review Decisions Using the *h* Index: Extent of and Reasons for Type I and Type II Errors." *Journal of Informetrics* 1, no. 3 (2007): 204–13. www .lutz-bornmann.de/icons/BornmannDanielRevised.pdf.

———. "What Do Citation Counts Measure? A Review of Studies on Citing Behavior." *Journal of Documentation* 64, no. 1 (2008): 45–80. www .lutz-bornmann.de/icons/BornmannLutzCitingBehavior.pdf.

Bornmann, Lutz, Rüdiger Mutz, and Hans-Dieter Daniel. "Are There Better Indices for Evaluation Purposes Than the *h* Index? A Comparison of Nine Different Variants of the *h* Index Using Data from Biomedicine." *Journal of the American Society for Information Science and Technology* 59, no. 5 (2008): 830–37. www.lutz-bornmann.de/icons/BornmannMutzDanielFinal.pdf.

———. "The *b* Index as a Measure of Scientific Excellence: A Promising Supplement to the *h* Index." *Cybermetrics* 11, no. 1, paper 6 (2007). www.cindoc.csic.es/cybermetrics/articles/v11i1p6.html.

Bornmann, Lutz, Gerlind Wallon, and Anna Ledin. "Is the *h* Index Related to (Standard) Bibliometric Measures and to the Assessments by Peers? An Investigation of the *h* Index by Using Molecular Life Sciences Data." *Research Evaluation* 17, no. 2 (2008): 149–56. www.lutz-bornmann.de/icons/BornmannWallonLedinRevised.pdf.

Bossy, Marcia J. "The Last of the Litter: 'Netometrics'." *Solaris* 2 (1995). biblio-fr.info.unicaen.fr/bnum/jelec/Solaris/d02/2bossy.html.

Bottle, Robert T., and Y. T. Gong. "A Bibliometric Study on the Ageing and Content Typology Relationship of the Biochemical Literature." *Journal of Information Science* 13, no. 1 (1987): 59–63.

Boudourides, Moses A., Sigrist Beatrice, and D. Alevizos Philippos. "Webometrics and the Self-Organizazion of the European Information Society." Paper presented at the Task 2.1 of the SOEIS Project, Rome Meeting, June 17–19, 1999. hyperion.math.upatras.gr/webometrics/.

Bourdieu, Pierre. *Homo Academicus*. Stanford, Calif.: Stanford University Press, 1988.

Bowden, Mary E., Trudi Bellardo Hahn, and Robert V. Williams, eds. *Proceedings of the 1998 Conference on the History and Heritage of Science Information Systems*. Medford, N.J.: Information Today, 1999.

Boyack, Kevin W., Richard Klavans, and Katy Börner. "Mapping the Backbone of Science." *Scientometrics* 64, no. 3 (2005): 351–74. ivl.slis.indiana.edu/km/pub/2005-boyack-mapbckbn.pdf.

Boyack, Kevin W., Brian N. Wylie, and George S. Davidson, "Domain Visualization Using Vxinsight for Science and Technology Management." *Journal of the American Society for Information Science and Technology* 53, no. 9 (2002): 764–74.

Braam, Robert R., Henk F. Moed, and Anthony F. J. van Raan. "Mapping of Science by Combined Co-Citation and Word Analysis. I: Structural Aspects." *Journal of the American Society for Information Science* 42, no. 4 (1991): 233–51.

———. "Mapping of Science by Combined Co-Citation and Word Analysis. II: Dynamical Aspects." *Journal of the American Society for Information Science* 42, no. 4 (1991): 252–66.

Bradford, Samuel Clement. *Documentation*. London: Crosby Lockwood, 1948.

———. "Sources of Information on Specific Subjects." *Engineering* 137, no. 3550 (1934): 85–86.

Braga, Gilda Maria. "Some Aspects of the Bradford Distribution." *Proceedings of the American Society for Information Science* 15 (1978): 51–54.

Brantle, Thomas F., and M. Hosein Fallah. "Complex Innovation Networks, Patent Citations and Power Laws." In *PICMET 2007 Proceedings,* 540–49. Portland, Ore.: PICMET, 2007.

Braun, Tibor. "Keeping the Gates of Science Journals: Gatekeeping Indicators of National Performance in the Sciences." In *Handbook of Quantitative Science and Technology Research: The Use of Publication and Patent Statistics in Studies of S&T Systems*, edited by Henk F. Moed, Wolfgang Glänzel, and Ulrich Schmoch, 95–111. Dordrecht: Kluwer Academic, 2004.

Braun, Tibor, Ernö Bujdosó, and Andreás Schubert. *Literature of Analytical Chemistry: A Scientometric Evaluation.* Boca Raton, Fla.: CRC Press, 1987.

Braun, Tibor, and Wolfgang Glänzel. "The Sweet and Sour of Journal Citation Rates." *The Chemical Intelligencer* 1 (1995): 31–32.

Braun, Tibor, Wolfgang Glänzel, and Andreás Schubert. "Assessing Assessments of British Science: Some Facts and Figures to Accept or Decline." *Scientometrics* 15, no. 3–4 (1989): 165–70.

———. "The Bibliometric Assessment of UK Scientific Performance—Some Comments on Martin's 'Reply'." *Scientometrics* 20, no. 2 (1991): 359–62.

———. "Publication and Cooperation Patterns of the Authors of Neuroscience Journals." *Scientometrics* 51, no. 3 (2001): 499–510.

Bridgstock, Martin. "The Quality of Single and Multiple Authored Papers: An Unresolved Problem." *Scientometrics* 21, no. 1 (1991): 37–48.

Brin, Sergey, and Lawrence Page. "The Anatomy of a Large-Scale Hypertextual Web Search Engine." *Computer Networks and ISDN Systems* 30, nos. 1–7 (1998): 107–17. www-db.stanford.edu/pub/papers/google.pdf.

Broadus, Robert N. "Toward a Definition of 'Bibliometrics'." *Scientometrics* 12, nos. 5–6 (1987): 373–79.

Broder, Andrei, Ravi Kumar, Farzin Maghoul, Prabhakar Raghavan, Sridhar Rajagopalan, Raymie Stata, Andrew Tomkins, and Janet Wiener. "Graph Structure in the Web." In *Proceedings of the 9th International World Wide Web Conference on Computer Networks: The International Journal of Computer and Telecommunications Networking,* 309–20. Amsterdam: North-Holland Publishing, 2000. www9.org/w9cdrom/160/160.html.

Brody, Tim. *Citation Analysis in the Open Access World.* 2004. eprints.ecs.soton.ac.uk/10000/.

———. "Citebase Search: Autonomous Citation Database for E-Print Archives." Paper presented at the Worldwide Coherent Workforce, Satisfied Users. New Services for Scientific Information, Oldenburg, Germany, September 17–19, 2003. physnet.physik.uni-oldenburg.de/projects/SINN/sinn03/proceedings/brody/brody_sinn03_paper.pdf.

Brody, Tim, Stevan Harnad, and Leslie Carr. "Earlier Web Usage Statistics as Predictors of Later Citation Impact." *Journal of the American Society for Information Science and Technology* 57, no. 8 (2006): 1060–72. eprints.ecs .soton.ac.uk/10713/.

Brookes, Bertram C. "Biblio-, Sciento-, Infor-Metrics??? What Are We Talking About?" In *Informetrics 89/90: Selection of Papers Submitted for the Second International Conference on Bibliometrics, Scientometrics and Informetrics*, edited by Leo Egghe and Ronald Rousseau, 31–43. Amsterdam: Elsevier Science, 1990. doclib.uhasselt.be/dspace/handle/1942/857.

———. "The Bradford Law: A New Calculus for the Social Sciences?" *Journal of the American Society for Information Science* 30, no. 4 (1979): 233–34.

———. "Bradford's Law and the Bibliography of Science." *Nature* 224, no. 5223 (1969): 953–56.

———. "The Complete Bradford-Zipf 'Bibliograph'." *Journal of Documentation* 25, no. 1 (1969): 58–60.

———. "The Growth, Utility and Obsolescence of Scientific Periodical Literature." *Journal of Documentation* 26, no. 4 (1970): 283–94.

———. "The Haitun Dichotomy and the Relevance of Bradford's Law." *Journal of Information Science* 8, no. 1 (1984): 19–24.

———. "Ranking Techniques and the Empirical Log Law." *Information Processing & Management* 20, nos. 1–2 (1984): 37–46.

———. "Theory of the Bradford Law." *Journal of Documentation* 33, no. 3 (1977): 180–209.

———. "Towards Informetrics: Haitun, Laplace, Zipf, Bradford and the Alvey Programme." *Journal of Documentation* 40, no. 2 (1984): 120–43.

Brookes, Bertram C., and José M.Griffith. "Frequency-Rank Distributions." *Journal of the American Society for Information Science* 29, no. 1 (1978): 5–13.

Brooks, Terrence A. "Private Acts and Public Objects: An Investigation of Citer Motivations." *Journal of the American Society for Information Science* 36, no. 4 (1985): 223–29.

Browman, Howard I., and Konstantinos I. Stergiou. "The Use and Misuse of Bibliometric Indices in Evaluating Scholarly Performance," *Ethics in Science and Environmental Politics* [Theme Section] 8, no. 1 (2008). www.int-res .com/abstracts/esep/v8/n1/.

Buchanan, Robert A. "Accuracy of Cited References: The Role of Citation Databases." *College & Research Libraries* 67, no. 4 (2006): 292–303. www.ala .org/ala/acrl/acrlpubs/crljournal/backissues2006a/julycrl/Buchanan06.pdf.

Burrell, Quentin L. "Ambiguity and Scientometric Measurement: A Dissenting View." *Journal of the American Society for Information Science and Technology* 52, no. 12 (2001): 1075–80.

——. "Defining a Core: Theoretical Observations on the Egghe-Rousseau Proposal." *Scientometrics* 57, no. 1 (2003): 75–92.

——. "The 80/20 Rule: Library Lore or Statistical Law?" *Journal of Documentation* 41, no. 1 (1985): 24–39.

——. "Hirsch's *h*-Index: A Stochastic Model." *Journal of Informetrics* 1, no. 1 (2007): 16–25.

——. "Hirsch Index or Hirsch Rate? Some Thoughts Arising from Liang's Data." *Scientometrics* 73, no. 1 (2007): 19–28.

——. "Measuring Similarity of Concentration Between Different Informetric Distributions: Two New Approaches." *Journal of the American Society for Information Science and Technology* 56, no. 7 (2005): 704–14.

——. "Modelling the Bradford Phenomenon." *Journal of Documentation* 44, no. 1 (1988): 1–18.

——. "A Note on Ageing in a Library Circulation Model." *Journal of Documentation* 41, no. 2 (1985): 100–15.

——. "On the Growth of Bibliographies with Time: An Exercise in Bibliometric Prediction." *Journal of Documentation* 45, no. 4 (1989): 302–17.

——. "Review of Virgil Diodato. Dictionary of Bibliometrics." *Journal of Documentation* 51, no. 4 (1995): 448–50.

——. "Scientostochastics?" *Scientometrics* 30, no. 2–3 (1994): 505–9.

——. "Stochastic Modelling of the First-Citation Distribution." *Scientometrics* 52, no. 1 (2001): 3–12.

Burrell, Quentin L., and Violet R. Cane. "The Analysis of Library Data." *Journal of the Royal Statistical Society. Series A (General)* 145, no. 4 (1982): 439–71.

Burrell, Quentin L., and Michael R. Fenton."Yes, the GIGP Really Does Work—and Is Workable!" *Journal of the American Society for Information Science* 44, no. 2 (1993): 61–69.

Burton, R. E., and R. W. Kebler. "The Half-Life of Some Scientific and Technical Literatures." *American Documentation* 11, no. 1 (1960): 18–22.

Bush, Vannevar. *Science: The Endless Frontier*. Washington, D.C.: U.S. Government Printing Office, 1945. www.nsf.gov/about/history/vbush1945.htm.

Callaham, Michael, Robert L. Wears, and Ellen Weber. "Journal Prestige, Publication Bias, and Other Characteristics Associated with Citation of Published Studies in Peer-Reviewed Journals." *JAMA* 287, no. 21 (2002): 2847–50. jama.ama-assn.org/cgi/content/full/287/21/2847.

Callon, Michel, Serge Bauin, Jean Pierre Courtial, and William A. Turner. "From Translation to Problematic Networks: An Introduction to Coword Analysis." *Social Science Information* 22, no. 2 (1983): 191–235.

Callon, Michel, John Law, and Arie Rip, eds. *Mapping the Dynamics of Science and Technology: Sociology of Science in the Real World*. Houndmills, UK: Macmillan Press, 1986.

Cameron, Robert D. "A Universal Citation Database as a Catalyst for Reform in Scholarly Communication." *First Monday* 2, no. 4 (1997). www.firstmonday.dk/issues/issue2_4/cameron/index.html.

Campanario, Juan Miguel. "Using Neural Networks to Study Networks of Scientific Journals." *Scientometrics* 33, no. 1 (1995): 23–40.

Cano, V. "Citation Behavior: Classification, Utility, and Location." *Journal of the American Society for Information Science* 40, no. 4 (1989): 284–90.

Carpenter, Mark P., and Francis Narin. "Validation Study: Patent Citations as Indicatoris of Science and Foreign Dependence." *World Patent Information* 5, no. 3 (1983): 180–85.

Case, Donald O., and Georgeann M. Higgins. "How Can We Investigate Citation Behavior? A Study of Reasons for Citing Literature in Communication." *Journal of the American Society for Information Science* 51, no. 7 (2000): 635–45.

Cason, Hulsey, and Marcella Lubotsky. "The Influence and Dependence of Psychological Journals on Each Other." *Psychological Bulletin* 33, no. 1 (1936): 95–103.

Cawkell, Anthony E. "Checking Research Progress on Image Retrieval by 'Shape-Matching' Using the Web of Science." *ASLIB Proceedings* 50, no. 2 (1998): 27–31.

———. "Methods of Information Retrieval Using Web of Science: Pulmonary Hypertension as a Subject Example." *Journal of Information Science* 26, no. 1 (2000): 66–70.

Chakrabarti, Soumen, Byron Dom, David Gibson, Jon M. Kleinberg, S. Ravi Kumar, Prabhakar Raghavan, Sridhar Rajagopalan, and Andrew Tomkins. "Hypersearching the Web." *Scientific American* 280, no. 6 (1999): 54–60. www.cs.cornell.edu/home/kleinber/sciam99.html.

Chen, Chaomei. *Mapping Scientific Frontiers: The Quest for Knowledge Visualization*. London: Springer, 2003.

Chen, Chaomei, Jian Zhang, Weizhong Zhu, and Michael Vogeley. "Delineating the Citation Impact of Scientific Discoveries." Paper presented at the IEEE/ACM Joint Conference on Digital Libraries (JCDL 2007), Vancouver, BC, Canada, June 17–22, 2007. hdl.handle .net/1860/1885.

Chen, Ye-Sho. "Analysis of Lotka's Law: The Simon-Yule Approach." *Information Processing & Management* 25, no. 5 (1989): 527–44.

Chen, Ye-Sho, P. Pete Chong, and Morgan Y. Tong. "Dynamic Behavior of Bradford's Law." *Journal of the American Society for Information Science* 46, no. 5 (1995): 370–83.

Chen, Ye-Sho, and Ferdinand F. Leimkuhler. "A Relationship Between Lotka's Law, Bradford's Law, and Zipf's Law." *Journal of the American Society for Information Science* 37, no. 5 (1999): 307–14.

Chubin, Daryl E. *Sociology of Sciences: An Annotated Bibliography on Invisible Colleges, 1972–1981*. New York: Garland, 1983.

Chubin, Daryl E., and Soumyo D. Moitra. "Content Analysis of References: Adjunct or Alternative to Citation Counting?" *Social Studies of Science* 5, no. 4 (1975): 423–41.

Cicchetti, Domenic V. "The Reliability of Peer Review for Manuscript and Grant Submissions: A Cross Disciplinary Investigation." *Behavioral and Brain Sciences* 14, no. 1 (1991): 119–86.

Clark, Kenneth E. "The APA Study of Psychologists." *American Psychologists* 9, no. 3 (1954): 117–20.

Cole, Francis Joseph, and Nellie B. Eales. "The History of Comparative Anatomy. Part I: A Statistical Analysis of the Literature." *Science Progress* 11, no. 43 (1917): 578–96.

Cole, Jonathan R., and Stephen Cole. "The Ortega Hypothesis." *Science* 178, no. 4059 (1972): 368–75.

——. "Scientific Output and Recognition: A Study in the Operation of the Reward System in Science." *American Sociological Review* 32, no. 3 (1967): 377–90.

——. *Social Stratification in Science*. Chicago: University of Chicago Press, 1973.

Coleman, S. R. "Disciplinary Variables That Affect the Shape of Bradford's Bibliograph." *Scientometrics* 29, no. 1 (1994): 59–81.

Costas, Rodrigo, and Maria Bordons. "The *h*-Index: Advantages, Limitations, and Its Relation with Other Bibliometric Indicators at the Micro Level." *Journal of Informetrics* 1, no. 3 (2007): 193–203.

Cothey, Viv. "Web-Crawling Reliability." *Journal of the American Society for Information Science and Technology* 55, no. 14 (2004): 1228–38. www.scit.wlv.ac.uk/~in7803/publications/cothey_2004.pdf.

Coughlin, James P., and Robert H. Baran. "Stochastic Models of Information Obsolescence." *Mathematical and Computer Modelling* 11 (1988): 760–65.

Councill, Isaac G., C. Lee Giles, Hui Han, and Eren Manavoglu. "Automatic Acknowledgment Indexing: Expanding the Semantics of Contribution in the CiteSeer Digital Library." In *Proceedings of the Third International Conference on Knowledge Capture (K-CAP 2005),* 19–26. New York: ACM Press, 2005. citeseer.ist.psu.edu/councill05automatic.html.

Courtial, Jean Pierre. "Comments on Leydesdorff's 'A Validation Study of Leximappe'." *Scientometrics* 25, no. 2 (1992): 313–16.

———. "Comments on Leydesdorff's Article." *Journal of the American Society for Information Science* 49, no. 1 (1998): 98.

Cozzens, Susan E. "Comparing the Sciences: Citation Context Analysis of Papers from Neuropharmacology and the Sociology of Science." *Social Studies of Science* 15, no. 1 (1985): 127–53.

———. "Using the Archive: Derek Price's Theory of Differences Among the Sciences." *Scientometrics* 7, nos. 3–6 (1985): 431–41.

———. "What Do Citations Count? The Rhetoric-First Model." *Scientometrics* 15, no. 5–6 (1989): 437–47.

Craig, Iain D., Andrew M. Plume, Marie E. McVeigh, James Pringle, and Mayur Amin. "Do Open Access Articles Have Greater Citation Impact? A Critical Review of the Literature." *Journal of Informetrics* 1, no. 3 (2007): 239–48. www.publishingresearch.net/Citations-SummaryPaper3_000.pdf.pdf.

Crane, Diana. *Invisible Colleges: Diffusion of Knowledge in Scientific Communication*. Chicago; London: University of Chicago Press, 1972.

"Crime Against Nature?" *Science Watch* 13, no. 1 (2002). www.sciencewatch.com/jan-feb2002/index.html.

Cronin, Blaise. "Agreement and Divergence on Referencing Practice." *Journal of Information Science* 3, no. 1 (1981): 27–33.

———. "Bibliometrics and Beyond: Some Thoughts on Web-Based Citation Analysis." *Journal of Information Science* 27, no. 1 (2001): 1–7.

———. *The Citation Process: The Role and Significance of Citations in Scientific Communications*. London: Taylor Graham, 1984. garfield.library.upenn.edu/cronin/citationprocess.pdf.

———. *The Hand of Science: Academic Writing and Its Rewards*. Lanham, Md.: Scarecrow Press, 2005.

———. "Hyperauthorship: A Postmodern Perversion or Evidence of a Structural Shift in Scholarly Communication Practices?" *Journal of the American Society for Information Science and Technology* 52, no. 7 (2001): 558–69.

———. "The Need for a Theory of Citing." *Journal of Documentation* 37, no. 1 (1981): 16–24.

Cronin, Blaise, and Helen Barsky Atkins, eds. *The Web of Knowledge: A Festschrift in Honor of Eugene Garfield*. Medford, N.J.: Information Today, 2000.

Cronin, Blaise, and Lokman I. Meho. "Using the *h*-Index to Rank Influential Information Scientists." *Journal of the American Society for Information Science and Technology* 57, no. 9 (2006): 1275–78. dlist.sir.arizona.edu/1717/.

Cui, Lei. "Rating Health Web Sites Using the Principles of Citation Analysis: A Bibliometric Approach." *Journal of Medical Internet Research* 1, no. 1 (1999): e4. www.jmir.org/1999/1/e4/.

Curry, Michael R. "On the Possibility of Ethics in Geography: Writing, Citing and the Construction of Intellectual Property." *Progress in Human Geography* 15, no. 2 (1991): 125–47.

Davis, Philip M. "Author-Choice Open Access Publishing in the Biological and Medical Literature: A Citation Analysis." *Journal of the American Society for Information Science and Technology* 59, no. 14 (2008): 1–6. arxiv.org/abs/0808.2428.

Davis, Philip M., and Michael J. Fromerth. "Does the ArXiv Lead to Higher Citations and Reduced Publisher Downloads for Mathematics Articles?" *Scientometrics* 71, no. 2 (2007): 203–15. arxiv.org/abs/cs.DL/0603056.

Davis, Philip M., Bruce V. Lewenstein, Daniel H. Simon, James G. Booth, and Matthew J. L. Connolly. "Open Access Publishing, Article Downloads, and Citations: Randomised Controlled Trial." *BMJ* [Online First] 337 (2008): a568. www.bmj.com/cgi/content/full/337/jul31_1/a568.

Davis, Philip M., and Jason S. Price. "Ejournal Interface Can Influence Usage Statistics: Implications for Libraries, Publishers, and Project COUNTER." *Journal of the American Society for Information Science and Technology* 57, no. 9 (2006): 1243–48. arxiv.org/abs/cs.IR/0602060.

Dean, Jeffrey, and Monika R. Henzinger. "Finding Related Pages in the World Wide Web." Paper presented at the 8th International World Wide Web Conference, May 1999, Toronto, Canada. www.cindoc.csic.es/cybermetrics/pdf/47.pdf.

Deng, Zhen, Baruch Lev, and Francis Narin. "Science & Technology as Predictors of Stock Performance." *Financial Analysts Journal* 55, no. 3 (1999): 20–32.

Dieks, Dennis, and Hans Chang. "Differences in Impact of Scientific Publications: Some Indices Derived from a Citation Analysis." *Social Studies of Science* 6, no. 2 (1976): 247–67.

Dietrich, Jörg P. "Disentangling Visibility and Self-Promotion Bias in the Arxiv: Astro-Ph Positional Citation Effect." *Publications of the Astronomical Society of the Pacific* 120, no. 869 (2008): 801–4. arxiv.org/abs/0805.0307v2.

Dietze, Erich von. *Paradigms Explained: Rethinking Thomas Kuhn's Philosophy of Science*. Westport, Conn.: Praeger, 2001.

Ding, Ying, Gobinda G. Chowdhury, and Schubert Foo. "Journals as Markers of Intellectual Space: Journal Co-Citation Analysis of Information Retrieval Area, 1987–1997." *Scientometrics* 47, no. 1 (2000): 55–73.

Diodato, Virgil Pasquale. *Dictionary of Bibliometrics*. New York: Haworth Press, 1994.

Doreian, Patrick, and Thomas J. Fararo. "Structural Equivalence in a Journal Network." *Journal of the American Society for Information Science* 36, no. 1 (1985): 28–37.

Drott, Carl M. "Bradford's Law: Theory, Empiricism and the Gaps Between." *Library Trends* 30, no. 1 (1981): 42–52. hdl.handle.net/2142/7189.

Dunlop, Lasley, and Charles Oppenheim. "The Effect of Recycling on a Patent Citation Network." *World Patent Information* 2, no. 3 (1980): 100–102.

Eagly, Robert V. "Economic Journals as a Communications Network." *Journal of Economic Literature* 13, no. 3 (1975): 878–88.

Eco, Umberto. *On Literature*. Orlando, Fla.: Harcourt, 2004.

Edge, David O. "Quantitative Measures of Communication in Science: A Critical Review." *History of Science* 17 (1979): 102–34.

———. "Why I Am Not a Cocitationist." *Society for the Social Studies of Science Newsletter* 2 (1977): 13–19. www.garfield.library.upenn.edu/essays/v3p240y1977-78.pdf.

Egghe, Leo. "The Duality of Informetric Systems with Applications to the Empirical Laws." *Journal of Information Science* 16, no. 1 (1990): 17–27.

———. "Dynamic *h*-Index: The Hirsch Index in Function of Time." *Journal of the American Society for Information Science and Technology* 58, no. 3 (2007): 452–54. hdl.handle.net/1942/980.

———. "Generalized Success-Breeds-Success Principle Leading to Time-Dependent Informetric Distributions." *Journal of the American Society for Information Science* 46, no. 6 (1995): 426–45.

———. "On the Influence of Growth on Obsolescence." *Scientometrics* 27, no. 2 (1993): 195–214.

———. *Power Laws in the Information Production Process: Lotkaian Informetrics*. Amsterdam: Elsevier Academic Press, 2005.

———. "Pratt's Measure for Some Bibliometric Distributions and Its Relation with the 80/20 Rule." *Journal of the American Society for Information Science* 38, no. 4 (1987): 288–97.

———. "Stochastic Processes Determined by a General Success-Breeds-Success Principle." *Mathematical and Computer Modelling* 23, no. 4 (1996): 93–104.

———. "Theory and Practise of the *g*-Index." *Scientometrics* 69, no. 1 (2006): 131–52. hdl.handle.net/1942/4536.

Egghe, Leo, and Inna K. Ravichandra Rao. "Citation Age Data and Obsolescence Function: Fits and Explanations." *Information Processing & Management* 28, no. 2 (1992): 201–17.

———. "Classification of Growth Models Based in Growth Rates and Its Implications." *Scientometrics* 25, no. 1 (1992): 5–46.

Egghe, Leo, and Ronald Rousseau. "Aging, Obsolescence, Impact, Growth and Utilization: Definitions and Relations." *Journal of the American Society for Information Science* 51, no. 11 (2000): 1004–17. http://hdl.handle.net/1942/786.

——. "A Characterization of Distributions Which Satisfy Price's Law and Consequences for the Laws of Zipf and Mandelbrot." *Journal of Information Science* 12, no. 4 (1986): 193–97.

——. "An *h*-Index Weighted by Citation Impact." *Information Processing & Management* 44, no. 2 (2008): 770–80.

——. "An Informetric Model for the Hirsch-Index." *Scientometrics* 69, no. 1 (2006): 121–29. hdl.handle.net/1942/943.

——. *Introduction to Informetrics: Quantitative Methods in Library, Documentation and Information Science*. Amsterdam: Elsevier, 1990. hdl.handle .net/1942/587.

Egghe, Leo, and Ronald Rousseau, eds. *Informetrics 87/88: Select Proceedings of the First International Conference on Bibliometrics and Theoretical Aspects of Information Retrieval*. Amsterdam: Elsevier, 1988. Individual articles are searchable and downloadable at doclib.uhasselt.be/dspace/.

——. *Informetrics 89/90: Selection of Papers Submitted for the Second International Conference on Bibliometrics, Scientometrics and Informetrics*. Amsterdam: Elsevier Science, 1990. Individual articles are searchable and downloadable at doclib.uhasselt.be/dspace/.

——. "A Proposal to Define a Core of a Scientific Subject: A Definition Using Concentration and Fuzzy Sets." *Scientometrics* 54, no. 1 (2002): 51–62. hdl .handle.net/1942/762.

——. "Transfer Principles and a Classification of Concentration Measures." *Journal of the American Society for Information Science* 42, no. 7 (1991): 479–89.

Egghe, Leo, Ronald Rousseau, and Guido Van Hooydonk. "Methods for Accrediting Publications to Authors or Countries: Consequence for Evaluation Studies." *Journal of the American Society for Information Science* 51, no. 2 (2000): 145–57. hdl.handle.net/1942/792.

Ellis, Philip, G. Hepburn, and Charles Oppenheim. "Studies on Patent Citation Networks." *Journal of Documentation* 34, no. 1 (1978): 12–20.

Elkana, Yehuda, Joshua Lederberg, Robert K. Merton, Arnold Thackray, and Harriet Zuckerman, eds. *Toward a Metric of Science: The Advent of Science Indicators*. New York: John Wiley & Sons, 1979.

Evans, James A. "Electronic Publication and the Narrowing of Science and Scholarship." *Science* 321, no. 5887 (2008): 395–99.

Evidence Ltd. *The Use of Bibliometrics to Measure Research Quality in UK Higher Education Institutions [Research Report]*. 2007. bookshop .universitiesuk.ac.uk/downloads/bibliometrics.pdf.

Eysenbach, Gunther. "Citation Advantage of Open Access Articles." *PLoS Biology* 4, no. 5 (2006): e157. biology.plosjournals.org/archive/1545-7885/4/5/ pdf/10.1371_journal.pbio.0040157-L.pdf.

Faba-Pérez, Cristina, Vicente P. Guerrero-Bote, and Felix Moya-Anegon. "Data Mining in a Closed Web Environment." *Scientometrics* 58, no. 3 (2003): 623–40.

Fairthorne, Robert A. "Empirical Hyperbolic Distributions (Bradford-Zipf-Mandelbrot) for Bibliometric Description and Prediction." *Journal of Documentation* 25, no. 4 (1969): 319–43.

Fedorowicz, Jane. "The Theoretical Foundation of Zipf's Law and Its Application to the Bibliographic Database Environment." *Journal of the American Society for Information Science* 33, no. 5 (1982): 285–93.

Fernandez-Cano, Antonio, Manuel Torralbo, and Monica Vallejo. "Reconsidering Price's Model of Scientific Growth: An Overview." *Scientometrics* 61, no. 3 (2004): 301–21.

Finocchiaro, Maurice A. *Galileo and the Art of Reasoning: Rhetorical Foundations of Logic and Scientific Method.* Dordrecht; Boston; London: Reidel, 1980.

Foucault, Michel. "What Is an Author?" In *Language, Counter-Memory, Practice. Selected Essays and Interviews by Michel Foucault*, edited by Donald F. Bouchard, 113–38. Ithaca, N.Y.: Cornell University Press, 1979.

Furner, Jonathan. "Little Book, Big Book: Before and after Little Science, Big Science; A Review Article, Part I." *Journal of Librarianship and Information Science* 35, no. 2 (2003): 115–25.

———. "Little Book, Big Book: Before and after Little Science, Big Science; A Review Article, Part II." *Journal of Librarianship and Information Science* 35, no. 3 (2003): 189–201.

Fussler, Herman H. "Characteristics of the Research Literature Used by Chemists and Physicists in the United States." *Library Quarterly* 19, no. 2 (1949): 19–35.

Gapen, D. Kaye, and Sigrid P. Milner. "Obsolescence." *Library Trends* 30, no. 1 (1981): 107–24. hdl.handle.net/2142/7189.

Garfield, Eugene. "The Articles Most Cited in the SCI, 1961–1982. 9: More Contemporary Classics of Science." In *Essays of an Information Scientist 1986*, 55–64. Philadelphia: ISI Press, 1988. www.garfield.library.upenn.edu/essays/v9p055y1986.pdf.

———. "Automatic Indexing and the Linguistics Connection." In *Essays of an Information Scientist 1981–1982*, 31–38. Philadelphia: ISI Press, 1983. www.garfield.library.upenn.edu/essays/v5p031y1981-82.pdf.

———. "Bradford's Law and Related Statistical Patterns." In *Essays of an Information Scientist 1979–1980*, 476–83. Philadelphia: ISI Press, 1981. www.garfield.library.upenn.edu/essays/v4p476y1979-80.pdf.

———. "Breaking the Subject Index Barrier. A Citation Index for Chemical Patents." *Journal of the Patent Office Society* 39, no. 8 (1957): 583–95. www.garfield.library.upenn.edu/essays/v6p472y1983.pdf.

——. "Can Citation Indexing Be Automated?" In *Essays of an Information Scientist 1962–1973,* 84–90. Philadelphia: ISI Press, 1977. www.garfield.library.upenn.edu/essays/V1p084y1962-73.pdf.

——. "Chemico-Linguistics: Computer Translation of Chemical Nomenclature." *Nature* 192, no. 4798 (1961): 192. www.garfield.library.upenn.edu/essays/v6p489y1983.pdf.

——. "Citation Analysis as a Tool in Journal Evaluation." *Science* 178, no. 4060 (1972): 471–79. www.garfield.library.upenn.edu/essays/V1p527y1962-73.pdf.

——. "Citation Consciousness: Interview with Eugene Garfield, Chairman Emeritus of ISI, Philadelphia, U.S.A." *Password,* no. 6 (2002): 22–25. www.garfield.library.upenn.edu/papers/passwordinterview062002.pdf.

——. "Citation Frequency as a Measure of Research Activity and Performance." In *Essays of an Information Scientist 1962–1973,* 406–8. Philadelphia: ISI Press, 1977. www.garfield.library.upenn.edu/essays/V1p406y1962-73.pdf.

——. "Citation Indexes—New Paths to Scientific Knowledge." *The Chemical Bulletin* 43, no. 4 (1956): 11. www.garfield.library.upenn.edu/papers/31.html.

——. "Citation Indexes for Science: A New Dimension in Documentation Through Association of Ideas." *Science* 122, no. 3159 (1955): 108–11. www.garfield.library.upenn.edu/essays/v6p468y1983.pdf.

——. *Citation Indexing: Its Theory and Application in Science, Technology, and Humanities.* Philadelphia: ISI Press, 1983. www.garfield.library.upenn.edu/ci/title.pdf.

——. "Do Nobel Prize Winners Write Citation Classics?" In *Essays of an Information Scientist 1986,* 182–87. Philadelphia: ISI Press, 1988. www.garfield.library.upenn.edu/essays/v9p182y1986.pdf.

——. "From Bibliographic Coupling to Co-Citation Analysis via Algorithmic Historio-Bibliography: A Citationist's Tribute to Belver C. Griffith." Paper presented at Drexel University, Philadelphia, Pa., November 27, 2001. www.garfield.library.upenn.edu/papers/drexelbelvergriffith92001.pdf.

——. "From Computational Linguistics to Algorithmic Historiography." Paper presented at the Lazerow Lecture held in conjunction with Panel on Knowledge and Language: Building Large-Scale Knowledge Bases for Intelligent Applications, University of Pittsburgh, September 19, 2001. www.garfield.library.upenn.edu/papers/pittsburgh92001.pdf.

——. "How ISI Selects Journals for Coverage: Quantitative and Qualitative Considerations." In *Essays of an Information Scientist 1990,* 185–93. Philadelphia: ISI Press, 1991. www.garfield.library.upenn.edu/essays/v13p185y1990.pdf.

———. "How to Use Citation Analysis for Faculty Evaluations, and When Is It Relevant? Part 1." In *Essays of an Information Scientist 1983,* 354–62. Philadelphia: ISI Press, 1984. www.garfield.library.upenn.edu/essays/v6p354y1983.pdf.

———. "How to Use Citation Analysis for Faculty Evaluations, and When Is It Relevant? Part 2." In *Essays of an Information Scientist 1983,* 363–72. Philadelphia: ISI Press, 1984. www.garfield.library.upenn.edu/essays/v6p363y1983.pdf.

———. "Introducing the ISI Atlas of Science: Biochemistry and Molecular Biology, 1978/80." In *Essays of an Information Scientist 1981–1982,* 279–87. Philadelphia: ISI Press, 1983. www.garfield.library.upenn.edu/essays/v5p279y1981-82.pdf.

———. "Is the Ratio Between Number of Citations and Publications Cited a True Constant?" In *Essays of an Information Scientist 1974–1976,* 419–25. Philadelphia: ISI Press, 1977. www.garfield.library.upenn.edu/essays/v2p419y1974-76.pdf.

———. "The ISI-VINITI Connection—Remarks on the 50th Anniversary of VINITI." Paper presented at the 6th International Conference of Information Society, Intelligent Information Processing and Information Technologies, October 16–18, 2002, Moscow. www.garfield.library.upenn.edu/papers/viniti101502.html.

———. "Long-Term vs. Short-Term Journal Impact: Does It Matter?" *The Scientist* 12, no. 3 (1998): 10–12. www.garfield.library.upenn.edu/commentaries/tsv12(03)p10y19980202.pdf.

———. "Long-Term vs. Short-Term Journal Impact, Part II: Cumulative Impact Factors." *The Scientist* 12, no. 14 (1998): 12–13. www.garfield.library.upenn.edu/commentaries/tsv12(14)p12y19980706.pdf.

———. "Mapping Science in the Third World." *Science and Pubic Policy* 10, no. 3 (1983): 112–27. www.garfield.library.upenn.edu/papers/mapscithirdworldp112y1983.pdf.

———. "The Mistery of the Transposed Journal Lists Wherein Bradford's Law of Scattering Is Generalized According to Garfield's Law of Concentration." In *Essays of an Information Scientist 1962–1973,* 222–23. Philadelphia: ISI Press, 1977. www.garfield.library.upenn.edu/essays/V1p222y1962-73.pdf.

———. "The Relationship Between Mechanical Indexing, Structural Linguistics and Information Retrieval." *Journal of Information Science* 18, no. 5 (1992): 343–54. www.garfield.library.upenn.edu/papers/jis18(5)p343y1992.html.

———. *Science Citation Index: An International Interdisciplinary Index to the Literature of Science.* Philadelphia: Institute for Scientific Information, 1963.

———. "A Unified Index to Science." In *Proceedings of the International Conference on Scientific Information Washington, D.C., Novermber, 16–21 1958,* 461–74. Washington, D.C.: NAS/NRC, 1959. www.garfield.library.upenn.edu/essays/v2p674y1974-76.pdf.

———. "What's in a Name? The Eponymic Route to Immortality." In *Essays of an Information Scientist 1983,* 384–95. Philadelphia: ISI Press, 1984. www.garfield.library.upenn.edu/essays/v6p384y1983.pdf.

———. "When to Cite?" *Library Quarterly* 66, no. 4 (1996): 449–58. www.garfield.library.upenn.edu/papers/libquart66(4)p449y1996.pdf.

Garfield, Eugene, Alexander Pudovkin, and Vladimir S. Istomin. "Algorithmic Citation-Linked Historiography—Mapping the Literature of Science." Paper presented at the ASIS&T 2002: Information, Connections and Community. 65th Annual Meeting of ASIST, Philadelphia, Pa., November 18–21, 2002. www.garfield.library.upenn.edu/papers/asis2002/asis2002presentation.html.

———. "Why Do We Need Algorithmic Historiography?" *Journal of the American Society for Information Science and Technology* 54, no. 5 (2003): 400–412. www.garfield.library.upenn.edu/papers/jasist54(5)400y2003.pdf.

Garfield, Eugene, and Irving H. Sher. "ISI's Experiences with ASCA—A Selective Dissemination System." *Journal of Chemical Documentation* 7, no. 3 (1967): 147–53. www.garfield.library.upenn.edu/essays/v6p533y1983.pdf.

———. "Keywords Plus™ Algorithmic Derivative Indexing." *Journal of the American Society for Information Science* 44, no. 5 (1993): 298–99. www.garfield.library.upenn.edu/papers/jasis44(5)p298y1993.html.

———. "New Factors in the Evaluation of Scientific Literature Through Citation Indexing." *American Documentation* 14, no. 3 (1963): 195–201. www.garfield.library.upenn.edu/essays/v6p492y1983.pdf.

Garfield, Eugene, Irving H. Sher, and Richard J. Thorpie. *The Use of Citation Data in Writing the History of Science*. Philadelphia: Institute for Scientific Information, 1964. www.garfield.library.upenn.edu/papers/useofcitdatawritinghistofsci.pdf.

Garfield, Eugene, and Alfred Welljams-Dorof. "Citation Data: Their Use as Quantitative Indicators for Science and Technology Evaluation and Policy-Making." In *Essays of an Information Scientist 1992–1993,* 190–98. Philadelphia: ISI Press, 1993. www.garfield.library.upenn.edu/essays/v15p188y1992-93.pdf.

Garvey, William D. *Communication: The Essence of Science*. Oxford: Pergamon Press, 1979.

Geisler, Eliezer. *The Metrics of Science and Technology*. Westport, Conn.; London: Quorum Books, 2000.

Geller, Nancy L., John S. De Cani, and Robert E. Davies. "Lifetime-Citation Rates: A Mathematical Model to Compare Scientists' Work." *Journal of the American Society for Information Science* 32, no. 1 (1981): 345–65.

Gibbons, Michael, Camille Limoges, Helga Nowotny, Simon Schwartzman, Peter Scott, and Martin Trow. *The New Production of Knowledge: The Dynamics of Science and Research in Contemporary Societies*. London: Sage, 1994.

Gilbert, G. Nigel. "Referencing as Persuasion." *Social Studies of Science* 7, no. 1 (1977): 113–22.

Gilbert, G. Nigel, and Michael Mulkay. *Opening Pandora's Box: A Sociological Analysis of Scientists' Discourse*. Cambridge: Cambridge University Press, 1984.

Gilbert, G. Nigel, and Steve Woolgar. "The Quantitative Study of Science: An Examination of the Literature." *Science Studies* 4, no. 3 (1974): 279–94.

Giles, C. Lee, and Isaac G. Councill. "Who Gets Acknowledged: Measuring Scientific Contributions Through Automatic Acknowledgment Indexing." *Proceedings of the National Academy of Sciences* 101, no. 51 (2004): 17599–604. www.pnas.org/cgi/content/abstract/0407743101v1.

Giles, C. Lee, Kurt D. Bollacker, and Steve Lawrence. "CiteSeer: An Automatic Citation Indexing System." In *Digital Libraries 98—Third ACM Conference on Digital Libraries*, edited by I. Witten, R. Akscyn and F. Shipman III, 89–98. New York: ACM Press, 1998. clgiles.ist.psu.edu/papers/DL-1998-citeseer.pdf.

Glänzel, Wolfgang. "Coauthorship Patterns and Trends in the Sciences: A Bibliometric Study with Implications for Database Indexing and Search Strategies—1980–1998." *Library Trends* 50, no. 3 (2002): 461–73. findarticles.com/p/articles/mi_m1387/is_3_50/ai_88582625.

———. "National Characteristics in International Scientific Co-Authorship Relations." *Scientometrics* 51, no. 1 (2001): 69–115.

———. "On the *h*-Index—A Mathematical Approach to a New Measure of Publication Activity and Citation Impact." *Scientometrics* 67, no. 2 (2006): 315–21.

Glänzel, Wolfgang, and Hans-Jürgen Czerwon. "A New Methodological Approach to Bibliographic Coupling and Its Application to the National, Regional and Institutional Level." *Scientometrics* 37, no. 2 (1996): 195–221.

Glänzel, Wolfgang, and Henk F. Moed. "Journal Impact Measures in Bibliometric Research." *Scientometrics* 53, no. 2 (2002): 171–93.

Glänzel, Wolfgang, Balázs Schlemmer, and Bart Thijs. "Better Late Than Never? On the Chance to Become Highly Cited Only Beyond the Standard Bibliometric Time Horizon." *Scientometrics* 58, no. 3 (2003): 571–86.

Glänzel, Wolfgang, and Urs Schoepflin. "A Bibliometric Study on Ageing and Reception Processes of Scientific Literature in the Science." *Journal of Information Science* 21, no. 1 (1995): 37–53.

———. "A Stochastic Model for the Ageing of Scientific Literature." *Scientometrics* 30, no. 1 (1994): 49–64.

———. "Little Scientometrics, Big Scientometrics . . . and Beyond?" *Scientometrics* 30, nos. 2–3 (1994): 375–84.

Glänzel, Wolfgang, and Andreás Schubert. "Analysing Scientific Networks Through Co-Authorships." In *Handbook of Quantitative Science and Technology Research: The Use of Publication and Patent Statistics in Studies of S&T Systems*, edited by Henk F. Moed, Wolfgang Glänzel, and Ulrich Schmoch, 257–76. Dordrecht: Kluwer Academic, 2004.

———. "Predictive Aspects of a Stochastic Model for Citation Processes." *Information Processing & Management* 31, no. 1 (1995): 69–80.

———. "Price Distribution: An Exact Formulation of Price's Square Root Law." *Scientometrics* 7, nos. 3–6 (1985): 211–19.

Glänzel, Wolfgang, Bart Thijs, Andreás Schubert. "A Concise Review on the Role of Author Self-Citations in Information Science, Bibliometrics and Science Policy." *Scientometrics* 67, no. 2 (2006): 263–77. www.cais-acsi.ca/proceedings/2005/glanzel_2_2005.pdf.

Glenisson, Patrick, Wolfgang Glänzel, Frizo Janssens, and Bart De Moor. "Combining Full Text and Bibliometric Information in Mapping Scientific Disciplines." *Information Processing & Management* 41, no. 6 (2005): 1548–72. ftp.esat.kuleuven.ac.be/sista/ida/reports/04-205.pdf.

Godin, Benoît. *From Eugenics to Scientometrics: Galton, Cattell and Men of Science*; Project on the History and Sociology of S&T Statistics. Working Paper No. 32. Montreal: Canadian Science and Innovation Indicators Consortium, 2006. www.csiic.ca/PDF/Godin_32.pdf.

———. "The Knowledge-Based Economy: Conceptual Framework or Buzzword." *Journal of Technology Transfer* 31, no. 1 (2006): 17–30.

———. *On the Origins of Bibliometrics*: Project on the History and Sociology of S&T Statistics, *Working Paper No. 33*. Montreal: Canadian Science and Innovation Indicators Consortium, 2006. www.csiic.ca/PDF/Godin_33.pdf.

Goffman, William, and Vaun A. Newill. "Generalization of Epidemic Theory: An Application to the Transmission of Ideas." *Nature* 204, no. 4955 (1964): 225–28.

Goldstein, Harvey, and David J. Spiegelhalter. "League Tables and Their Limitations: Statistical Issues in Comparisons of Institutional Performance." *Journal of the Royal Statistical Society, Series A* 159, no. 3 (1996): 385–443. www

.cmm.bristol.ac.uk/team/HG_Personal/limitations-of-league-tables.pdf (Preprint devoid of figures).

Gomperts, M. C. "The Law of Constant Citation for Scientific Literature." *Journal of Documentation* 24, no. 2 (1968): 113–17.

Gosnell, Charles F. "Obsolescence of Books in College Libraries." *College & Research Libraries* 5, no. 2 (1944): 115–25.

Grafton, Anthony. *The Footnote: A Curious History*. London: Faber and Faber, 1997.

Granovetter, Mark. "The Strength of Weak Ties." *American Journal of Sociology* 78, no. 6 (1973): 1360–80.

Granovsky, Yuri V. "Is It Possible to Measure Science? V. V. Nalimov's Research in Scientometrics." *Scientometrics* 52, no. 2 (2001): 127–50.

Gray, W. A., and A. J. Harley. "Computer Assisted Indexing." *Information Storage & Retrieval* 7, no. 4 (1971): 167–74.

Green, Gary S. "A Test of the Ortega Hypothesis in Criminology." *Criminology* 19, no. 1 (1981): 45–52.

Greenwood, Darren C. "Reliability of Journal Impact Factor Rankings." *BMC Medical Research Methodology* 7, no. 48 (2007). www.biomedcentral.com/1471-2288/7/48.

Gregory, Derek. "Editorial." *Environment and Planning. D: Society and Space* 8, no. 1 (1990): 1–6.

Griffith, Belver C. "Derek Price's Puzzles: Numerical Metaphors for the Operation of Science." *Science, Technology & Human Values* 13, nos. 3–4 (1988): 351–60.

Griffith, Belver C., Patricia N. Servi, Anita L. Anker, and M. Carl Drott. "The Aging of Scientific Literature: A Citation Analysis." *Journal of Documentation* 35, no. 3 (1979): 179–96.

Griffith, Belver C., Henry G. Small, Judith A. Stonehill, and Sandra Dey. "The Structure of Scientific Literature. II: Toward a Macro- and Microstructure for Science." *Science Studies* 4, no. 4 (1974): 339–65.

Griliches, Zvi. "Patent Statistics as Economic Indicators: A Survey." *Journal of Economic Literature* 28, no. 4 (1990): 1661–1707.

Groos, Ole V. "Bradford's Law and the Keenan-Atherton Data." *American Documentation* 18, no. 1 (1967): 46.

Gross, P. L. K., and E. M. Gross. "College Libraries and Chemical Education." *Science* 66, no. 1713 (1927): 385–89.

Gupta, B. M., and C. R. Karisiddappa. "Modelling the Growth of Literature in the Area of Theoretical Population Genetics." *Scientometrics* 49, no. 2 (2000): 321–55.

Gurjeva, Lyubov G. "Early Soviet Scientometrics and Scientometricians." Thesis for the degree of MSc in science dynamics, Universiteit van Amsterdam, 1992. www.chstm.man.ac.uk/people/gurjeva/thesis-msc.pdf.

Gurjeva, Lyubov G., and Paul Wouters. "Scientometrics in the Context of Probabilistic Philosophy." *Scientometrics* 52, no. 2 (2001): 111–26.

Habibzadeh, Farrokh, and Mahboobeh Yadollahie. "Journal Weighted Impact Factor: A Proposal." *Journal of Informetrics* 2, no. 2 (2008): 164–72.

Hagstrom, Warren O. *The Scientific Community*. New York: Basic Books, 1965.

Hahn, Roger. *A Bibliography of Quantitative Studies on Science and Its History*. Berkeley: Office for History of Science and Technology, University of California, 1980.

Hahn, Trudi Bellardo, and Michael K. Buckland, eds. *Historical Studies in Information Science*. Medford, N.J.: Information Today, 1998.

Haitun, Sergej D. "Stationary Scientometric Distributions, Part I: Different Approximations." *Scientometrics* 4, no. 1 (1982): 5–25.

———. "Stationary Scientometric Distributions, Part II: Non-Gaussian Nature of Scientific Activities." *Scientometrics* 4, no. 2 (1982): 89–104.

———. "Stationary Scientometric Distributions, Part III: The Role of the Zipf Distribution." *Scientometrics* 4, no. 3 (1982): 181–94.

Hajjem, Chawki, Yves Gingras, Tim Brody, Les Carr, and Stevan Harnad. *Open Access to Research Increases Citation Impact [Technical Report]*. Montréal: Université du Québec, 2005. eprints.ecs.soton.ac.uk/11687/.

Hajjem, Chawki, Stevan Harnad, and Yves Gingras. "Ten-Year Cross-Disciplinary Comparison of the Growth of Open Access and How It Increases Research Citation Impact." *Bulletin of the IEEE Computer Society Technical Committee on Data Engineering* 28, no. 4 (2005): 39–47. eprints.ecs.soton.ac.uk/11688/.

Hall, Bronwyn H., Adam B. Jaffe, and Manuel Trajtenberg. "Market Value and Patent Citations." *RAND Journal of Economics* 36, no. 1 (2005): 16–38. repositories.cdlib.org/postprints/2454/.

Hand, Jeff. "Feasibility of Using Citations as Document Summaries." Thesis in partial fulfillment of the requirements for the degree of Doctor of Philosophy, Drexel University, December 2003. dspace.library.drexel.edu/bitstream/1860/288/8/hand_jeff_thesis.pdf.

Harley, John B. "Maps, Knowledge and Power." In *The Iconography of Landscape: Essays on the Symbolic Representation, Design and Use of Past Environments*, edited by Denis Cosgrove and Stephen Daniels, 277–312. Cambridge, Mass.: Cambridge University Press, 1988.

———. *The New Nature of Maps: Essays in the History of Cartography*. Baltimore: The Johns Hopkins University Press, 2001.

Harnad, Stevan. *Confirmation Bias and the Open Access Advantage: Some Methodological Suggestions for Davis's Citation Study.* August 25, 2008. openaccess.eprints.org/index.php?/archives/451-Confirmation-Bias-and-the-Open-Access-Advantage-Some-Methodological-Suggestions-for-Daviss-Citation-Study.html.

———. *Davis et Al.'s 1-Year Study of Self-Selection Bias: No Self-Archiving Control, No OA Effect, No Conclusion.* July 31, 2008. openaccess.eprints.org/index.php?/archives/441–Davis-et-als-1–year-Study-of-Self-Selection-Bias-No-Self-Archiving-Control,-No-OA-Effect,-No-Conclusion.html.

———. "Open Access Scientometrics and the UK Research Assessment Exercise." Paper presented at the 11th Annual Meeting of the International Society for Scientometrics and Informetrics, Madrid, June 25–27, 2007. arxiv.org/abs/cs/0703131.

Harnad, Stevan, and Tim Brody. "Comparing the Impact of Open Access (OA) vs. Non-OA Articles in the Same Journals." *D-Lib Magazine* 10, no. 6 (2004). www.dlib.org/dlib/june04/harnad/06harnad.html.

Harnad, Stevan, Tim Brody, François Vallières, Les Carr, Steve Hitchcock, Yves Gingras, Charles Oppenheim, Heinrich Stamerjohanns, and Eberhard R. Hilf. "The Access/Impact Problem and the Green and Gold Roads to Open Access." *Serials Review* 30, no. 4 (2004): 310–14. eprints.ecs.soton.ac.uk/10209/.

———. "The Access/Impact Problem and the Green and Gold Roads to Open Access: An Update." *Serials Review* 34, no. 1 (2008): 36–40. eprints.ecs.soton.ac.uk/15852/.

Harsanyi, Martha A. "Multiple Authors, Multiple Problems: Bibliometrics and the Study of Scholarly Collaboration, A Literature Review." *Library & Information Science Research* 15, no. 4 (1993): 325–54.

Harter, Stephen P., and Charlotte E. Ford. "Web-Based Analysis of E-Journal Impact: Approaches, Problems, and Issues." *Journal of the American Society for Information Science* 51, no. 13 (2000): 1159–76.

Hartley, James. "What Do We Know About Footnotes? Opinions and Data." *Journal of Information Science* 25, no. 3 (1999): 205–12.

Haslam, Nick, Lauren Ban, Leah Kaufmann, Stephen Loughnan, Kim Peters, Jennifer Whelan, and Sam Wilson. "What Makes an Article Influential? Predicting Impact in Social and Personality Psychology." *Scientometrics* 76, no. 1 (2008): 169–85.

Havemann, Frank, Roland Wagner-Döbler, and Hildrun Kretschmer, eds. *Collaboration in Science and in Technology, Proceedings of the Second Berlin Workshop on Scientometrics and Informetrics, September 1–3, 2000.* Berlin:

Gesellschaft für Wissenschaftsforschung, 2001. www.collnet.de/workshop/Berlin2000.pdf.

He, Oin. "Knowledge Discovery Through Co-Word Analysis." *Library Trends* 48, no. 1 (1999): 133–59. findarticles.com/p/articles/mi_m1387/is_1_48/ai_57046530.

Heifetz, Stephen R. "Blue in the Face: The Bluebook, the Bar Exam, and the Paradox of Our Legal Culture." *Rutgers Law Review* 51, no. 3 (1999): 695–711.

Henkle, Hermann H. "The Periodical Literature of Biochemistry." *Bulletin of the Medical Library Association* 27 (1938): 139–47.

Hertzel, Dorothy H. "Bibliometrics History." In *Encyclopedia of Library and Information Science,* 2nd ed., edited by Miriam A. Drake, 288–328. New York: M. Dekker, 2003. books.google.it/books?id=tPnkxsklgXoC.

Hicks, Diana. "Limitations of Co-Citation Analysis as a Tool for Science Policy." *Social Studies of Science* 17, no. 2 (1987): 295–316.

Hicks, Diana, Peter Kroll, Francis Narin, Patrick Thomas, Rosalie Ruegg, Hiroyuki Tomizawa, Yoshiko Saitoh, and Shinichi Kobayashi. *Quantitative Methods of Research Evaluation Used by the U.S. Federal Government.* Japan: National Institute of Science and Technology Police, 2002. www.nistep.go.jp/achiev/ftx/eng/mat086e/pdf/mat086e.pdf.

Hirsch, Jorge E. "Does the *h* Index Have Predictive Power?" *Proceedings of the National Academy of Sciences* 104, no. 49 (2007): 19193–98. arxiv.org/abs/0708.0646.

———. "An Index to Quantify an Individual's Scientific Research Output." *Proceedings of the National Academy of Sciences* 102, no. 46 (2005): 16569–72. rxiv.org/abs/physics/0508025.

Hirst, Graeme. "Discipline Impact Factors: A Method for Determining Core Journal Lists." *Journal of the American Society for Information Science* 29, no. 4 (1978): 171–72.

Hitchcock, Steve, Donna Bergmark, Tim Brody, Cristopher Gutteridge, Les Carr, Wendy Hall, Carl Lagoze, and Stevan Harnad. "Open Citation Linking: The Way Forward." *D-Lib Magazine* 8, no. 10 (2002). eprints.ecs.soton.ac.uk/7717/.

Hjerppe, Roland. *A Bibliography of Bibliometrics and Citation Indexing & Analysis.* Stockholm: Royal Institute of Technology, 1980.

Hoerman, Heidi Lee, and Carole Elizabeth Nowicke. "Secondary and Tertiary Citing: A Study of Referencing Behavior in the Literature of Citation Analysis Deriving from the Ortega Hypothesis of Cole and Cole." *The Library Quarterly* 65, no. 4 (1995): 415–34.

Hood, William W., and Concepción S. Wilson. "The Literature of Bibliometrics, Scientometrics, and Informetrics." *Scientometrics* 52, no. 2 (2001): 291–314.

Huang, Mu-hsuan, and Yu-wei Chang. "Characteristics of Research Output in Social Sciences and Humanities: From a Research Evaluation Perspective." *Journal of the American Society for Information Science and Technology* 59, no. 11 (2008): 1819–28.

Huber, John C. "Cumulative Advantage and Success-Breeds-Success: The Value of Time Pattern Analysis." *Journal of the American Society for Information Science* 49, no. 5 (1998): 471–76. topology.eecs.umich.edu/archive/huber_99.pdf.

———. "Invention and Inventivity Is a Random, Poisson Process: A Potential Guide to Analysis of General Creativity." *Creativity Research Journal* 11, no. 3 (1998): 231–41.

———. "Inventive Productivity and the Statistics of Exceedances." *Scientometrics* 45, no. 1 (1999): 33–53.

———. "A New Method for Analyzing Scientific Productivity." *Journal of the American Society for Information Science and Technology* 52, no. 13 (2001): 1089–99.

Hubert, John J. "Bibliometric Models for Journal Productivity." *Social Indicators Research* 4, no. 4 (1977): 441–73.

———. "General Bibliometric Models." *Library Trends* 30, no. 1 (1981): 65–81. hdl.handle.net/2142/7189.

Hulme, Edward Wyndham. *Statistical Bibliography in Relation to the Growth of Modern Civilization*. London: Grafton, 1923.

Hummon, Norman P., and Patrick Doreian. "Connectivity in a Citation Network: The Development of DNA Theory." *Social Networks* 11, no. 1 (1989): 39–63. www.garfield.library.upenn.edu/papers/hummondoreian1989.pdf.

Iglesias, Juan, and Carlos Pecharromán. "Scaling the *h*-Index for Different Scientific ISI Fields." *Scientometrics* 73, no. 3 (2007): 303–20. arxiv.org/abs/physics/0607224.

Ingwersen, Peter. "The Calculation of Web Impact Factors." *Journal of Documentation* 54, no. 2 (1998): 236–43.

Ingwersen, Peter, and Kalervo Järvelin. *The Turn: Integration of Information Seeking and Retrieval in Context*. Dordrecht: Springer, 2005.

Ingwersen, Peter, Birger Larsen, Ronald Rousseau, and Jane Russel. "The Publication-Citation Matrix and Its Derived Quantities." *Chinese Science Bulletin* 46, no. 6 (2001): 524–28, 219.238.6.200/getfile?category=article&code=00WC0565&file-name=01ky0524.pdf.

International Committee of Medical Journal Editors. *Uniform Requirements for Manuscripts Submitted to Biomedical Journals: Writing and Editing for Biomedical Publication*. 2007. www.icmje.org/.

Irvine, John, and Ben R. Martin. "Assessing Basic Research: The Case of the Isaac Newton Telescope." *Social Studies of Science* 13, no. 1 (1983): 49–86.

——. "Basic Research in the East and West: A Comparison of the Scientific Performance of High-Energy Physics Accelerators." *Social Studies of Science* 15, no. 2 (1985): 293–341.

——. "Evaluating Big Science: CERN's Past Performance and Future Prospects." *Scientometrics* 7, nos. 3–6 (1985): 281–308.

Irvine, John, Ben R. Martin, John Abraham, and Tim Peacock. "Assessing Basic Research: Reappraisal and Update of an Evaluation of Four Radio Astronomy Observatories." *Research Policy* 16, nos. 2–4 (1987): 213–27.

Irvine, John, Ben R. Martin, Tim Peacock, and Roy Turner. "Charting the Decline in British Science." *Nature* 316, no. 6029 (1985): 587–90.

Ivancheva, Ludmila E. "The Non-Gaussian Nature of Bibliometric and Scientometric Distributions: A New Approach to Interpretation." *Journal of the American Society for Information Science and Technology* 52, no. 13 (2001): 1100–1105.

Jackson, Michele H. "Assessing the Structure of Communication on the World Wide Web." *Journal of Computer-Mediated Communication* 3, no. 1 (1997). www.nicoladoering.de/Hogrefe/jackson.htm.

Jaffe, Adam B., and Manuel Trajtenberg. *Patents, Citations & Innovations: A Window on the Knowledge Economy*. Cambridge, Mass.: The MIT Press, 2002.

Jaffe, Adam B., Manuel Trajtenberg, and Rebecca Henderson. "Geographic Localization of Knowledge Spillovers as Evidenced by Patent Citations." *The Quarterly Journal of Economics* 108, no. 3 (1993): 577–98.

Jarneving, Bo. "Bibliographic Coupling and Its Application to Research-Front and Other Core Documents." *Journal of Informetrics* 1, no. 4 (2007): 287–307.

Järvelin, Kalervo, and Olle Persson. "The DCI Index: Discounted Cumulated Impact-Based Research Evaluation." *Journal of the American Society for Information Science and Technology* 59, no. 9 (2008): 1433–40. www.info.uta.fi/tutkimus/fire/archive/2008/DCF-JASIST-08.pdf.

Jin, BiHui, LiMing Liang, Ronald Rousseau, and Leo Egghe. "The *R*- and *AR*-Indices: Complementing the *h*-Index," *Chinese Science Bulletin* 52, no. 6 (2007): 855–63. users.pandora.be/ronald.rousseau/CSB_Jin_et_al.pdf.

Johnson, Ben, and Charles Oppenheim. "How Socially Connected Are Citers to Those That They Cite?" *Journal of Documentation* 63, no. 5 (2007): 609–37.

Journal Citation Reports® on the Web V. 4.0. Philadelphia: The Thomson Corporation, 2005. scientific.thomson.com/media/scpdf/jcr4_sem_0305.pdf.

Judge, Timothy A., Daniel M. Cable, Amy E. Colbert, and Sara L. Rynes. "What Causes a Management Article to Be Cited—Article, Author, Or Journal?" *Academy of Management Journal* 50, no. 3 (2007): 491–506. journals.aomonline.org/amj/editorials/Judge.etal.Citations.2007.pdf.

Kantor, Paul B. "A Note on Cumulative Advantage Distributions." *Journal of the American Society for Information Science* 29, no. 4 (1978): 202–4.

Kaplan, Norman. "The Norms of Citation Behavior: Prolegomena to the Footnote." *American Documentation* 16, no. 3 (1965): 179–84.

Katz, J. Sylvan. "Scale Independent Indicators and Research Evaluation." *Science and Public Policy* 27, no. 1 (2000): 23–36. www.sussex.ac.uk/Users/sylvank/pubs/SI-RE.pdf.

——. "The Self-Similar Science System." *Research Policy* 28, no. 5 (1999): 501–17. www.sussex.ac.uk/Users/sylvank/pubs/SSSS.pdf.

Katz, J. Sylvan, and Ben R. Martin. "What Is Research Collaboration?" *Research Policy* 26, no. 1 (1997): 1–18. www.sussex.ac.uk/Users/sylvank/pubs/Res_col9.pdf.

Kealey, Terence. "Government-Funded Academic Science Is a Consumer Good, Not a Producer Good: A Comparative Reassessment of Britain's Scientific and Technological Achievements Since 1794 and a Comment on the Bibliometry of B. Martin and J. Irvine." *Scientometrics* 20, no. 2 (1991): 369–94.

Kendall, Maurice G. "The Bibliography of Operational Research." *Operational Research Quarterly* 11, nos. 1–2 (1960): 31–36.

——. "Natural Law in the Social Sciences." *Journal of the Royal Statistical Society. Series A (General)* 124, no. 1 (1961): 1–16.

Kessler, Myer M. "Bibliographic Coupling Between Scientific Papers." *American Documentation* 14, no. 1 (1963): 10–25.

——. "Bibliographic Coupling Extended in Time: Ten Case Histories." *Information Storage & Retrieval* 1, no. 4 (1963): 169–87.

Knorr-Cetina, Karin D., and Michael Mulkay, eds. *Science Observed: Perspectives in the Social Study of Science*. London: Sage, 1983.

Kochen, Manfred. *Principles of Information Retrieval.* Los Angeles: Melville Publishing, 1974.

——. "Stability in the Growth of Knowledge." *American Documentation* 20, no. 3 (1969): 186–97.

Kochen, Manfred, ed. *The Small World*. Norwood, N.J.: Ablex, 1989.

Koehler, Wallace. "A Longitudinal Study of Web Pages Continued: A Consideration of Document Persistence." *Information Research* 9, no. 2 (2004). informationr.net/ir/9–2/paper174.html.

Koenig, Michael E.D. "Bibliometric Indicators Versus Expert Opinion in Assessing Research Performance." *Journal of the American Society for Information Science* 34, no. 2 (1983): 136–45.

Kosmulski, Marek. "A New Hirsch-Type Index Saves Time and Works Equally Well as the Original *h*-Index." *ISSI Newsletter* 2, no. 3 (2006): 4–6.

Kostoff, Ronald N. "Citation Analysis of Research Performer Quality." *Scientometrics* 53, no. 1 (2002): 49–71.

———. *The Handbook of Research Impact Assessment*. Arlington, Va.: Office of Naval Research, 1997. handle.dtic.mil/100.2/ADA296021.

Kretschmer, Hildrun. "Coauthorship Networks of Invisible Colleges and Institutionalized Communities." *Scientometrics* 30, no. 1 (1994): 363–69.

———. "Patterns of Behaviour in Coauthorship Networks of Invisible Colleges." *Scientometrics* 40, no. 3 (1997): 579–91.

Kretschmer, Hildrun, and Ronald Rousseau. "Author Inflation Leads to a Breakdown of Lotka's Law." *Journal of the American Society for Information Science and Technology* 52, no. 8 (2001): 610–14.

Kuhn, Thomas S. *The Essential Tension: Selected Studies in Scientific Tradition and Change*. Chicago and London: The University of Chicago Press, 1977.

———. *The Structure of Scientific Revolutions*. 2nd. ed. Chicago: University of Chicago Press, 1970.

Kumar, Ravi, P. Raghauam, Sridhar Rajagopalan, and Andrew Tomkins. "Trawling the Web for Emerging Cyber-Communities." *Computer Networks* 31, no. 11 (1999): 1481–93. www8.org/w8-papers/4a-search-mining/trawling/trawling.html.

Kurtz, Michael J., Guenther Eichhorn, Alberto Accomazzi, Carolyn Grant, Markus Demleitner, Edwin Henneken, and Stephen S. Murray. "The Effect of Use and Access on Citations." *Information Processing & Management* 41, no. 6 (2005): 1395–1402. arxiv.org/abs/cs/0503029.

Kwok, Kui-Lam. "A Probabilistic Theory of Indexing and Similarity Measure Based on Cited and Citing Documents." *Journal of the American Society for Information Science* 36, no. 5 (1985): 342–51.

Laherrere, Jean, and Didier Sornette. "Stretched Exponential Distributions in Nature and Economy: Fat Tails with Characteristic Scales." *European Physical Journal B* 2, no. 4 (1998): 525–39. arxiv.org/abs/cond-mat/9801293.

Lakatos, Imre, and Alan Musgrave, eds. *Criticism and the Growth of Knowledge: Proceedings of the International Colloquium in the Philosophy of Science, London, 1965*. Cambridge: Cambridge University Press, 1970.

Lancaster, Frederick W. *Indexing and Abstracting in Theory and Practice*. Champaign: University of Illinois—School of Library and Information Science, 1998.

Larivière, Vincent, Éric Archambault, and Yves Gingras. "Long-Term Variations in the Aging of Scientific Literature: From Exponential Growth to Steady-State Science (1900–2004)." *Journal of the American Society for Information Science and Technology* 59, no. 2 (2008): 288–96. www.ost.uqam.ca/Portals/0/docs/articles/2008/JASIST_Aging.pdf.

Larsen, Birger. "References and Citations in Automatic Indexing and Retrieval Systems: Experiments with the Boomerang Effect." Ph.D. dissertation, Royal School of Library and Information Science, Department of Information Studies, 2004. www.db.dk/dbi/samling/phd/birgerlarsen-phd.htm.

Larson, Ray R. "Bibliometrics of the World Wide Web: An Exploratory Analysis of the Intellectual Structure of Cyberspace." In *Global Complexity: Information, Chaos and Control; Proceedings of the ASIS 1996 Annual Meeting*. 1996. sherlock.berkeley.edu/asis96/asis96.html.

Latour, Bruno. *Reassembling the Social: An Introduction to Actor-Network-Theory*. Oxford: Oxford University Press, 2005.

———. *Science in Action: How to Follow Scientists and Engineers Through Society*. Cambridge, Mass.: Harvard University Press, 1987.

Latour, Bruno, and Steve Woolgar. *Laboratory Life*. Princeton, N.J.: Princeton University Press, 1986.

Laudel, Grit. "What Do We Measure by Co-Authorships?" *Research Evaluation* 11, no. 1 (2002): 3–15. www.ingentaconnect.com/content/beech/rev/2002/00000011/00000001/art00001.

Law, John. "Notes on the Theory of the Actor-Network: Ordering, Strategy, and Heterogeneity." *Systemic Practice and Action Research* 5, no. 4 (1992): 379–93. www.lancs.ac.uk/fass/sociology/papers/law-notes-on-ant.pdf.

Lawrence, Steve. "Free Online Availability Substantially Increases a Paper's Impact." *Nature* 411, no. 6837 (2001): 521. citeseer.ist.psu.edu/online-nature01/.

Lawrence, Steve, C. Lee Giles, and Kurt D. Bollacker. "Digital Libraries and Autonomous Citation Indexing." *IEEE Computer* 32, no. 6 (1999): 67–71. citeseer.ist.psu.edu/lawrence99digital.html.

Leach, Edmund Ronald. *Culture and Communication: The Logic by Which Symbols Are Connected: An Introduction to the Use of Structuralist Analysis in Social Anthropology*. Cambridge, Mass.: Cambridge University Press, 1976.

Lee, Dongwon, Jaewoo Kang, Prasenjit Mitra, C. Lee Giles, and Byung-Won On. "Are Your Citations Clean? New Scenarios and Challenges in Maintaining Digital Libraries." *Communications of the ACM* 50, no. 12 (2007): 33–38. www.cse.psu.edu/~on/cacm06.pdf.

Lehmann, Sune, Andrew D. Jackson, and Benny E. Lautrup. "Measures for Measures." *Nature* 444, no. 7122 (2006): 1003–4.

Leimkuhler, Ferdinand F. "The Bradford Distribution." *Journal of Documentation* 23, no. 3 (1967): 197–207.

Leimu, Roosa, and Julia Koricheva. "What Determines the Citation Frequency of Ecological Papers?" *Trends in Ecology and Evolution* 20, no. 1 (2005): 28–32.

Le Pair, Cornelius. "The Citation Gap of Applicable Science." In *Handbook of Quantitative Studies of Science and Technology*, edited by Anthony F. J. van Raan, 537–53. Amsterdam: Elsevier, 1988.

Lewison, Grant. "Beyond Outputs: New Measures of Biomedical Research Impact." *ASLIB Proceedings* 55, nos. 1/2 (2003): 32–42.

Leydesdorff, Loet. "Can Scientific Journals Be Classified in Terms of Aggregated Journal-Journal Citation Relations Using the Journal Citation Reports?" *Journal of the American Society for Information Science and Technology* 57, no. 5 (2006): 601–13. www.leydesdorff.net/classif03/index.htm.

———. "*Caveats* for the Use of Citation Indicators in Research and Journal Evaluations." *Journal of the American Society for Information Science and Technology* 59, no. 2 (2008): 278–87. www.leydesdorff.net/cit_indicators/cit_indicators.pdf.

———. *The Challenge of Scientometrics: The Development, Measurement, and Self-Organization of Scientific Communications*. Parkland, Fla.: Universal Publishers, 2001.

———. "Clusters and Maps of Science Journals Based on Bi-Connected Graphs in the *Journal Citation Reports*." *Journal of Documentation* 60, no. 4 (2004): 371–427. users.fmg.uva.nl/lleydesdorff/jcr01/art/index.htm.

———. "Indicators of Structural Change in the Dynamics of Science: Entropy Statistics of the *SCI Journal Citation Reports*." *Scientometrics* 53, no. 1 (2002): 131–59. users.fmg.uva.nl/lleydesdorff/jcr/index.htm.

———. *The Knowledge Based Economy: Modeled, Measured, Simulated*. Boca Raton, Fla.: Universal Publishers, 2006.

———. "On the 'Scientometric Decline' of British Science: One Additional Graph in Reply to Ben Martin." *Scientometrics* 20, no. 2 (1991): 363–67.

———. "Problems with the 'Measurement' of National Scientific Performance." *Science and Public Policy* 15, no. 3 (1988): 149–52.

———. "A Reply to Courtial's Comments." *Scientometrics* 25, no. 2 (1992): 317–19.

———. "The Science Citation Index and the Measurement of National Performance in Terms of Numbers of Scientific Publications." *Scientometrics* 17, no. 1 (1989): 111–20.

———. "Theories of Citation?" *Scientometrics* 43, no. 1 (1998): 5–25.

———. "A Validation Study of Leximappe." *Scientometrics* 25, no. 2 (1992): 295–312.

———. "Various Methods for the Mapping of Science." *Scientometrics* 11, nos. 5–6 (1987): 295–324.

———. "Why Words and Co-Words Cannot Map the Development of the Sciences." *Journal of the American Society for Information Science* 48, no. 5 (1997): 418–27. users.fmg.uva.nl/lleydesdorff/cowords/index.htm.

———. "Words and Co-Words as Indicators of Intellectual Organization." *Research Policy* 18, no. 4 (1989): 209–23.

Leydesdorff, Loet, and Olga Amsterdamska. "Dimensions of Citation Analysis." *Science, Technology & Human Values* 15, no. 3 (1990): 305–35.

Leydesdorff, Loet, and Stephen Bensman. "Classification and Powerlaws: The Logarithmic Transformation." *Journal of the American Society for Information Science and Technology* 58, no. 11 (2006): 1470–86. users.fmg.uva.nl/lleydesdorff/log05/.

Leydesdorff, Loet, and Susan E. Cozzens. "The Delineation of Specialties in Terms of Journals Using the Dynamic Journal Set of the *SCI*." *Scientometrics* 26, no. 1 (1993): 135–56.

Leydesdorff, Loet, and Michael Curran. "Mapping University-Industry-Government Relations on the Internet: The Construction of Indicators for a Knowledge-Based Economy." *Cybermetrics* 4, no. 1, paper 2 (2000). www.cindoc.csic.es/cybermetrics/articles/v4i1p2.html.

Leydesdorff, Loet, and Ismael Rafols. *A Global Map of Science Based on the ISI Subject Categories.* 2008. http://users.fmg.uva.nl/lleydesdorff/map06/texts/index.htm.

Li, Xin, Hsinchun Chen, Zan Huang, and Mihail C. Roco. "Patent Citation Network in Nanotechnology (1976–2004)." *Journal of Nanoparticle Research* 9, no. 3 (2007): 337–52.

Li, Xuemei. "A Review of the Development and Application of the Web Impact Factor." *Online Information Review* 27, no. 6 (2003): 407–17.

Liang, Liming. "*h*-Index Sequence and *h*-Index Matrix: Constructions and Applications." *Scientometrics* 69, no. 1 (2006): 153–59.

Lievrouw, Leah A. "Reconciling Structure and Process in the Study of Scholarly Communication." In *Scholarly Communication and Bibliometrics*, edited by Christine L. Borgman, 59–69. Newbury Park, Calif.: Sage, 1990.

Lin, Bou-Wen, Chung-Jen Chen, and Hsueh-Liang Wu. "Predicting Citations to Biotechnology Patents Based on the Information from the Patent Documents." *International Journal of Technology Management* 40, no. 1–3 (2007): 87–100.

Lindsey, Duncan. "Production and Citation Measures in the Sociology of Science: The Problem of Multiple Authorship." *Social Studies of Science* 10, no. 2 (1980): 145–62.

Line, Maurice B. "Changes in the Use of Literature with Time—Obsolescence Revisited." *Library Trends* 41, no. 4 (1993): 665–83. hdl.handle.net/2142/7869.

———. "'Half Life' of Periodical Literature: Apparent and Real Obsolescence." *Journal of Documentation* 26, no. 1 (1970): 46–52.

Line, Maurice B., and Alexander Sandison. "'Obsolescence' and Changes in the Use of Literature with Time." *Journal of Documentation* 30, no. 3 (1974): 283–350.

Liu, Mengxiong. "The Complexities of Citation Practice: A Review of Citation Studies." *Journal of Documentation* 49, no. 4 (1993): 370–408.

———. "A Study of Citing Motivations of Chinese Scientists." *Journal of Information Science* 19, no. 1 (1993): 13–23.

Liu, Zao. "Visualizing the Intellectual Structure in Urban Studies: A Journal Co-Citation Analysis (1992–2002)." *Scientometrics* 62, no. 3 (2005): 385–402.

Lokker, Cynthia, Ann K. McKibbon, R. James McKinkay, Nancy L. Wilczynski, and R. Brian Haynes. "Prediction of Citation Counts for Clinical Articles at Two Years Using Data Available Within Three Weeks of Publication: Retrospective Cohort Study." *BMJ* 336, no. 7645 (2008): 655–57. bmj.com/cgi/content/full/336/7645/655.

Lotka, Alfred J. "Statistics—The Frequency Distribution of Scientific Productivity." *Journal of the Washington Academy of Sciences* 16, no. 12 (1926): 317–25.

Lovaglia, Michael J. "Predicting Citations to Journal Articles: The Ideal Number of References." *American Sociologist* 22, no. 1 (1991): 49–64.

Luhn, Hans Peter. "The Automatic Creation of Literature Abstracts." *IBM Journal of Research and Development* 2, no. 2 (1958): 159–65. courses.ischool.berkeley.edu/i256/f06/papers/luhn58.pdf.

———. "A Statistical Approach to Mechanized Encoding and Searching of Literary Information." *IBM Journal of Research and Development* 1, no. 4 (1957): 309–17. www.research.ibm.com/journal/rd/014/ibmrd0104D.pdf.

Lundberg, Jonas. "Lifting the Crown—*Citation z-score*." *Journal of Informetrics* 1, no. 2 (2007): 145–54.

Luukkonen, Terttu. "Why Has Latour's Theory of Citations Been Ignored by the Bibliometric Community? Discussion of Sociological Interpretations of Citation Analysis." *Scientometrics* 38, no. 1 (1997): 27–37.

MacRoberts, Michael H., and Barbara R. MacRoberts. "Testing the Ortega Hypothesis: Facts and Artifacts." *Scientometrics* 12, no. 5 (1987): 293–95.

Madigan, Robert, Susan Johnson, and Patricia Linton. "The Language of Psychology: APA Style as Epistemology." *American Psychologist* 50, no. 6 (1995): 428–36.

Mandelbrot, Benoît. "Final Note on a Class of Skew Distribution Functions: Analysis and Critique of a Model Due to H. A. Simon." *Information and Control* 4, nos. 2–3 (1959): 198–216.

———. "An Information Theory of the Statistical Structure of Language." In *Communication Theory*, edited by Willis Jackson, 486–502. London: Butterworths Scientific Publications, 1953.

———. "New Methods in Statistical Economics." *Journal of Political Economics* 71, no. 5 (1963): 421–40.

———. "A Note on a Class of Skew Distribution Functions: Analysis and Critique of a Paper by H. A. Simon." *Information and Control* 2, no. 1 (1959): 90–99.

——— "The Pareto-Levy Law and the Distribution of Income." *International Economic Review* 1, no. 2 (1960): 79–106.

———. "Structure Formelle Des Textes Et Communication." *Word* 10 (1954): 1–27.

Marion, Linda S., Concepción S. Wilson, and Mari Davis. "Intellectual Structure and Subject Themes in Information Systems Research: A Journal Cocitation Study." *Proceedings of the American Society for Information Science and Technology* 42, no. 1 (2005). eprints.rclis.org/archive/00005217/.

Martello, Angela. *Selection of Content for the Web Citation Index (Tm): Institutional Repositories and Subject-Specific Archives*. Thomson Scientific, 2005. www.thomsonreuters.com/business_units/scientific/free/essays/wci-selection/.

Martin, Ben R. "The Bibliometric Assessment of UK Scientific Performance: Reply to Braun, Glänzel and Schubert." *Scientometrics* 20, no. 2 (1991): 333–57.

———. "British Science in the 1980s—Has the Relative Decline Continued?" *Scientometrics* 29, no. 1 (1994): 27–56.

Martin, Ben R., and John Irvine. "Assessing Basic Research: Some Partial Indicators of Scientific Progress in Radio Astronomy." *Research Policy* 12, no. 2 (1983): 61–90.

———. "The Position of British Science." *Nature* 355, no. 6363 (1992): 760.

Martin, Ben R., John Irvine, Francis Narin, and Chris Sterritt. "The Continuing Decline of British Science." *Nature* 330, no. 6144 (1987): 123–26.

Martin, Ben R., John Irvine, and Roy Turner. "The Writing on the Wall for British Science." *New Scientist* 104 (1984): 25–29.

Matter, George A., and Henri Broms. "The Myth of Garfield and Citation Indexing." *Tidskrift for Dokumentation* 39, no. 1 (1983): 1–8.

May, Kenneth O. "Growth and Quality of the Mathematical Literature." *ISIS* 59, no. 4 (1968): 363–71.

McCain, Katherine W. "Core Journal Networks and Cocitation Maps: New Bibliometric Tools for Serials Research and Management." *Library Quarterly* 61, no. 3 (1991): 311–36.

———. "Descriptor and Citation Retrieval in the Medical Behavioral Sciences Literature: Retrieval Overlaps and Novelty Distribution." *Journal of the American Society for Information Science* 40, no. 2 (1989): 110–14.

———. "Mapping Authors in Intellectual Space: A Technical Overview." *Journal of the American Society for Information Science* 41, no. 6 (1990): 433–43.

———. "Mapping Economics Through the Journal Literature: An Experiment in Journal Cocitation Analysis." *Journal of the American Society for Information Science* 42, no. 4 (1991): 290–96.

——. "Neural Networks Research in Context: A Longitudinal Journal Cocitation Analysis of an Emerging Interdisciplinary Field." *Scientometrics* 41, no. 3 (1998): 389–410.

McDonald, John D. "Understanding Online Journal Usage: A Statistical Analysis of Citation and Use." *Journal of the American Society for Information Science and Technology* 58, no. 1 (2007): 39–50. resolver.caltech.edu/CaltechLIB:2006.001.

McVeigh, Marie E. *Open Access Journals in the ISI Citation Databases: Analysis of Impact Factors and Citation Patterns; A Citation Study from Thomson Scientific*. Stamford, Conn.: Thomson Corporation, 2004. www.isinet.com/media/presentrep/essayspdf/openaccesscitations2.pdf.

Meadow, Charles T., Jiabin Wang, and Manal Stamboulie. "An Analysis of Zipf-Mandelbrot Language Measures and Their Application to Artificial Languages." *Journal of Information Science* 19, no. 4 (1993): 247–58.

Meadows, A. Jack. "The Citation Characteristics of Astronomical Research Literature." *Journal of Documentation* 23, no. 1 (1967): 28–33.

——. *Communication in Science*. London: Butterworths, 1974.

——. "The Scientific Paper as an Archaeological Artefact." *Journal of Information Science* 11, no. 1 (1985): 27–30.

Medawar, Peter B. "Is the Scientific Paper Fraudulent? Yes; It Misrepresents Scientific Thought." *Saturday Review*, 1 August (1964): 42–43.

Menard, Henry W. *Science: Growth and Change*. Cambridge, Mass.: Harvard University Press, 1971.

Merton, Robert K. "The Matthew Effect in Science." *Science* 159, no. 3810 (1968): 56–63. www.garfield.library.upenn.edu/merton/matthew1.pdf.

——. "The Matthew Effect in Science, 2: Cumulative Advantage and the Symbolism of Intellectual Property." *ISIS* 79, no. 299 (1988): 606–23. www.garfield.library.upenn.edu/merton/matthewii.pdf.

——. "The Normative Structure of Science." In *The Sociology of Science: Theoretical and Empirical Investigations,* 267–78. Chicago: University of Chicago Press, 1973. books.google.it/books?id=zPvcHuUMEMwC.

——. "Science and Technology in a Democratic Order." *Journal of Legal and Political Sociology* 1 (1942): 115–26.

Metcalfe, Travis S. "The Rise and Citation Impact of Astro-Ph in Major Journals." *Bulletin of the American Astronomical Society* 37, no. 2 (2005): 555–57. arxiv.org/pdf/astro-ph/0503519.

Meyer, Martin. "What Is Special About Patent Citations? Differences Between Scientific and Patent Citations." *Scientometrics* 49, no. 1 (2000): 93–123.

Milgram, Stanley. "The Small World Problem." *Psychology Today* 1, no. 1 (1967): 61–67.

Moed, Henk F. *Bibliometric Ranking of World Universities (CWTS Report 2006–01)*. Leiden: CWTS, 2006. www.cwts.nl/hm/bibl_rnk_wrld_univ_full.pdf.

——. *Citation Analysis in Research Evaluation*. Dordrecht: Springer, 2005.

——. "The Effect of 'Open Access' upon Citation Impact: An Analysis of ArXiv's Condensed Matter Section." *Journal of the American Society for Information Science and Technology* 58, no. 13 (2007): 2047–54. arxiv.org/abs/cs/0611060.

——. "UK Research Assessment Exercises: Informed Judgments on Research Quality Or Quantity?" *Scientometrics* 74, no. 1 (2008): 153–61.

Moed, Henk F., Renger E. De Bruin, and Thed N. van Leeuwen. "New Bibliometric Tools for the Assessment of National Research Performance: Database Description." *Scientometrics* 33, no. 3 (1995): 381–422.

Moed, Henk F., Wolfgang Glänzel, and Ulrich Schmoch, eds. *Handbook of Quantitative Science and Technology Research: The Use of Publication and Patent Statistics in Studies of S&T Systems*. Dordrecht: Kluwer Academic, 2004.

Moed, Henk F., and Thed N. van Leeuwen. "Improving the Accuracy of Institute for Scientific Information's Journal Impact Factors." *Journal of the American Society for Information Science* 46, no. 6 (1995): 461–67.

Moed, Henk F., Thed N. van Leeuwen, and Jan Reedijk. "A Critical Analysis of the Journal Impact Factors of Angewandte Chemie and the Journal of the American Chemical Society Inaccuracies in Published Impact Factors Based on Overall Citations Only." *Scientometrics* 37, no. 1 (1996): 105–16.

——. "A New Classification System to Describe the Ageing of Scientific Journals and Their Impact Factors." *Journal of Documentation* 54, no. 4 (1998): 387–419.

Moed, Henk F., and Martijn S. Visser. *Developing Bibliometric Indicators of Research Performance in Computer Science: An Exploratory Study (CWTS Report 2007–01)*. Leiden: CWTS, 2007. www.socialsciences.leidenuniv.nl/cwts/publications/nwo_indcompsciv.jsp.

Moed, Henk F., and M. Vriens. "Possible Inaccuracies Occurring in Citation Analysis." *Journal of Information Science* 15, no. 2 (1989): 95–107.

Moravcsik, Michael J. "Assessing the Methodology for Finding a Methodology for Assessment." *Social Studies of Science* 16, no. 3 (1986): 534–39.

——. "Life in a Multidimensional World." *Scientometrics* 6, no. 2 (1984): 75–86.

——. "Measures of Scientific Growth." *Research Policy* 2, no. 3 (1973): 266–75.

Moravcsik, Michael J., and Poovanalingam Murugesan. "Citation Patterns in Scientific Revolutions." *Scientometrics* 1, no. 2 (1979): 161–69.

———. "Some Results on the Quality and Function of Citations." *Social Studies of Science* 5, no. 1 (1975): 86–92.

Morris, Steven A., and Betsy Van der Veer Martens. "Mapping Research Specialties." *Annual Review of Information Science and Technology* 42 (2008): 213–95.

Muddiman, Dave. "Red Information Science: J. D. Bernal and the Nationalization of Scientific Information in Britain from 1930 to 1949." In *Conference on the History and Heritage of Scientific and Technological Information Systems: Proceedings of the 2002 Conference*, edited by W. Boyd Rayward and Mary E. Bowden, 258–66. Medford, N.J.: Information Today, for the American Society of Information Science and Technology and the Chemical Heritage Foundation, 2004. www.chemheritage.org/pubs/asist2002/21–muddiman.pdf.

———. "Red Information Scientist: The Information Career of J. D. Bernal." *Journal of Documentation* 59, no. 4 (2003): 387–409.

Nakamoto, Hideshiro. "Synchronous and Diachronous Citation Distributions." In *Informetrics 87/88: Select Proceedings of the First International Conference on Bibliometrics and Theoretical Aspects of Information Retrieval*, edited by Leo Egghe and Ronald Rousseau, 157–63. Amsterdam: Elsevier, 1988. doclib .uhasselt.be/dspace/handle/1942/837.

Nalimov, V. V., and B. M. Mulchenko. *Measurement of Science: Study of the Development of Science as an Information Process*. Washington, D.C.: Foreign Technology Division, U.S. Air Force System Command, 1971. www.garfield .library.upenn.edu/nalimovmeasurementofscience/book.pdf.

Nanba, Hidetsugu, and Manabu Okumura. "Towards Multi-Paper Summarization Using Reference Information." In *Proceedings of the 16th International Joint Conferences on Artificial Intelligence (IJCAI-99), Stockolm, Sweden*, vol. 2, edited by Thomas Dean, 926–31. San Francisco: Morgan Kaufmann, 1999. citeseer.ist.psu.edu/204343.html.

Naranan, Sundaresan. "Bradford's Law of Bibliography of Science: An Interpretation." *Nature* 227, no. 5258 (1970): 631–32.

Narin, Francis. *Evaluative Bibliometrics: The Use of Publication and Citation Analysis in the Evaluation of Scientific Activity*. Cherry Hill, N.J.: Computer Horizons, 1976. www.conceptsymbols.com/narin/narin_1975_ eval-bibliometrics_images.pdf.

———. "Patent Bibliometrics." *Scientometrics* 30, no. 1 (1994): 147–55.

Narin, Francis, Kimberly S. Hamilton, and Dominic Olivastro. "The Increasing Linkage Between U.S. Technology and Public Science." *Research Policy* 26, no. 3 (1997): 317–30.

Narin, Francis, and Joy K. Moll. "Bibliometrics." *Annual Review of Information Science and Technology* 12 (1977): 35–58.

Narin, Francis, Kimberly Stevens, and Edith S. Whitlow. "Scientific Co-Operation in Europe and the Citation of Multinationally Authored Papers." *Scientometrics* 21, no. 3 (1990): 313–23.

Nederhof, Anton J. "Evaluating Research Output Through Life Work Citation Counts." *Scientometrics* 7, nos. 1–2 (1985): 23–28.

Nederhof, Anton J., and Anthony F. J. van Raan. "Citation Theory and the Ortega Hypothesis." *Scientometrics* 12, nos. 5–6 (1987): 325–28.

Newman, Mark E. J. "The Structure and Function of Complex Networks." *SIAM Review* 45, no. 2 (2003): 167–256. citeseer.ist.psu.edu/newman03structure.html.

Nicholls, Paul Travis. "Bibliometric Modelling Processes and the Empirical Validity of Lotka's Law." *Journal of the American Society for Information Science* 40, no. 6 (1989): 379–85.

———. "Price's Square Root Law: Empirical Validity and Relation to Lotka's Law." *Information Processing & Management* 24, no. 4 (1988): 469–77.

Nicolaisen, Jeppe. "Citation Analysis." *Annual Review of Information Science and Technology* 41 (2007): 609–41.

Nicolaisen, Jeppe, and Birger Hjørland. "Practical Potentials of Bradford's Law: A Critical Examination of the Received View." *Journal of Documentation* 63, no. 3 (2007): 359–77. dlist.sir.arizona.edu/2123/.

Nicolini, Claudio, Sergei Vakula, M. Italo Balla, and Ezio Gandini. "Can the Assignment of University Chairs Be Automated?" *Scientometrics* 32, no. 2 (1995): 93–107.

Nieminen, Pentti, Gerta Rucker, Jouko Miettunen, James Carpenter, and Martin Schumacher. "Statistically Significant Papers in Psychiatry Were Cited More Often Than Others." *Journal of Clinical Epidemiology* 60, no. 9 (2007): 939–46.

Norris, Michael, Charles Oppenheim, and Fytton Rowland. "The Citation Advantage of Open-Access Articles." *Journal of the American Society for Information Science and Technology* 59, no. 12 (2008): 1963–72.

Noruzi, Alireza. "The Web Impact Factor: A Critical Review." *The Electronic Library* 24, no. 4 (2006): 490–500. eprints.rclis.org/archive/00005543/.

Noyons, Ed. "Science Maps Within a Science Policy Context." In *Handbook of Quantitative Science and Technology Research: The Use of Publication and Patent Statistics in Studies of S&T Systems,* edited by Henk F. Moed, Wolfgang Glänzel, and Ulrich Schmoch, 237–55. Dordrecht: Kluwer Academic, 2004.

Nunn, Hilary, and Charles Oppenheim. "A Patent-Journal Citation Network on Prostaglandins." *World Patent Information* 2, no. 2 (1980): 57–60.

Oberski, J. E. J. "Mapping of Science: Possibilities and Limitations." In *Handbook of Quantitative Studies of Science and Technology*, edited by Anthony F.J. van Raan, 431–62. Amsterdam: Elsevier, 1988.

Odlyzko, Andrew. "The Rapid Evolution of Scholarly Communication." *Learned Publishing* 15, no. 1 (2002): 7–19. www.si.umich.edu/PEAK-2000/odlyzko.pdf.

Ogden, Patti. "'Mastering the Lawless Science of Our Law': A Story of Legal Citation Indexes." *Law Library Journal* 85, no. 1 (1993): 1–47.

Oluić-Vuković, Vesna. "Bradford's Distribution: From the Classical Bibliometric 'Law' to the More General Stochastic Models." *Journal of the American Society for Information Science* 48, no. 9 (1998): 833–42.

———. "Impact of Productivity Increase on the Distribution Patterns of Journals." *Scientometrics* 17, no. 1–2 (1989): 97–109.

———. "Journal Productivity Distribution: Quantitative Study of Dynamic Behavior." *Journal of the American Society for Information Science* 43, no. 6 (1992): 412–21.

OpCit Project. *The Effect of Open Access and Downloads ("Hits") on Citation Impact: A Bibliography of Studies*. opcit.eprints.org/oacitation-biblio.html.

Oppenheim, Charles. "Do Patent Citations Count?" In *The Web of Knowledge: A Festschrift in Honor of Eugene Garfield*, edited by Blaise Cronin and Helen Barsky Atkins, 405–32. Medford, N.J.: Information Today, 2000.

Oppenheim, Charles, and Susan P. Renn. "Highly Cited Old Papers and the Reasons Why They Continue to Be Cited." *Journal of the American Society for Information Science* 29, no. 5 (1978): 225–31.

Oppenheim, Charles, and Mark A. C. Summers. "Citation Counts and the Research Assessment Exercise, Part VI: Unit of Assessment 67 (Music)." *Information Research* 13, no. 2 (2008). informationr.net/ir/13–2/paper342.html.

Oromaner, Mark. "The Ortega Hypothesis and Influential Articles in American Sociology." *Scientometrics* 7, nos. 1–2 (1985): 3–10.

Ortega y Gasset, José. *The Revolt of the Masses*. New York & London: W.W. Norton & Company, 1999. www.19.5degs.com/ebook/revolt-of-the-masses/1419/read.

Osareh, Farideh. "Bibliometrics, Citation Analysis and Co-Citation Analysis: A Review of Literature I." *Libri: International Library Review* 46, no. 3 (1996): 149–58.

———. "Bibliometrics, Citation Analysis and Co-Citation Analysis: A Review of Literature II." *Libri: International Library Review* 46, no. 4 (1996): 217–25.

Otlet, Paul. *Traitée De Documentation: Le Livre Sur Le Livre. Theorie Et Pratique*. Bruxelles: Van Keerberghen, 1934. https://archive.ugent.be/handle/1854/5612?mode=full.

Otte, Evelien, and Ronald Rousseau. "Social Network Analysis: A Powerful Strategy; Also for the Information Sciences." *Journal of Information Science* 28, no. 6 (2002): 441–53.

Pao, Miranda Lee. "An Empirical Examination of Lotka's Law." *Journal of the American Society for Information Science* 37, no. 1 (1986): 26–33.

Pao, Miranda Lee, and Dennis B. Worthen. "Retrieval Effectiveness by Semantic and Citation Searching." *Journal of the American Society for Information Science* 40, no. 4 (1989): 226–35.

Park, Han Woo. "Hyperlink Network Analysis: A New Method for the Study of Social Structure on the Web." *Connections* 25, no. 1 (2003): 49–61.

Park, Han Woo, and Mike Thelwall. "Hyperlink Analyses of the World Wide Web: A Review." *Journal of Computer-Mediated Communication* 8, no. 4 (2003). jcmc.indiana.edu/vol8/issue4/park.html.

Paul, Danette. "In Citing Chaos: A Study of the Rhetorical Use of Citations." *Journal of Business and Technical Communication* 14, no. 2 (2000): 185–222.

Payne, Nigel, and Mike Thelwall. "Longitudinal Trends in Academic Web Links." *Journal of Information Science* 34, no. 1 (2008): 3–14. citeseerx.ist.psu.edu/viewdoc/summary?doi=10.1.1.83.3769.

Peritz, Bluma C. "A Bradford Distribution for Bibliometrics." *Scientometrics* 18, no. 5–6 (1990): 323–29.

———. "Predictors of Citation Impact: An Exploratory Study." *Scripta Hierosolymitana* 29 (1989): 205–16.

Perneger, Thomas V. "Relations Between Online 'Hit Counts' and Subsequent Citations: Prospective Study of Research Papers in the BMJ." *BMJ* 329, no. 7465 (2004): 546–47. www.bmj.com/cgi/content/full/329/7465/546.

Peters, Harry P. F., and Anthony F. J. van Raan. "On Determinants of Citation Scores: A Case Study in Chemical Engineering." *Journal of the American Society for Information Science* 45, no. 1 (1994): 39–49.

Pinski, Gabriel, and Francis Narin. "Citation Influence for Journal Aggregates of Scientific Publications." *Information Processing & Management* 12, no. 5 (1976): 297–312.

Plomp, Reinier. "The Highly Cited Papers of Professors as an Indicator of a Research Groups Scientific Performance." *Scientometrics* 29, no. 3 (1994): 377–93.

———. "The Significance of the Number of Highly Cited Papers as an Indicator of Scientific Prolificacy." *Scientometrics* 19, no. 3 (1990): 185–97.

Polanco, Xavier, Moses A. Boudourides, Dominique Besagni, and Ivana Roche. "Clustering and Mapping Web Sites for Displaying Implicit Associations and Visualizing Networks." Paper presented at the NTTS&ETK 2001 Conference, Hersonissos, 2001. www.math.upatras.gr/~mboudour/articles/web_clustering&mapping.pdf.

Polanco, Xavier, Ivana Roche, and Dominique Besagni. "User Science Indicators in the Web Context and Co-Usage Analysis." *Scientometrics* 66, no. 1 (2006): 171–82.

Pollmann, Thus. "Forgetting and the Ageing of Scientific Publications." *Scientometrics* 47, no. 1 (2000): 43–54.

Pool, Ithiel de Sola, and Manfred Kochen. "Contacts and Influence." *Social Networks* 1, no. 1 (1978): 5–51.

Potter, William Gray. "Lotka's Law Revisited." *Library Trends* 30, no. 1 (1981): 21–39. hdl.handle.net/2142/7189.

Prabha, Chandra G. "Some Aspects of Citation Behavior: A Pilot Study in Business Administration." *Journal of the American Society for Information Science* 34, no. 3 (1983): 202–6.

Prabowo, Rudy, and Mike Thelwall. "Finding and Tracking Subjects Within an Ongoing Debate." *Journal of Informetrics* 2, no. 2 (2008): 107–27.

Price, Derek John de Solla. "Citation Measures of Hard Science, Soft Science, Technology, and Non-Science." In *Communication Among Scientists and Engineers*, edited by Carnot E. Nelson and Donald K. Pollock, 3–22. Lexington, Mass.: Heath Lexington Books, 1970.

———. "Cumulative Advantage Urn Games Explained: A Reply to Kantor." *Journal of the American Society for Information Science* 29, no. 4 (1978): 204–6.

———. "A General Theory of Bibliometric and Other Cumulative Advantage Processes." *Journal of the American Society for Information Science* 27, no. 2 (1976): 292–306. www.garfield.library.upenn.edu/price/pricetheory1976.pdf.

———. *Little Science, Big Science . . . and Beyond.* New York: Columbia University Press, 1986.

———. "Networks of Scientific Papers." *Science* 149, no. 3683 (1965): 510–15. www.garfield.library.upenn.edu/papers/pricenetworks1965.pdf.

———. *Science Since Babylon.* Enlarged ed. New Haven, Conn. and London: Yale University Press, 1975.

———. "Toward a Model for Science Indicators." In *Toward a Metric of Science: The Advent of Science Indicators*, edited by Yehuda Elkana, Joshua Lederberg, Robert K. Merton, Arnold Thackray, and Harriet Zuckerman, 69–95. New York: John Wiley & Sons, 1979.

Price, Derek John de Solla, and Donald DeB. Beaver. "Collaboration in an Invisible College." *American Psychologist* 21, no. 11 (1966): 1011–18.

Prime, Camille, Elise Bassecoulard, and Michel Zitt. "Co-Citations and Co-Sitations: A Cautionary View on an Analogy." *Scientometrics* 54, no. 2 (2002): 291–308.

Prime-Claverie, Camille, Michael Beigbeder, and Thierry Lafouge. "Transposition of the Cocitation Method with a View to Classifying Web Pages." *Journal of the American Society for Information Science and Technology* 55, no. 14 (2004): 1282–89. citeseerx.ist.psu.edu/viewdoc/summary?doi=10.1.1.76.336.

Pritchard, Alan. "Statistical Bibliography or Bibliometrics?" *Journal of Documentation* 25, no. 4 (1969): 348–49.

Qiu, Junping, Yejun Li, Jiang Li, and Quane Ren. "An Exploratory Study on Substantive Co-Link Analysis: A Modification to Total Co-Link Analysis." *Scientometrics* 76, no. 2 (2008): 327–41.

Qiu, Liwen. "An Empirical Examination of the Existing Models for Bradford's Law." *Information Processing & Management* 26, no. 5 (1990): 655–72.

Qiu, Liwen, and Jean Tague. "Complete or Incomplete Data Sets: The Groos Droop Investigated." *Scientometrics* 19, no. 3 (1990): 223–37.

Radicchi, Filippo, Santo Fortunato, and Claudio Castellano. *Universality of Citation Distributions: Towards an Objective Measure of Scientific Impact*. 2008. arxiv.org/abs/0806.0974.

Ravetz, Jerome R. *Scientific Knowledge and Its Social Problems*. Oxford: Clarendon Press, 1971.

Ravichandra Rao, Inna K. "Classical Laws of Informetrics: An Overview." Paper presented at the DRTC Workshop on Informetrics and Scientometrics, Bangalore, March 16–19, 1998. https://drtc.isibang.ac.in/bitstream/1849/143/2/AB.pdf.

———. "The Distribution of Scientific Productivity and Social Change." *Journal of the American Society for Information Science* 31, no. 2 (1980): 111–22.

———. *Quantitative Methods for Library and Information Science*. New Delhi: Wiley Eastern Ltd., 1983.

Redner, Sidney. "How Popular Is Your Paper? An Empirical Study of the Citation Distribution." *European Physical Journal B* 4, no. 2 (1998): 131–34. arxiv.org/abs/cond-mat/9804163.

Reedijk, Jan, and Henk F. Moed. "Is the Impact of Journal Impact Factor Decreasing?" *Journal of Documentation* 64, no. 2 (2008): 183–92.

Reisner, Phyllis. "A Machine Stored Citation Index to Patent Literature Experimentation and Planning." In *Automation and Scientific Communication: Short Papers Contributed to the Theme Session of the 26th Annual Meeting of the American Documentation Institute*, edited by Hans Peter Luhn, 71–72. Washington, D.C.: American Documentation Institute, 1963.

Rodríguez i Gairín, Josep Manuel. "Valorando el Impacto de la Información en Internet: Altavista, El 'Citation Index' de la Red." *Revista Española de Documentación Científica* 20, no. 2 (1997): 175–81. www.cindoc.csic.es/cybermetrics/pdf/120.pdf.

Ross, Nancy C. M., and Dietmar Wolfram. "End User Searching on the Internet: An Analysis of Term Pair Topics Submitted to the Excite Search Engine." *Journal of the American Society for Information Science* 51, no. 10 (2000): 949–58.

Rossner, Mike, Heather Van Hepps, and Emma Hill. "Show Me the Data." *The Journal of Cell Biology* 179, no. 6 (2007): 1091–92. www.jcb.org/cgi/content/full/179/6/1091.

Rousseau, Ronald. "Breakdown of the Robustness Property of Lotka's Law: The Case of Adjusted Counts for Multiauthorship Attribution." *Journal of the American Society for Information Science* 43, no. 10 (1992): 645–47.

———. "Sitations: An Exploratory Study." *Cybermetrics* 1, no. 1, paper 1 (1997). www.cindoc.csic.es/cybermetrics/vol1iss1.html.

Rousseau, Ronald, and Sandra Rousseau. "Informetric Distributions: A Tutorial Review." *Canadian Journal of Information and Library Science* 18, no. 2 (1993): 51–63.

Ruane, Frances, and Richard S. J. Tol. *Refined (Successive) h-Indices: An Application to Economics in the Republic of Ireland.* 2007. ideas.repec.org/p/sgc/wpaper/130.html.

Ruiz-Baños, Rosario, Rafael Bailón-Moreno, Evaristo Jiménez-Contreras, and Jean Pierre Courtial. "Structure and Dynamics of Scientific Networks. Part I: Fundamentals of the Quantitative Model of Translation." *Scientometrics* 44, no. 2 (1999): 217–34.

———. "Structure and Dynamics of Scientific Networks. Part II: The New Zipf's Law, the Clusters of Co-Citations and the Model of the Descriptor Presence." *Scientometrics* 44, no. 2 (1999): 235–65.

Rusmevichientong, Paat, David M. Pennock, Steve Lawrence, and Lee C. Giles. "Methods for Sampling Pages Uniformly from the World Wide Web." Paper presented at the AAAI Fall Symposium on Using Uncertainty Within Computation, 2001. citeseer.ist.psu.edu/455785.html.

Salton, Gerard. "Automatic Indexing Using Bibliographic Citations." *Journal of Documentation* 27, no. 2 (1971): 98–110.

———. "Automatic Text Analysis." *Science* 168, no. 3929 (1970): 335–43.

———. "Developments in Automatic Text Retrieval." *Science* 253, no. 5023 (1991): 974–80.

Salton, Gerard, and Michael J. McGill. *Introduction to Modern Information Retrieval.* New York: McGraw-Hill, 1983.

Samoylenko, I., T. C. Chao, W. C. Liu, and C. M. Chen. "Visualizing the Scientific World and Its Evolution." *Journal of the American Society for Information Science and Technology* 57, no. 11 (2006): 1461–69. phy.ntnu.edu.tw/~cchen/pdf/jasist1461.pdf.

Sanderson, Mark. "Revisiting *h* Measured on UK LIS and IR Academics." *Journal of the American Society for Information Science and Technology* 59, no. 7 (2008): 1184–90. dis.shef.ac.uk/mark/publications/my_papers/2008–JASIST.pdf.

Saracevic, Tefko. "Relevance: A Review of and a Framework for the Thinking on the Notion in Information Science." *Journal of the American Society for Information Science* 26, no. 6 (1975): 321–43.

———. "Relevance: A Review of the Literature and a Framework for the Thinking on the Notion in Information Science. Part II: Nature and Manifestations of Relevance." *Journal of the American Society for Information Science and Technology* 58, no. 13 (2007): 1915–33.

———. "Relevance: A Review of the Literature and a Framework for the Thinking on the Notion in Information Science. Part III: Behavior and Effects of Relevance." *Journal of the American Society for Information Science and Technology* 58, no. 13 (2007): 2126–44.

Savoy, Jacques. "Citation Schemes in Hypertext Information Retrieval." In *Information Retrieval and Hypertext*, edited by Maristella Agosti and Alan F. Smeaton, 99–120. Boston: Kluwer, 1996.

Scharnhorst, Andrea. "Complex Networks and the Web: Insights from Nonlinear Physics." *Journal of Computer-Mediated Communication* 8, no. 4 (2003). jcmc.indiana.edu/vol8/issue4/scharnhorst.html.

Scharnhorst, Andrea, and Paul Wouters. "Web Indicators—a New Generation of S&T Indicators?" *Cybermetrics* 10, no. 1, paper 6 (2006). www.cindoc.csic.es/cybermetrics/articles/v10i1p6.html.

Scheff, Thomas J. *Peer Review: An Iron Law of Disciplines?* 2002. www.soc.ucsb.edu/faculty/scheff/23.html.

Schmoch, Ulrich, and Torben Schubert. "Are International Co-Publications an Indicator for Quality of Scientific Research?" *Scientometrics* 74, no.3 (2008): 361–77.

Schreiber, Michael. "An Empirical Investigation of the *g*-Index for 26 Physicists in Comparison with the *h*-Index, the *A*-Index, and the *R*-Index." *Journal of the American Society for Information Science and Technology* 59, no. 9 (2008): 1513–22. arxiv.org/abs/0802.1820.

———. "Self-Citation Corrections for the Hirsch Index." *Europhysics Letters* 78, no. 3 (2007): 30002. arxiv.org/abs/physics/0701231.

———. "To Share the Fame in a Fair Way, h_m Modifies *h* for Multi-Authored Papers." *New Journal of Physics* 10, no. 4 (2008): 040201. iopscience.iop.org/1367–2630/10/4/040201/fulltext.

Schubert, Andreás, and Tibor Braun. "Cross-Field Normalization of Scientometric Indicators." *Scientometrics* 36, no. 3 (1986): 311–24.

———. "Reference Standards for Citation Based Assessments." *Scientometrics* 26, no. 1 (1993): 21–35.

———. "Relative Indicators and Relational Charts for Comparative Assessment of Publication Output and Citation Impact." *Scientometrics* 9, nos. 5–6 (1986): 281–91.

Schubert, Andreás, and Wolfgang Glänzel. "A Dynamic Look at a Class of Skew Distributions: A Model with Scientometric Applications." *Scientometrics* 6, no. 3 (1984): 149–67.

———. "Statistical Reliability of Comparisons Based on the Citation Impact of Scientific Publications." *Scientometrics* 5, no. 1 (1983): 59–73.

Schubert, Andreás, Wolfgang Glänzel, and Tibor Braun. "World Flash on Basic Research: Scientometric Datafiles; A Comprehensive Set of Indicators on 2649 Journals and 96 Countries in All Major Science Fields and Subfields 1981–1985." *Scientometrics* 16, nos. 1–6 (1989): 3–478.

Schwarz, Greg J., and Robert C. Kennicutt Jr. "Demographic and Citation Trends in Astrophysical Journal Papers and Preprints." *Bulletin of the American Astronomical Society* 36, no. 5 (2004): 1654–63. arxiv.org/abs/astro-ph/0411275.

Seglen, Per O. "Causal Relationship Between Article Citedness and Journal Impact." *Journal of the American Society for Information Science* 45, no. 1 (1994): 1–11.

———. "The Skewness of Science." *Journal of the American Society for Information Science* 43, no. 9 (1999): 628–38.

———. "Why the Impact Factor of Journals Should Not Be Used for Evaluating Research." *BMJ* 314, no. 7079 (1997): 498–502. www.bmj.com/cgi/content/full/314/7079/497.

Sellen, Mary K. *Bibliometrics: An Annotated Bibliography, 1970–1990.* New York: G.K. Hall, 1993.

Sen, S. K., and S. K. Chatterjee. "Bibliographic Scattering and Time: An Empirical Study Through Temporal Partitioning of Bibliographies." *Scientometrics* 41, nos. 1–2 (1998): 135–54.

Sengupta, I. N. "Bibliometrics, Informetrics, Scientometrics and Librametrics: An Overview." *Libri: International Library Review* 42, no. 2 (1992): 75–98.

Shadish, William R., Donna Tolliver, Maria Gray, and Sunil K. Sen Gupta. "Author Judgements About Works They Cite: Three Studies from Psychology Journals." *Social Studies of Science* 25, no. 3 (1995): 477–98.

Shapiro, Fred R. "Origins of Bibliometrics, Citation Indexing, and Citation Analysis: The Neglected Legal Literature." *Journal of the American Society for Information Science* 43, no. 5 (1992): 337–39.

Shaw, William M., Jr. "Subject and Citation Indexing, Part 1: The Clustering Structure of Composite Representations in the Cystic-Fibrosis Document Collection." *Journal of the American Society for Information Science* 42, no. 9 (1991): 669–75.

———. "Subject and Citation Indexing, Part 2: The Optimal, Cluster-Based Retrieval Performance of Composite Representations." *Journal of the American Society for Information Science* 42, no. 9 (1991): 675–84.

Shin, Eun-Ja. "Do Impact Factors Change with a Change of Medium? A Comparison of Impact Factors When Publication Is by Paper and Through Parallel Publishing." *Journal of Information Science* 29, no. 6 (2003): 527–33.

Sichel, Herbert S. "Anatomy of the Generalized Inverse Gaussian-Poisson Distribution with Special Applications to Bibliometric Studies." *Information Processing & Management* 28, no. 1 (1992): 5–17.

———. "A Bibliometric Distribution Which Really Works." *Journal of the American Society for Information Science* 36, no. 5 (1985): 314–21.

Sidiropoulos, Antonis, Dimitrios Katsaros, and Yannis Manolopoulos. "Generalized Hirsch *h*-Index for Disclosing Latent Facts in Citation Networks." *Scientometrics* 72, no. 2 (2007): 253–80. arxiv.org/abs/cs/0607066.

Simon, Herbert A. "On a Class of Skew Distribution Functions." *Biometrika* 42, nos. 3/4 (1955): 425–40.

Small, Henry G. "Citations and Consilience in Science." *Scientometrics* 43, no. 1 (1998): 143–48.

———. "Cited Documents as Concept Symbols." *Social Studies of Science* 8, no. 3 (1978): 327–40.

———. "Co-Citation in the Scientific Literature: A New Measure of the Relationship Between Two Documents." *Journal of the American Society for Information Science* 24, no. 4 (1973): 265–69. www.garfield.library.upenn.edu/essays/v2p028y1974–76.pdf.

———. "A Co-Citation Model of a Scientific Specialty: A Longitudinal Study of Collagen Research." *Social Studies of Science* 7, no. 2 (1977): 139–66.

———. "On the Shoulders of Robert Merton: Towards a Normative Theory of Citation." *Scientometrics* 60, no. 1 (2004): 71–79.

———. "Paradigms, Citations and Maps of Science: A Personal History." *Journal of the American Society for Information Science and Technology* 54, no. 5 (2003): 394–99.

———. "A Passage Through Science: Crossing Disciplinary Boundaries." *Library Trends* 48, no. 1 (1999): 72–108. findarticles.com/p/articles/mi_m1387/is_1_48/ai_57046528.

———. "The Synthesis of Specialty Narratives from Co-Citation Clusters." *Journal of the American Society for Information Science* 37, no. 3 (1986): 97–110. www.asis.org/Publications/JASIS/Best_Jasist/1987Small.pdf.

Small, Henry G., and Belver C. Griffith. "The Structure of Scientific Literature. I: Identifying and Graphing Specialties." *Science Studies* 4, no. 1 (1974): 17–40.

Small, Henry G., and E. Sweeney. "Clustering the Science Citation Index Using Co-Citations, 1: A Comparison of Methods." *Scientometrics* 7, no. 3–6 (1985): 391–409.

Small, Henry G., E. Sweeney, and Edwin Grenlee. "Clustering the Science Citation Index Using Co-Citations, 2: Mapping Science." *Scientometrics* 8, nos. 5–6 (1985): 321–40.

Smart, John C., and Alan E. Bayer. "Author Collaboration and Impact: A Note on Citation Rates of Single and Multiple Authored Articles." *Scientometrics* 10, no. 5 (1986): 297–305.

Smith, David C., Peter M. D. Collins, Diana Hicks, and Suzanne Wyatt. "National Performance in Basic Research." *Nature* 323, no. 6090 (1986): 681–84.

Snizek, William E. "A Re-Examination of the Ortega Hypothesis: The Dutch Case." *Scientometrics* 9, no. 1 (1986): 3–11.

Snizek, William E., Kay Oehler, and Nicholas C. Mullins. "Textual and Nontextual Characteristics of Scientific Papers: Neglected Science Indicators." *Scientometrics* 20, no. 1 (1991): 25–35.

Snyder, Herbert W., and Howard Rosenbaum. "Can Search Engines Be Used as Tools for Web-Link Analysis? A Critical View." *Journal of Documentation* 55, no. 4 (1999): 375–84.

Soyibo, Adedoyin, and W. Olabode Aiyepeku. "On the Categorization, Exactness and Probable Utility of Bibliometric Laws and Their Extension." *Journal of Information Science* 14, no. 4 (1988): 243–51.

Sokal, Michael. "Stargazing: James McKeen Cattell, American Men of Science, and the Reward Structure of the American Scientific Community, 1906–1944." In *Psychology, Science, and Human Affairs: Essays in Honor of William Bevan*, edited by Frank Kessel, 64–86. Boulder, Colo.: Westview Press, 1995.

Solomon, Sorin, and Peter Richmond. "Stability of Pareto-Zipf Law in Non-Stationary Economies." Paper presented at the 5th Workshop on Economics and Heterogeneous Interacting Agents, Marseille, June 15–17, 2001. arxiv .org/abs/cond-mat/0012479v2.

Sparck Jones, Karen. "Statistics and Retrieval: Past and Future." In *International Conference in Computing: Theory and Applications,* 396–405. Kolkata, India: IEEE Computer Society, 2007.

Sternberg, Robert J., and Tamara Gordeeva. "The Anatomy of Impact: What Makes an Article Influential?" *Psychological Science* 7, no. 2 (1996): 69–75.

Stewart, John A. "Achievement and Ascriptive Processes in the Recognition of Scientific Articles." *Social Forces* 62, no. 1 (1983): 166–89.

Stigler, Stephen M. "Precise Measurement in the Face of Error: A Comment on MacRoberts and MacRoberts." *Social Studies of Science* 17, no. 2 (1987): 332–34.

Stowe, Robert C. *Annotated Bibliography of Publications Dealing with Qualitative and Quantitative Indicators of the Quality of Science*. Cambridge, Mass.: Jefferson Physical Lab., Harvard University, 1986. eric.ed.gov/ERICDocs/data/ericdocs2sql/content_storage_01/0000019b/80/2f/77/bf.pdf.

Stremersch, Stefan, Isabel Verniers, and Peter C. Verhoef. "The Quest for Citations: Drivers of Article Impact." *Journal of Marketing* 71, no. 3 (2007): 171–93.

Studer, Kenneth E., and Daryl E. Chubin. *The Cancer Mission: Social Contexts of Biomedical Research*. Beverly Hills, Calif.: Sage, 1980.

Subramanyam, K. "Bibliometric Studies of Research Collaboration: A Review." *Journal of Information Science* 6, no. 1 (1983): 33–38.

Swales, John M. *Genre Analysis: English in Academic and Research Settings*. Cambridge: Cambridge University Press, 1990.

Szabó, A. T. "Alphonse De Candolle's Early Scientometrics (1883, 1885) with References to Recent Trends in the Field." *Scientometrics* 8, nos. 1–2 (1985): 13–33.

Száva-Kováts, Endre. "The False 'Ortega Hypothesis': A Literature Science Study." *Journal of Information Science* 30, no. 6 (2004): 496–508.

———. "Unfounded Attribution of the 'Half-Life' Index-Number of Literature Obsolescence to Burton and Kebler: A Literature Science Study." *Journal of the American Society for Information Science* 53, no. 13 (2002): 1098–1105.

Tabah, Albert N. "Literature Dynamics: Studies on Growth, Diffusion, and Epidemics." *Annual Review of Information Science and Technology* 34 (1999): 249–76.

Tagliacozzo, Renata. "Self-Citations in Scientific Literature." *Journal of Documentation* 33, no. 4 (1977): 251–65.

Tague, Jean. "The Success-Breeds-Success Phenomena and Bibliometric Processes." *Journal of the American Society for Information Science* 32, no. 4 (1981): 280–86.

Tague, Jean, Jamshid Beheshti, and Lorna Rees-Potter. "The Law of Exponential Growth: Evidence, Implications and Forecasts." *Library Trends* 30, no. 1 (1981): 125–49. hdl.handle.net/2142/7189.

Tague-Sutcliffe, Jean. "An Introduction to Informetrics." *Information Processing & Management* 28, no. 1 (1992): 1–3.

———. *Measuring Information: An Information Services Perspective*. San Diego: Academic Press, 1995.

Tang, Rong, and Martin A. Safer. "Author-Rated Importance of Cited References in Biology and Psychology Publications." *Journal of Documentation* 64, no. 2 (2008): 246–72.

Tenopir, Carol, and Donald W. King. *Towards Electronic Journals: Realities for Scientists, Librarians, and Publishers*. Washington, D.C.: Special Libraries Association, 2000.

Testa, James, and Marie E. McVeigh. *The Impact of Open Access Journals: A Citation Study from Thomson ISI*. Stamford, Conn.: Thomson Corporation, 2004. www.isinet.com/media/presentrep/acropdf/impact-oa-journals.pdf.

Thelwall, Mike. "Bibliometrics to Webometrics." *Journal of Information Science* 34, no. 4 (2008): 605–21. www.scit.wlv.ac.uk/~cm1993/papers/JIS-0642–v4–Bibliometrics-to-Webometrics.pdf.

——. "Conceptualising Documentation on the Web: An Evaluation of Different Heuristic-Based Models for Counting Links Between University Web Sites." *Journal of the American Society for Information Science and Technology* 53, no. 12 (2002): 995–1005. citeseerx.ist.psu.edu/viewdoc/summary?doi=10.1.1.65.1637.

——. "Extracting Macroscopic Information from Web Links." *Journal of the American Society for Information Science and Technology* 52, no. 13 (2001): 1157–68. citeseerx.ist.psu.edu/viewdoc/summary?doi=10.1.1.64.7031.

——. "A Free Database of University Web Links: Data Collection Issues." *Cybermetrics* 6/7, no. 1, paper 2 (2002). www.cindoc.csic.es/cybermetrics/articles/v6i1p2.html.

——. *Link Analysis: An Information Science Approach*. Amsterdam: Elsevier, 2004.

——. "A Web Crawler Design for Data Mining." *Journal of Information Science* 27, no. 5 (2001): 319–25.

——. "Web Impact Factors and Search Engine Coverage." *Journal of Documentation* 56, no. 2 (2000): 185–89. wlv.openrepository.com/wlv/handle/2436/3141.

——. "What Is This Link Doing Here? Beginning a Fine-Grained Process of Identifying Reasons for Academic Hyperlink Creation." *Information Research* 8, no. 3 (2003). informationr.net/ir/8–3/paper151.html.

Thelwall, Mike, and Gareth Harries. "The Connection Between the Research of a University and Counts of Links to Its Web Pages: An Investigation Based Upon a Classification of the Relationships of Pages to the Research of the Host University." *Journal of the American Society for Information Science and Technology* 54, no. 7 (2003): 594–602. citeseerx.ist.psu.edu/viewdoc/summary?doi=10.1.1.20.4172.

Thelwall, Mike, Liwen Vaughan, and Lennart Björneborn. "Webometrics." *Annual Review of Information Science and Technology* 39 (2005): 81–135.

Thelwall, Mike, and David Wilkinson. "Finding Similar Academic Web Sites with Links, Bibliometric Couplings and Colinks." *Information Processing &*

Management 40, no. 3 (2004): 515–26. citeseerx.ist.psu.edu/viewdoc/summary ?doi=10.1.1.65.95.

———. "Three Target Document Range Metrics for University Web Sites." *Journal of the American Society for Information Science and Technology* 54, no. 6 (2003): 489–96. citeseerx.ist.psu.edu/viewdoc/summary?doi=10.1.1.64.2423.

Thompson, Peter. "Patent Citations and the Geography of Knowledge Spillovers: Evidence from Inventor- and Examiner-Added Citations." *The Review of Economics and Statistics* 88, no. 2 (2006): 383–88. www.fiu.edu/~economic/wp2004/04–05.pdf.

Tijssen, Robert J. W. "Africa's Contribution to the Worldwide Research Literature: New Analytical Perspectives, Trends, and Performance Indicators." *Scientometrics* 71, no. 2 (2007): 303–27. www.springerlink.com/content/e401 26m448gv0127/?p=7f6f962118034e1d8e6449a2632c6eb9&pi=9.

———. *Cartography of Science: Scientometric Mapping with Multidimensional Scaling Methods*. Leiden: DSWO Press, 1992.

Tijssen, Robert J. W., Jan De Leeuw, and Anthony F. J. van Raan. "A Method for Mapping Bibliometric Relations Based on Field-Classifications and Citations of Articles." In *Informetrics 87/88: Select Proceedings of the First International Conference on Bibliometrics and Theoretical Aspects of Information Retrieval*, edited by Leo Egghe and Ronald Rousseau, 279–92. Amsterdam: Elsevier, 1988. hdl.handle.net/1942/847.

———. "Quasi-Correspondence Analysis on Scientometric Transaction Matrices." *Scientometrics* 11, no. 5–6 (1987): 351–66.

Tijssen, Robert J. W., Anthony F. J. van Raan, W. J. Heisser, and L. Wachmann. "Integrating Multiple Sources of Information in Literature-Based Maps of Science." *Journal of Information Science* 16, no. 4 (1990): 217–27.

Tijssen, Robert J. W., Martijn S.Visser, and Thed N. van Leeuwen. "Benchmarking International Scientific Excellence: Are Highly Cited Research Papers an Appropriate Frame of Reference?" *Scientometrics* 54, no. 3 (2002): 381–97.

Trajtenberg, Manuel, Rebecca Henderson, and Adam Jaffe. "University Versus Corporate Patents: A Window on the Basicness of Invention." *Economics of Innovation and New Technology* 5, no. 1 (1997): 19–50.

Tsay, Ming-Yueh, Hong Xu, and Chia-Wen Wu. "Journal Co-Citation Analysis of Semiconductor Literature." *Scientometrics* 57, no. 1 (2003): 7–25.

Turnbull, Don. *Bibliometrics and the World Wide Web*. Technical Report FIS-12 –19–1996–1. Toronto: Faculty of Information Studies, University of Toronto, 1996. www.ischool.utexas.edu/~donturn/research/bibweb.html.

van Dalen, Hendrik P., and Kène Henkens. "What Makes a Scientific Article Influential? The Case of Demographers." *Scientometrics* 50, no. 3 (2001): 455–82. ssrn.com/abstract=250885.

Van Hooydonk, Guido. “Fractional Counting of Multiauthored Publications: Consequences for the Impact of Authors.” *Journal of the American Society for Information Science* 48, no. 10 (1997): 944–45.

van Leeuwen, Thed N. “Testing the Validity of the Hirsch-Index for Research Assessment Purposes.” *Research Evaluation* 17, no. 2 (2008): 157–60.

van Leeuwen, Thed N., and Henk F. Moed. “Development and Application of Journal Impact Measures in the Dutch Science System.” *Scientometrics* 53, no. 2 (2002): 249–66.

van Raan, Anthony F. J. “Comparison of the Hirsch-Index with Standard Bibliometric Indicators and with Peer Judgment for 147 Chemistry Research Groups.” *Scientometrics* 67, no. 3 (2006): 491–502. front.math.ucdavis.edu/0511.7206.

———. “Fatal Attraction: Conceptual and Methodological Problems in the Ranking of Universities by Bibliometric Methods.” *Scientometrics* 62, no. 1 (2005): 133–43. www.cwts.nl/TvR/documents/AvR-RankingScientom.pdf.

———, ed. *Handbook of Quantitative Studies of Science and Technology*. Amsterdam: Elsevier, 1988.

———. “In Matters of Quantitative Studies of Science the Fault of Theorists Is Offering Too Little and Asking Too Much.” *Scientometrics* 43, no. 1 (1998): 129–39.

———. “Measuring Science: Capita Selecta of Current Main Issues.” In *Handbook of Quantitative Science and Technology Research: The Use of Publication and Patent Statistics in Studies of S&T Systems*, edited by Henk F. Moed, Wolfgang Glänzel, and Ulrich Schmoch, 19–50. Dordrecht: Kluwer, 2004. www.cwts.nl/TvR/documents/AvR-HandbChKluw.pdf.

———. “On Growth, Ageing, and Fractal Differentiation of Science.” *Scientometrics* 47, no. 2 (2000): 347–62. www.cwts.nl/TvR/documents/AvR-Growth-Scientometrics.pdf.

———. “The Pandora’s Box of Citation Analysis: Measuring Scientific Excellence—The Last Evil?” In *The Web of Knowledge: A Festschrift in Honor of Eugene Garfield*, edited by Blaise Cronin and Helen Barsky Atkins, 301–19. Medford, N.J.: Information Today Inc, 2000.

———. “Self-Citation as an Impact-Reinforcing Mechanism in the Science System.” *Journal of the American Society for Information Science and Technology* 59, no. 10 (2008): 1631–43. arxiv.org/abs/0801.0524.

———.”Statistical Properties of Bibliometric Indicators: Research Group Indicator Distributions and Correlations.” *Journal of the American Society for Information Science and Technology* 57, no. 3 (2006): 408–30.

———. “The Use of Bibliometric Analysis in Research Performance Assessment and Monitoring of Interdisciplinary Scientific Developments.” *Technikfol-*

genabschätzung: Theorie und Praxis 12, no. 1 (2003): 20–29. www.itas.fzk.de/tatup/031/raan03a.htm#[1].

van Raan, Anthony F. J., Henk F. Moed, and Thed N. van Leeuwen. *Scoping Study on the Use of Bibliometric Analysis to Measure the Quality of Research in UK Higher Education Institutions [Report to HEFCE]*. 2007. www.hefce.ac.uk/pubs/rdreports/2007/rd18_07/rd18_07.pdf.

Vargas-Quesada, Benjamin, and Félix Móya-Anegon. *Visualizing the Structure of Science*. Berlin: Springer, 2007.

Vaughan, Liwen, and Kathy Hysen. "Relationship Between Links to Journal Web Sites and Impact Factors." *ASLIB Proceedings* 54, no. 6 (2002): 356–61.

Vaughan, Liwen, and Mike Thelwall. "Scholarly Use on the Web: What Are the Key Inducers of Links to Journal Web Sites?" *Journal of the American Society for Information Science and Technology* 54, no. 1 (2003): 29–38. citeseerx.ist.psu.edu/viewdoc/summary?doi=10.1.1.20.3055.

Vaughan, Liwen, and Justin You. "Comparing Business Competition Positions Based on Web Co-Link Data: The Global Market vs. the Chinese Market." *Scientometrics* 68, no. 3 (2006): 611–28.

Vickery, Brian C. "Bradford's Law of Scattering." *Journal of Documentation* 4, no. 3 (1948): 198–203.

Vidal-Arbona, Carlos. "Comparing the Retrieval Effectiveness of Free-Text and Citation Search Strategies in the Subject of Technology Planning." Ph.D. dissertation, Case Western Reserve University, 1986.

Vinkler, Paul. "Characterization of the Impact of Sets of Scientific Papers: The Garfield (Impact) Factor." *Journal of the American Society for Information Science and Technology* 55, no. 5 (2004): 431–35.

———. "The Origin and Features of Information Referenced in Pharmaceutical Patents." *Scientometrics* 30, no. 1 (1994): 283–302.

———. "A Quasi-Quantitative Citation Model." *Scientometrics* 12, nos. 1–2 (1987): 47–72.

———. "Research Contribution, Authorship and Team Cooperativeness." *Scientometrics* 26, no. 1 (1993): 213–30.

———. "Some Aspects of the Evaluation of Scientific and Related Performances of Individuals." *Scientometrics* 32, no. 2 (1995): 109–16.

Virgo, Julie A. "A Statistical Procedure for Evaluating the Importance of Scientific Papers." *Library Quarterly* 47, no. 4 (1977): 415–30.

Vlachý, Jan. "Frequency Distributions of Scientific Performance: A Bibliography of Lotka's Law and Related Phenomena." *Scientometrics* 1 (1978): 109–30.

———. "Time Factor in Lotka's Law." *Probleme de Informare si Documentare* 10, no. 2 (1976): 44–87.

Voos, Henry, and Katherine S. Dagaev. "Are All Citations Equal? Or Did We Op. Cit. Your Idem.?" *Journal of Academic Librarianship* 1, no. 6 (1976): 19–21.

Wagner-Döbler, Roland. "Time Dependencies of Bradford Distributions: Structures of Journal Output in 20th-Century Logic and 19th-Century Mathematics." *Scientometrics* 39, no. 3 (1997): 231–52.

———. "Where Has the Cumulative Advantage Gone? Some Observations About the Frequency Distribution of Scientific Productivity, of Duration of Scientific Participation, and of Speed of Publication." *Scientometrics* 32, no. 2 (1995): 123–32.

Walters, Glenn D. "Predicting Subsequent Citations to Articles Published in Twelve Crime-Psychology Journals: Author Impact Versus Journal Impact." *Scientometrics* 69, no. 3 (2006): 499–510.

Wan, Jin-kun, Ping-huan Hua, and Ronald Rousseau. *The Pure h-Index: Calculating an Author's h-Index by Taking Co-Authors into Account.* 2007. eprints .rclis.org/archive/00011401/.

Warner, Julian. "A Critical Review of the Application of Citation Studies to the Research Assessment Exercises." *Journal of Information Science* 26, no. 6 (2000): 453–60.

Watts, Duncan J., and Steven H. Strogatz. "Collective Dynamics of 'Small World' Networks." *Nature* 393, no. 6684 (1998): 440–42. www.tam.cornell .edu/tam/cms/manage/upload/SS_nature_smallworld.pdf.

Weinberg, Bella Hass. "Bibliographic Coupling: A Review." *Information Storage & Retrieval* 10, no. 5/6 (1974): 189–96.

Weinstock, Melvin. "Citation Indexes." In *Encyclopaedia of Library and Information Science*, Vol. 5, 16–40. New York: Marcel Dekker, 1971.

White, Howard D. "Authors as Citers over Time." *Journal of the American Society for Information Science and Technology* 52, no. 2 (2001): 87–108.

———. "'Bradfordizing' Search Output: How It Would Help Online Users." *Online Review* 5, no. 1 (1981): 47–54.

White, Howard D., and Belver C. Griffith. "Author Cocitation: A Literature Measure of Intellectual Structure." *Journal of the American Society for Information Science* 32, no. 3 (1981): 163–71. garfield.library.upenn.edu/hwhite/white jasist1981.pdf.

White, Howard D., and Katherine W. McCain. "Bibliometrics." *Annual Review of Information Science and Technology* 24 (1989): 119–65.

———. "Visualizing a Discipline: An Author Co-Citation Analysis of Information Science, 1972–1995." *Journal of the American Society for Information Science* 49, no. 4 (1998): 327–55. www.asis.org/Publications/JASIS/Best_ Jasist/1998WhiteandMcCain.pdf.

———. "Visualization of Literatures." *Annual Review of Information Science and Technology* 32 (1997): 99–168.

White, Marylin Domas, and Peiling Wang. "A Qualitative Study of Citing Behavior: Contributions, Criteria, and Metalevel Documentation Concern." *Library Quarterly* 67, no. 2 (1997): 122–54.

Whitley, Richard. *The Intellectual and Social Organization of the Sciences*: Oxford: Clarendon Press, 1984.

Whittaker, John, Jean Pierre Courtial, and John Law. "Creativity and Conformity in Science: Titles, Keywords and Co-Word Analysis." *Social Studies of Science* 19, no. 3 (1989): 473–96.

Wilkinson, David, Gareth Harries, Mike Thelwall, and Liz Price. "Motivations for Academic Web Site Interlinking: Evidence for the Web as a Novel Source of Information on Informal Scholarly Communication." *Journal of Information Science* 29, no. 1 (2003): 49–56.

Wilkinson, Elizabeth A. "The Ambiguity of Bradford's Law." *Journal of Documentation* 28, no. 2 (1972): 122–30.

Willinsky, John. *The Access Principle. The Case for Open Access to Research and Scholarship*. Cambridge, Mass.: MIT Press, 2006. mitpress.mit.edu/catalog/item/default.asp?ttype=2&tid=10611.

Wilson, Concepción S. "Informetrics." *Annual Review of Information Science and Technology* 34 (1999): 107–247.

Wilson, Perry W., and Edwin B. Fred. "The Growth Curve of a Scientific Literature: Nitrogen Fixation by Plants." *Scientific Monthly* 41, no. 3 (1935): 240–50.

Wolfram, Dietmar. *Applied Informetrics for Information Retrieval Research*. Westport, Conn.: Libraries Unlimited, 2003.

Wouters, Paul. "The Citation Culture." Ph.D. thesis, University of Amsterdam, 1999. www.garfield.library.upenn.edu/wouters/wouters.pdf.

Wren, Jonathan D. "Open Access and Openly Accessible: A Study of Scientific Publications Shared Via the Internet." *BMJ* 330, no. 7500 (2005): 1128. www.bmj.com/cgi/reprint/330/7500/1128.

Wu, Qiang. *The w-Index: A Significant Improvement of the h-Index*. 2008. arxiv.org/abs/0805.4650.

Wyllys, Ronald E. "Empirical and Theoretical Bases of Zipf's Law." *Library Trends* 30, no. 1 (1981): 53–64. hdl.handle.net/2142/7189.

Yablonsky, Anatolity I. "On Fundamental Regularities of the Distribution of Scientific Productivity." *Scientometrics* 2, no. 1 (1980): 3–34.

———. "Stable Non-Gaussian Distributions in Scientometrics." *Scientometrics* 7, no. 3–6 (1985): 459–70.

Yoon, Lanju Lee. "The Performance of Cited References as an Approach to Information Retrieval." *Journal of the American Society for Information Science* 45, no. 5 (1994): 287–99.

Zipf, George Kingsley. *The Psycho-Biology of Language: An Introduction to Dynamic Philology*. London: Routledge, 1936.

Zitt, Michel, and Elise Bassecoulard. "Delineating Complex Scientific Fields by an Hybrid Lexical-Citation Method: An Application to Nanosciences." *Information Processing & Management* 42, no. 6 (2006): 1513–31. www.nanodistrict.org/scientific-output/papers/delinnano-mz-eb.PDF.

Zitt, Michel, Suzy Ramanana-Rahary, and Elise Bassecoulard. "Relativity of Citation Performance and Excellence Measures: From Cross-Field to Cross-Scale Effects of Field-Normalisation." *Scientometrics* 63, no. 2 (2005): 373–401.

Zitt, Michel, and Henry Small. "Modifying the Journal Impact Factor by Fractional Citation Weighting: The Audience Factor." *Journal of the American Society for Information Science and Technology* 59, no. 11 (2008): 1856–60.

Zuccala, Alesia. "Modeling the Invisible College." *Journal of the American Society for Information Science and Technology* 57, no. 2 (2006): 152–68. individual.utoronto.ca/azuccala_web/InvisibleCollege.pdf.

Zuckerman, Harriet. "Citation Analysis and the Complex Problem of Intellectual Influence." *Scientometrics* 12, no. 5 (1987): 329–38.

———. *Scientific Elite: Nobel Laureates in the United States*. New Brunswick and London: Transaction Publishers, 1996.

Index

About the Author

Nicola De Bellis is currently a medical librarian at the University of Modena and Reggio Emilia. In 1998, he earned a Ph.D. in the history of science, a discipline that currently energizes his interest in information systems.